Path Integrals in Quantum Mechanics

Path Integrals in Quantum Mechanics

J. Zinn-Justin

Dapnia, CEA/Saclay, France

and

Institut de Mathématiques de Jussieu–Chevaleret,
University of Paris VII

OXFORD
UNIVERSITY PRESS

Great Clarendon Street, Oxford OX2 6DP

Oxford University Press is a department of the University of Oxford.
It furthers the University's objective of excellence in research, scholarship,
and education by publishing worldwide in

Oxford New York

Auckland Bangkok Buenos Aires Cape Town Chennai
Dar es Salaam Delhi Hong Kong Istanbul Karachi Kolkata
Kuala Lumpur Madrid Melbourne Mexico City Mumbai Nairobi
São Paulo Shanghai Taipei Tokyo Toronto

Oxford is a registered trade mark of Oxford University Press
in the UK and in certain other countries

Published in the United States
by Oxford University Press Inc., New York

© Oxford University Press 2005

The moral rights of the author have been asserted

Database right Oxford University Press (maker)

First published 2005

All rights reserved. No part of this publication may be reproduced,
stored in a retrieval system, or transmitted, in any form or by any means,
without the prior permission in writing of Oxford University Press,
or as expressly permitted by law, or under terms agreed with the appropriate
reprographics rights organization. Enquiries concerning reproduction
outside the scope of the above should be sent to the Rights Department,
Oxford University Press, at the address above

You must not circulate this book in any other binding or cover
and you must impose this same condition on any acquirer

British Library Cataloguing in Publication Data

Data available

Library of Congress Cataloging in Publication Data

Data available

ISBN 0-19-856674-3 (Hbk)

10 9 8 7 6 5 4 3 2 1

Printed in Great Britain
on acid-free paper by
Biddles Ltd, Kings Lynn, Norfolk

INTRODUCTION

This work originates from a course on *Advanced Quantum Mechanics* taught in Paris since fall 1996. It is inspired in part by several chapters of the textbook

J. Zinn-Justin, *Quantum Field Theory and Critical Phenomena*, Clarendon Press 1989 (Oxford 4th ed. 2002), chapters 1–6, 39–41, to which it may also serve as an introduction.

The first edition has been published in the French language:

J. Zinn-Justin, *Intégrale de chemin en mécanique quantique: introduction*, EDP Sciences (Les Ulis, 2003).

The main goal of this work is to familiarize the reader with a tool, the path integral, that offers an alternative point of view on quantum mechanics, but more importantly, under a generalized form, has become the key to a deeper understanding of quantum field theory and its applications, which extend from particle physics to phase transitions or properties of quantum gases.

Path integrals are mathematical objects that can be considered as generalizations to an infinite number of variables, represented by paths, of usual integrals. They share the algebraic properties of usual integrals, but have new properties from the viewpoint of analysis.

Path integrals are powerful tools for the study of quantum mechanics, because they emphasize very explicitly the correspondence between classical and quantum mechanics. Physical quantities are expressed as averages over all possible paths but, in the semi-classical limit $\hbar \to 0$, the leading contributions come from paths close to classical paths. Thus, path integrals lead to an intuitive understanding and simple calculations of physical quantities in the semi-classical limit. We will illustrate this observation with scattering processes, spectral properties or barrier penetration.

The formulation of quantum mechanics based on path integrals, which, even if it seems mathematically more complicated than the usual formulation based on partial differential equations, is well adapted to systems with many degrees of freedom, where a formalism of Schrödinger type is much less useful. Therefore, it allows an easy transition from quantum mechanics with a small number of particles to quantum field theory or statistical physics. In particular, generalized path integrals (field integrals) lead to an understanding of the deep relations between quantum field theory and critical phenomena in continuous phase transitions.

In this work, we first discuss *Euclidean* path integrals. This means that we construct a path integral representation for the matrix elements of the quantum statistical operator, or matrix density at thermal equilibrium $e^{-\beta H}$, H being the quantum hamiltonian and β the inverse temperature (measured in a unit where the Boltzmann constant k_B is 1). In this way, we are able to describe quantum statistical physics in terms of path integrals, but also, perhaps more surprisingly, to exhibit a relation between classical and quantum statistical mechanics.

The Euclidean formulation has one serious advantage: it is in general easier to rigorously define path integrals representing the operator $\mathrm{e}^{-\beta H}$ (the Feynman–Kac formula) than the quantum evolution operator $\mathrm{e}^{-iHt/\hbar}$.

Formally, the quantum statistical operator (or density matrix), whose trace is the quantum partition function

$$\mathcal{Z}(\beta) = \operatorname{tr} \mathrm{e}^{-\beta H},$$

describes 'evolution' in imaginary time since it is related to the evolution operator by the substitution $\beta \mapsto it/\hbar$. From this relation originates a drawback of the Euclidean formalism: the associated classical expressions have somewhat unusual forms since time is imaginary. We will speak of *Euclidean* actions or Lagrangians, as well as of Euclidean time.

Nevertheless, algebraic properties of Euclidean path integrals generalize easily to the case of the quantum evolution operator in real time, explicit expressions being obtained by analytic continuation. This will be illustrated by the calculation of scattering amplitudes.

Let us point out one important specific property of the statistical operator: it provides a tool to determine the structure of the ground state of a quantum system. For example, when H is bounded from below, the ground state energy E_0 is given by

$$E_0 = \lim_{\beta \to \infty} \left(-\frac{1}{\beta} \ln \operatorname{tr} \mathrm{e}^{-\beta H} \right).$$

If, in addition, the ground state is unique and isolated, $\mathrm{e}^{-\beta H}$ projects, when $\beta \to +\infty$, on to the ground state $|0\rangle$:

$$\mathrm{e}^{-\beta H} = \mathrm{e}^{-\beta E_0} \left[|0\rangle \langle 0| + O\left(\mathrm{e}^{-\beta(E_1 - E_0)}\right) \right].$$

Euclidean path integrals, thus, often lead to a simple and intuitive understanding of the structure of the ground state for systems with a large number of degrees of freedom.

It has been noticed that quantum barrier penetration effects, in the semi-classical limit, have an interpretation in terms of classical trajectories in imaginary time. The Euclidean path integral is thus naturally adapted to this problem.

Finally, the Euclidean integral is directly related to diffusion processes; for example, the Fokker–Planck equation has the form of a Schrödinger equation in imaginary time. This class of problems contains, as its simplest example, brownian motion, which has motivated the construction of the first path integral, the Wiener integral.

The set-up of the work is then the following:

The goal of Chapter 1 is to introduce, in the case of ordinary integrals, concepts and methods that can be generalized to path integrals.

The first part is devoted to the calculation of ordinary gaussian integrals, gaussian expectation values and the proof of the corresponding Wick's theorem. The notion of connected contributions is introduced. It is shown that expectation values of monomials can conveniently be associated with graphs, called Feynman diagrams.

Expectation values corresponding to measures that deviate slightly from gaussian measures can be reduced to sums of infinite series of gaussian expectation values, a method called perturbation theory.

Finally, Chapter 1 contains a short presentation of the steepest descent method, a method that allows evaluating a class of integrals by reducing them, in some limit, to gaussian integrals.

In Chapter 2, we construct the path integral associated with the statistical operator $\mathrm{e}^{-\beta H}$ in the case of hamiltonians of the simple form $p^2/2m + V(q)$. We then calculate the path integral corresponding to a harmonic oscillator coupled an external, time-dependent force explicitly. This result allows a perturbative evaluation of path integrals with general analytic potentials. We apply these results to the calculation of the partition function $\operatorname{tr} \mathrm{e}^{-\beta H}$, by perturbative and semi-classical methods.

The integrand for this class of path integrals defines a positive measure on paths. It is thus natural to introduce the corresponding expectation values, called correlation functions. Moments of such a distribution can be generated by a generating functional, and recovered by functional differentiation.

These results of Chapter 2 can be applied to the determination of the spectrum of a class of hamiltonians in several approximation schemes, as Chapter 3 tries to illustrate. We also show how path integrals lead to a variational principle and allow us to solve some $O(N)$ symmetric problems in the large N limit.

In Chapter 4, comparing the classical statistical mechanics of one-dimensional systems and the quantum statistical mechanics of the particle, we give an interpretation to the correlation functions introduced in Chapter 2 from the view point of classical statistical physics.

In Chapter 5, we construct path integrals for general hamiltonians with potentials linear in the velocities, like hamiltonians of particles in a magnetic field. We relate the new problem of the choice of the ordering of quantum operators to ambiguities in the definition of the path integral. We discuss some diffusion processes described by a Fokker–Planck equation, which lead to similar path integrals.

In Chapter 6, we introduce the holomorphic representation of quantum mechanics, because it allows a study of the properties of boson systems both from the point of view of evolution and of quantum statistical physics (in the so-called second quantization formalism). Path integrals then take the form of integrals over trajectories in phase space in a complex parametrization. A parallel formalism, based on integration over anti-commuting type or Grassmann variables that we describe in Chapter 7, allows us then to handle fermions in a way quite analogous to bosons.

Chapter 8 is devoted to barrier penetration in the semi-classical limit, a problem for which the Euclidean formalism is especially well suited. We consider the quantum lifting of degeneracy between classical degenerate energy states and the decay of metastable states. The concept of *instanton* plays there a major role.

In Chapter 9, introducing quantum evolution (i.e. in real time), we construct a path integral representation of the scattering or S-matrix, from which we recover the standard perturbative expansion in powers of the potential. We also define an S-matrix in the example of boson and fermion systems. Various other approximation schemes of semi-classical type are then discussed.

Finally, Chapter 10 contains a few additional results such as a definition of path integrals over phase space trajectories, and the problems generated by the quantization of lagrangians with potentials quadratic in the velocities.

We want to emphasize that we take in this work a rather empirical point of view and thus mathematical rigour will not be an essential issue. Our goal is more pedagogical: to understand the path integral, its properties and its usefulness from the physics viewpoint. It is also practical and, thus, much space is devoted to calculation methods.

Finally, this work assumes a minimal knowledge of concepts in quantum mechanics like the superposition principle, Hilbert spaces and operators acting on vectors in Hilbert space or the Schrödinger equation. We often use Dirac's bra and ket notation to indicate vectors in Hilbert space and their complex conjugate. Therefore, for completeness, we recall some basic notions of quantum mechanics in aAppendix A.

Brief historical remarks and bibliography. The first path integral seems to have been proposed by Wiener [1], as a tool to describe the statistical properties of the brownian motion, inspired by the well-known work of Einstein. A brownian motion can be considered as a continuum limit of a markovian random walk with discrete time steps. The motion at time t (t is an integer) is entirely determined by the position x at time t and the probability $p(x' - x)$ to go from a point x to another point x'. As a consequence, if the walker is at time 0 at a point x_0, the probability $P_n(x_n, x_0)$ to reach the point x_n at time n is given by

$$P_n(x_n, x_0) = \int dx_1\, dx_2 \ldots dx_{n-1}\, p(x_n - x_{n-1}) \ldots p(x_2 - x_1) p(x_1 - x_0).$$

The set of integrations over all variables x_i can be interpreted as a weighted average over all paths $\{x_i\}$ that go from x_0 to x_n in a time that takes the integer values $0, 1, \ldots, n$.

Moreover, asymptotically for $n \to \infty$, the discrete nature of time plays no role anymore. Also, as a consequence of the central limit theorem of probabilities, the distribution $p(x' - x)$ can be replaced by a gaussian distribution of the form $e^{-(x-x')^2/2D}$, which yields the same asymptotic limit. This limiting procedure then leads to a path integral: The statistical properties of the brownian motion can be derived from averages over a class of continuous paths, function of a continuous time and weighted by a gaussian distribution.

If Wiener's work is rather well known, a less-known article of Wentzel of about the same period [2] introduces, in the framework of quantum optics, the notions of sums over paths weighted by a phase factor, of destructive interference between paths that do not satisfy classical equations of motion, and the interpretation of the sum as a transition probability amplitude. Dirac [3] has written the first expression of the quantum evolution operator that resembles a path integral, but he did not go beyond the approximate version with discrete time intervals. Nevertheless, his observation was very important from the viewpoint of relativistic physics: he showed that the matrix elements of the evolution operator for an infinitesimal time interval δt could be calculated in terms of the lagrangian $\mathcal{L}$ as $e^{i\mathcal{L}\delta t/\hbar}$, an obviously covariant expression.

Introduction

Of course, the modern history of path integrals begins with the articles of Feynman [4] who formulates quantum evolution in terms of sums over a set of trajectories weighted by $e^{i\mathcal{S}/\hbar}$, where $\mathcal{S}$ is the value of the classical action (time integral of the lagrangian). This leads, in particular to an interpretation of the classical equations of motion as resulting from an evaluation of the path integral by the stationary phase method. When, in a physical system, the typical values of the action are large compared to $\hbar$, only paths close to the classical paths contribute to the integral.

The path integral formulation of quantum mechanics has led to a new and deeper understanding of the relation between quantum and classical mechanics. Path integrals have also provided new tools to study quantum mechanics in the semi-classical limit. However, beyond the conceptual interest of such a reformulation of quantum evolution, the study of path integrals has become unavoidable in modern physics through its generalization to systems with a large number of degrees of freedom, as in quantum field theory. In particular, the quantization of non-abelian gauge theories by Faddeev and Popov (1967) and DeWitt would have been extremely difficult in the usual formulation of the quantum theory in terms of operators and quantum field equations. Since non-abelian gauge theories form the basic ingredient of the Standard Model that describes all fundamental interactions except gravitation, one realizes the importance of such a result.

Moreover, path integrals have revealed the deep mathematical relations between quantum field theory and the statistical physics of phase transitions, relations that would been much more difficult to perceive otherwise. These relations have played a major role in our understanding of critical phenomena since Wilson.

From the mathematical point of view, path integrals related to quantum evolution are often more difficult to define because $e^{i\mathcal{S}/\hbar}$ has modulus one for all paths and thus the variable contribution of paths is a consequence of interferences. Kac [5] noticed that if one replaces the evolution operator $e^{-itH/\hbar}$ by the statistical operator $e^{-\beta H}$ (and thus the Schrödinger equation by a kind of diffusion or heat equation), one obtains a path integral where the integrand becomes a positive measure (at least in the simplest examples), generalizing Wiener's integral, which can be more easily defined [6]. A strategy that has been then followed by many authors has been to define path integrals corresponding to quantum evolution by an analytic continuation in the time variable [7]. This is the viewpoint we also adopt throughout this work.

In physics, several generalizations of the initial concept of path integral have proved useful. The integral over phase space (space and momentum) trajectories in a complex parametrization, in correspondence with the holomorphic representation of quantum mechanics [8], appears naturally in the study of the properties of boson systems in the so-called second quantization formalism. The integral over Grassmann paths [9,10] allows a parallel treatment of fermion and boson systems. The integral over real phase space trajectories [11–13] leads to an intuitive understanding of the integration measure when paths belong to curved riemannian manifolds [14] like, for example, spheres.

Finally, we do not discuss in this work the interesting problem of the quantization of constrained systems, but refer to the corresponding literature [15].

Many authors have emphasized that in general cases where the quantum hamiltonian cannot be inferred from its classical counterpart because a problem of ordering operators arises, the path integral does not quantize either, though formally it seems to depend only on classical quantities, because it is no longer uniquely defined [16,17].

Path integrals have allowed recovering many semi-classical approximations by simpler and more intuitive methods. Examples are provided by the semi-classical calculation of the evolution operator [14], from which semi-classical approximation of scattering amplitudes follows [18], or by the calculation of spectra of hamiltonians [19].

The imaginary time version (the quantum statistical operator, Feynman–Kac) allows studying barrier penetration in the semi-classical limit [20]. Path integrals are then dominated by classical solutions of *instantons* type [21] and the evaluation of their contributions requires the introduction of collectives coordinates [22]. The behaviour of perturbative expansions around the harmonic approximation at large orders is also obtained from an analogous barrier penetration calculation [23].

Finally, several books devoted to path integrals are of historical interest or present other viewpoints and contain additional references [16,24].

Contents

1 Gaussian integrals 1
  1.1 Generating function 1
  1.2 Gaussian expectation values. Wick's theorem 2
  1.3 Perturbed gaussian measure. Connected contributions 6
  1.4 Expectation values. Generating function. Cumulants 9
  1.5 Steepest descent method 12
  1.6 Steepest descent method: Several variables, generating functions . . . 18
  1.7 Gaussian integrals: Complex matrices 20
  Exercises 23
2 Path integrals in quantum mechanics 27
  2.1 Local markovian processes 28
  2.2 Solution of the evolution equation for short times 31
  2.3 Path integral representation 34
  2.4 Explicit calculation: gaussian path integrals 38
  2.5 Correlation functions: generating functional 41
  2.6 General gaussian path integral and correlation functions . . . 44
  2.7 Harmonic oscillator: the partition function 48
  2.8 Perturbed harmonic oscillator 52
  2.9 Perturbative expansion in powers of $\hbar$ 54
  2.10 Semi-classical expansion 55
  Exercises 59
3 Partition function and spectrum 63
  3.1 Perturbative calculation 63
  3.2 Semi-classical or WKB expansion 66
  3.3 Path integral and variational principle 73
  3.4 $O(N)$ symmetric quartic potential for $N \to \infty$ 75
  3.5 Operator determinants 84
  3.6 Hamiltonian: structure of the ground state 85
  Exercises 87
4 Classical and quantum statistical physics 91
  4.1 Classical partition function. Transfer matrix 91
  4.2 Correlation functions 94

    4.3 Classical model at low temperature: an example 97
    4.4 Continuum limit and path integral 98
    4.5 The two-point function: perturbative expansion, spectral representation   102
    4.6 Operator formalism. Time-ordered products 105
    Exercises 107

5 Path integrals and quantization 111
    5.1 Gauge transformations 111
    5.2 Coupling to a magnetic field: gauge symmetry 113
    5.3 Quantization and path integrals 116
    5.4 Magnetic field: direct calculation 120
    5.5 Diffusion, random walk, Fokker–Planck equation 122
    5.6 The spectrum of the $O(2)$ rigid rotator 126
    Exercises 131

6 Path integrals and holomorphic formalism 135
    6.1 Complex integrals and Wick's theorem 135
    6.2 Holomorphic representation 140
    6.3 Kernel of operators 143
    6.4 Path integral: the harmonic oscillator 146
    6.5 Path integral: general hamiltonians 150
    6.6 Bosons: second quantization 156
    6.7 Partition function 159
    6.8 Bose–Einstein condensation 161
    6.9 Generalized path integrals: the quantum Bose gas 164
    Exercises 169

7 Path integrals: fermions 179
    7.1 Grassmann algebras 179
    7.2 Differentiation in Grassmann algebras 181
    7.3 Integration in Grassmann algebras 182
    7.4 Gaussian integrals and perturbative expansion 184
    7.5 Fermion vector space and operators 190
    7.6 One-state hamiltonian 195
    7.7 Many-particle states. Partition function 197
    7.8 Path integral: one-state problem 200
    7.9 Path integrals: Generalization 203
    7.10 Quantum Fermi gas 206
    7.11 Real gaussian integrals. Wick's theorem 210
    7.12 Mixed change of variables: Berezinian and supertrace 212
    Exercises 214

8 Barrier penetration: semi-classical approximation 225
    8.1 Quartic double-well potential and instantons 225
    8.2 Degenerate minima: semi-classical approximation 229
    8.3 Collective coordinates and gaussian integration 232
    8.4 Instantons and metastable states 238
    8.5 Collective coordinates: alternative method 243
    8.6 The jacobian 245

    8.7 Instantons: the quartic anharmonic oscillator 247
    Exercises 251
9 Quantum evolution and scattering matrix 257
    9.1 Evolution of the free particle and S-matrix 257
    9.2 Perturbative expansion of the S-matrix 260
    9.3 S-matrix: bosons and fermions 266
    9.4 S-matrix in the semi-classical limit 269
    9.5 Semi-classical approximation: one dimension 270
    9.6 Eikonal approximation 272
    9.7 Perturbation theory and operators 276
    Exercises 277
10 Path integrals in phase space 279
    10.1 A few elements of classical mechanics 279
    10.2 The path integral in phase space 284
    10.3 Harmonic oscillator. Perturbative calculations 289
    10.4 Lagrangians quadratic in the velocities 290
    10.5 Free motion on the sphere or rigid rotator 294
    Exercises 299
Appendix Quantum mechanics: minimal background 301
    A1 Hilbert space and operators 301
    A2 Quantum evolution, symmetries and density matrix . . . 303
    A3 Position and momentum. Schrödinger equation 305
Bibliography 311
Index 315

1 GAUSSIAN INTEGRALS

Gaussian measures play a central role in many fields: in probability theory as a consequence of the central limit theorem, in quantum mechanics as we will show and, thus, in quantum field theory, in the theory of phase transitions in statistical physics. Therefore, as a preliminary to the discussion of path integrals, we recall in this chapter a few useful mathematical results about gaussian integrals, and properties of gaussian expectation values. In particular, we prove the corresponding Wick's theorem, a simple result but of major practical significance.

To discuss properties of expectation values with respect to some measure or probability distribution, it is always convenient to introduce a generating function of the moments of the distribution. This also allows defining the generating function of the cumulants of a distribution, which have simpler properties.

The steepest descent method provides an approximation scheme to evaluate certain types of complex integrals. Successive contributions take the form of gaussian expectation values. We explain the method here, for real and complex, simple and multiple integrals, with the aim of eventually applying it to path integrals.

Notation. In this work we use, in general, boldface characters to indicate matrices or vectors, while we use italics (then with indices) to indicate the corresponding matrix elements or vector components.

1.1 Generating function

We consider a positive measure or probability distribution $\Omega(x_1, x_2, \ldots, x_n)$ defined on $\mathbb{R}^n$ and properly normalized. We denote by

$$\langle F \rangle \equiv \int d^n x \, \Omega(\mathbf{x}) F(\mathbf{x}),$$

where $d^n x \equiv \prod_{i=1}^{n} dx_i$, the expectation value of a function $F(x_1, \ldots, x_n)$. The normalization is chosen such that $\langle 1 \rangle = 1$.

It is generally convenient to introduce the Fourier transform of the distribution. In this work, we consider mainly a special class of distributions such that the Fourier transform is an analytic function that also exists for imaginary arguments. We thus introduce the function

$$\mathcal{Z}(\mathbf{b}) = \langle e^{\mathbf{b} \cdot \mathbf{x}} \rangle = \int d^n x \, \Omega(\mathbf{x}) \, e^{\mathbf{b} \cdot \mathbf{x}} \quad \text{where } \mathbf{b} \cdot \mathbf{x} = \sum_{i=1}^{n} b_i x_i. \tag{1.1}$$

The advantage of this definition is that the integrand remains a positive measure for all real values of $\mathbf{b}$.

Expanding the integrand in powers of the variables b_k, one recognizes that the coefficients are expectation values, moments of the distribution:

$$\mathcal{Z}(\mathbf{b}) = \sum_{\ell=0}^{\infty} \frac{1}{\ell!} \sum_{k_1,k_2,\ldots,k_\ell=1}^{n} b_{k_1} b_{k_2} \ldots b_{k_\ell} \langle x_{k_1} x_{k_2} \ldots x_{k_\ell} \rangle.$$

The function $\mathcal{Z}(\mathbf{b})$ thus is a *generating function* of the moments of the distribution, that is, of the expectation values of monomial functions. The expectation values can be recovered by differentiating the function $\mathcal{Z}(\mathbf{b})$. Quite directly, differentiating both sides of equation (1.1) with respect to b_k, one obtains

$$\frac{\partial}{\partial b_k} \mathcal{Z}(\mathbf{b}) = \int \mathrm{d}^n x \, \Omega(\mathbf{x}) x_k \, \mathrm{e}^{\mathbf{b} \cdot \mathbf{x}}. \tag{1.2}$$

Repeated differentiation then yields, in the limit $\mathbf{b} = 0$,

$$\langle x_{k_1} x_{k_2} \ldots x_{k_\ell} \rangle = \left[\frac{\partial}{\partial b_{k_1}} \frac{\partial}{\partial b_{k_2}} \cdots \frac{\partial}{\partial b_{k_\ell}} \mathcal{Z}(\mathbf{b}) \right]_{\mathbf{b}=0}. \tag{1.3}$$

This notion of generating functions is very useful and will be extended in Section 2.5.3 to the limit where the number of variables becomes infinite.

1.2 Gaussian expectation values. Wick's theorem

As a consequence of the central limit theorem of probabilities, gaussian distributions play an important role in all stochastic phenomena and, therefore, also in physics. We recall here some algebraic properties of gaussian integrals and gaussian expectation values. Since most algebraic properties generalize to complex gaussian integrals, we consider also below this more general situation.

The gaussian integral

$$\mathcal{Z}(\mathbf{A}) = \int \mathrm{d}^n x \, \exp\left(-\sum_{i,j=1}^{n} \tfrac{1}{2} x_i A_{ij} x_j \right), \tag{1.4}$$

converges if the matrix $\mathbf{A}$ with elements A_{ij} is a symmetric complex matrix such that the real part of the matrix is non-negative (this implies that all eigenvalues of $\mathrm{Re}\,\mathbf{A}$ are non-negative) and no eigenvalue a_i of $\mathbf{A}$ vanishes:

$$\mathrm{Re}\,\mathbf{A} \geq 0, \quad a_i \neq 0.$$

Several methods then allow us to prove

$$\mathcal{Z}(\mathbf{A}) = (2\pi)^{n/2} (\det \mathbf{A})^{-1/2}. \tag{1.5}$$

When the matrix is complex, the meaning of the square root and thus the determination of the global sign requires, of course, some special care.

We derive below this result for real positive matrices. In Section 1.7, we give a proof for complex matrices. Another independent proof is indicated in the exercises.

1.2.1 Real matrices: a proof

The general one-dimensional gaussian integral can easily be calculated and one finds ($a > 0$)

$$\int_{-\infty}^{+\infty} dx\, e^{-ax^2/2+bx} = \sqrt{2\pi/a}\, e^{b^2/2a}. \tag{1.6}$$

More generally, any real symmetric matrix can be diagonalized by an orthogonal transformation and the matrix $\mathbf{A}$ in (1.4) can thus be written as

$$\mathbf{A} = \mathbf{ODO}^T, \tag{1.7}$$

where the matrix $\mathbf{O}$ is orthogonal and the matrix $\mathbf{D}$ with elements D_{ij} diagonal:

$$\mathbf{O}^T\mathbf{O} = \mathbf{1},\ D_{ij} = a_i \delta_{ij}.$$

We thus change variables, $\mathbf{x} \mapsto \mathbf{y}$, in the integral (1.4):

$$x_i = \sum_{j=1}^n O_{ij} y_j \Rightarrow \sum_{i,j} x_i A_{ij} x_j = \sum_{i,j} x_i O_{ik} a_k O_{jk} x_j = \sum_i a_i y_i^2.$$

The corresponding jacobian is $J = |\det \mathbf{O}| = 1$.

The integral then factorizes:

$$\mathcal{Z}(\mathbf{A}) = \prod_{i=1}^n \int dy_i\, e^{-a_i y_i^2/2}.$$

The matrix $\mathbf{A}$ is positive; all eigenvalues a_i are thus positive and each integral converges. From the result (1.6), one then infers

$$\mathcal{Z}(\mathbf{A}) = (2\pi)^{n/2} (a_1 a_2 \ldots a_n)^{-1/2} = (2\pi)^{n/2} (\det \mathbf{A})^{-1/2}.$$

1.2.2 General gaussian integral

We now consider a general gaussian integral

$$\mathcal{Z}(\mathbf{A}, \mathbf{b}) = \int d^n x\, \exp\left(-\sum_{i,j=1}^n \tfrac{1}{2} x_i A_{ij} x_j + \sum_{i=1}^n b_i x_i\right). \tag{1.8}$$

To calculate $\mathcal{Z}(\mathbf{A}, \mathbf{b})$, one looks for the minimum of the quadratic form:

$$\frac{\partial}{\partial x_k}\left(\sum_{i,j=1}^n \tfrac{1}{2} x_i A_{ij} x_j - \sum_{i=1}^n b_i x_i\right) = \sum_{j=1}^n A_{kj} x_j - b_k = 0.$$

Introducing the inverse matrix
$$\mathbf{\Delta} = \mathbf{A}^{-1},$$
one can write the solution as
$$x_i = \sum_{j=1}^{n} \Delta_{ij} b_j \,. \tag{1.9}$$

After the change of variables $x_i \mapsto y_i$,
$$x_i = \sum_{j=1}^{n} \Delta_{ij} b_j + y_i \,, \tag{1.10}$$

the integral becomes
$$\mathcal{Z}(\mathbf{A},\mathbf{b}) = \exp\left[\sum_{i,j=1}^{n} \tfrac{1}{2} b_i \Delta_{ij} b_j\right] \int d^n y \, \exp\left(-\sum_{i,j=1}^{n} \tfrac{1}{2} y_i A_{ij} y_j\right). \tag{1.11}$$

The change of variables has reduced the calculation to the integral (1.4). One concludes
$$\mathcal{Z}(\mathbf{A},\mathbf{b}) = (2\pi)^{n/2} (\det \mathbf{A})^{-1/2} \exp\left[\sum_{i,j=1}^{n} \tfrac{1}{2} b_i \Delta_{ij} b_j\right]. \tag{1.12}$$

Remark. Gaussian integrals have a remarkable property: after integration over one variable, one finds again a gaussian integral. This structural stability explains the stability of gaussian probability distributions and is also related to some properties of the harmonic oscillator, which will be discussed in Section 2.6.

1.2.3 Gaussian expectation values and Wick theorem

When the matrix $\mathbf{A}$ is real and positive, the gaussian integrand can be considered as a positive measure or a probability distribution on $\mathbb{R}^n$, which can be used to calculate expectation values of functions of the n variables x_i:
$$\langle F(\mathbf{x}) \rangle \equiv \mathcal{N} \int d^n x \, F(\mathbf{x}) \exp\left(-\sum_{i,j=1}^{n} \tfrac{1}{2} x_i A_{ij} x_j\right), \tag{1.13}$$

where the normalization $\mathcal{N}$ is determined by the condition $\langle 1 \rangle = 1$:
$$\mathcal{N} = \mathcal{Z}^{-1}(\mathbf{A},0) = (2\pi)^{-n/2} (\det \mathbf{A})^{1/2} \,.$$

The function
$$\mathcal{Z}(\mathbf{A},\mathbf{b})/\mathcal{Z}(\mathbf{A},0) = \langle e^{\mathbf{b}\cdot\mathbf{x}} \rangle, \tag{1.14}$$

where $\mathcal{Z}(\mathbf{A},\mathbf{b})$ is the function (1.8), is then a *generating function* of the moments of the distribution, that is, of the gaussian expectation values of monomials (see

Section 1.1). Expectation values are then obtained by differentiating equation (1.14) with respect to the variables b_i:

$$\langle x_{k_1} x_{k_2} \ldots x_{k_\ell} \rangle = (2\pi)^{-n/2} (\det \mathbf{A})^{1/2} \left[\frac{\partial}{\partial b_{k_1}} \frac{\partial}{\partial b_{k_2}} \cdots \frac{\partial}{\partial b_{k_\ell}} \mathcal{Z}(\mathbf{A}, \mathbf{b}) \right]\bigg|_{\mathbf{b}=0}$$

and, replacing $\mathcal{Z}(\mathbf{A}, \mathbf{b})$ by the explicit expression (1.12),

$$\langle x_{k_1} \ldots x_{k_\ell} \rangle = \left\{ \frac{\partial}{\partial b_{k_1}} \cdots \frac{\partial}{\partial b_{k_\ell}} \exp\left[\sum_{i,j=1}^{n} \tfrac{1}{2} b_i \Delta_{ij} b_j \right] \right\}\bigg|_{\mathbf{b}=0} . \tag{1.15}$$

More generally, if $F(\mathbf{x})$ is a power series in the variables x_i, its expectation value is given by the identity

$$\langle F(\mathbf{x}) \rangle = \left\{ F\left(\frac{\partial}{\partial \mathbf{b}}\right) \exp\left[\sum_{i,j} \tfrac{1}{2} b_i \Delta_{ij} b_j \right] \right\}\bigg|_{\mathbf{b}=0} . \tag{1.16}$$

Wick's theorem. Identity (1.15) leads to Wick's theorem. Each time one differentiates the exponential in the r.h.s., one generates one factor b. One must differentiate this b factor later, otherwise the corresponding contribution vanishes when $\mathbf{b}$ is set to zero. One concludes that the expectation value of the product $x_{k_1} \ldots x_{k_\ell}$ with a gaussian weight proportional to $\exp(-\tfrac{1}{2} x_i A_{ij} x_j)$ is given by the following expression: one pairs in all possible ways the indices $k_1, \ldots, k_\ell$ (ℓ must be even, otherwise the moment vanishes). To each pair $k_p k_q$, one associates the element $\Delta_{k_p k_q}$ of the matrix $\boldsymbol{\Delta} = \mathbf{A}^{-1}$. Then,

$$\langle x_{k_1} \ldots x_{k_\ell} \rangle = \sum_{\substack{\text{all possible pairings} \\ P \text{ of } \{k_1 \ldots k_\ell\}}} \Delta_{k_{P_1} k_{P_2}} \cdots \Delta_{k_{P_{\ell-1}} k_{P_\ell}} , \tag{1.17}$$

$$= \sum_{\substack{\text{all possible pairings} \\ P \text{ of } \{k_1 \ldots k_\ell\}}} \langle x_{k_{P_1}} x_{k_{P_2}} \rangle \cdots \langle x_{k_{P_{\ell-1}}} x_{k_{P_\ell}} \rangle . \tag{1.18}$$

Equations (1.17, 1.18) are characteristic properties of all centred (i.e. $\langle x_i \rangle = 0$) gaussian measures. They are known under the name of Wick's theorem. Suitably adapted to quantum mechanics or quantum field theory, they form the basis of perturbation theory. Note that the simplicity of this result should not hide its *major practical significance*. Note also that, since the derivation is purely algebraic, it generalizes to complex integrals. Only the interpretation of gaussian functions as positive measures or probability distributions disappears.

Examples. One finds, successively,

$$\langle x_{i_1} x_{i_2} \rangle = \Delta_{i_1 i_2} ,$$
$$\langle x_{i_1} x_{i_2} x_{i_3} x_{i_4} \rangle = \Delta_{i_1 i_2} \Delta_{i_3 i_4} + \Delta_{i_1 i_3} \Delta_{i_2 i_4} + \Delta_{i_1 i_4} \Delta_{i_3 i_2} .$$

More generally, the gaussian expectation value of a product of $2p$ variables is the sum of $(2p-1)(2p-3)\ldots 5 \times 3 \times 1$ contributions (a simple remark that provides a useful check).

A useful identity. We consider the gaussian expectation value of the product $x_i F(\mathbf{x})$:

$$\langle x_i\, F(\mathbf{x})\rangle = \mathcal{N} \int \mathrm{d}^n x\, x_i\, F(\mathbf{x}) \exp\left(-\sum_{j,k=1}^n \tfrac{1}{2} x_j A_{jk} x_k\right). \tag{1.19}$$

Using the identity

$$x_i \exp\left(-\sum_{j,k=1}^n \tfrac{1}{2} x_j A_{jk} x_k\right) = -\sum_\ell \Delta_{i\ell} \frac{\partial}{\partial x_\ell} \exp\left(-\sum_{j,k=1}^n \tfrac{1}{2} x_j A_{jk} x_k\right)$$

inside (1.19), and integrating by parts, one obtains the relation

$$\langle x_i F(\mathbf{x})\rangle = \mathcal{N} \sum_\ell \Delta_{i\ell} \int \mathrm{d}^n x \exp\left(-\sum_{j,k=1}^n \tfrac{1}{2} x_j A_{jk} x_k\right) \frac{\partial F}{\partial x_\ell},$$

which can also be written as

$$\langle x_i F(x)\rangle = \sum_\ell \langle x_i x_\ell\rangle \left\langle \frac{\partial F}{\partial x_\ell}\right\rangle, \tag{1.20}$$

and which can also be derived by applying Wick's theorem.

1.3 Perturbed gaussian measure. Connected contributions

Even in favourable situations where the central limit theorem applies, the gaussian measure is only an asymptotic distribution. Therefore, it is useful to also evaluate expectation values with perturbed gaussian distributions.

1.3.1 Perturbed gaussian measure

We consider a more general normalized distribution $\mathrm{e}^{-A(\mathbf{x},\lambda)}/\mathcal{Z}(\lambda)$, where the function $A(\mathbf{x}, \lambda)$ can be written as the sum of a quadratic part and a polynomial $\lambda V(\mathbf{x})$ in the variables x_i:

$$A(\mathbf{x}, \lambda) = \frac{1}{2} \sum_{i,j=1}^n x_i A_{ij} x_j + \lambda V(\mathbf{x}), \tag{1.21}$$

the parameter λ characterizing the amplitude of the deviation from the gaussian distribution.

The normalization $\mathcal{Z}(\lambda)$ is given by the integral

$$\mathcal{Z}(\lambda) = \int \mathrm{d}^n x\, \mathrm{e}^{-A(\mathbf{x},\lambda)}. \tag{1.22}$$

1.3 Gaussian integrals

To evaluate it, we expand the integrand in a formal power series in λ and integrate term by term:

$$\mathcal{Z}(\lambda) = \sum_{k=0}^{\infty} \frac{(-\lambda)^k}{k!} \int d^n x\, V^k(\mathbf{x}) \exp\left(-\sum_{i,j=1}^{n} \tfrac{1}{2} x_i A_{ij} x_j\right)$$

$$= \mathcal{Z}(0) \sum_{k=0}^{\infty} \frac{(-\lambda)^k}{k!} \langle V^k(\mathbf{x}) \rangle_0, \quad (1.23)$$

where the symbol $\langle \bullet \rangle_0$ refers to the expectation value with respect to the normalized gaussian measure $\exp[-\sum_{i,j} x_i A_{ij} x_j/2]/\mathcal{Z}(0)$. Each term in the expansion, which is a gaussian expectation value of a polynomial, can then be evaluated with the help of Wick's theorem (1.17).

Using equation (1.16) with $F = e^{-\lambda V}$, one also infers a formal representation of the function (1.22):

$$\mathcal{Z}(\lambda)/\mathcal{Z}(0) = \left\{ \exp\left[-\lambda V\left(\frac{\partial}{\partial b}\right)\right] \exp\left[\sum_{i,j=1}^{n} \tfrac{1}{2} b_i \Delta_{ij} b_j\right] \right\}\bigg|_{\mathbf{b}=0}. \quad (1.24)$$

Example. In the example of the perturbation

$$V(\mathbf{x}) = \frac{1}{4!} \sum_{i=1}^{n} x_i^4, \quad (1.25)$$

the expansion to order λ^2 is ($\mathbf{\Delta A} = \mathbf{1}$)

$$\mathcal{Z}(\lambda)/\mathcal{Z}(0) = 1 - \frac{1}{4!}\lambda \sum_i \langle x_i^4 \rangle_0 + \frac{1}{2!(4!)^2}\lambda^2 \sum_i \sum_j \langle x_i^4 x_j^4 \rangle_0 + O(\lambda^3)$$

$$= 1 - \tfrac{1}{8}\lambda \sum_i \Delta_{ii}^2 + \tfrac{1}{128}\lambda^2 \sum_i \Delta_{ii}^2 \sum_j \Delta_{jj}^2$$

$$+ \lambda^2 \sum_{i,j} \left(\tfrac{1}{16}\Delta_{ii}\Delta_{jj}\Delta_{ij}^2 + \tfrac{1}{48}\Delta_{ij}^4\right) + O(\lambda^3). \quad (1.26)$$

A simple verification of the factors is obtained by specializing to the case of only one variable. Then,

$$\mathcal{Z}(\lambda)/\mathcal{Z}(0) = 1 - \tfrac{1}{8}\lambda + \tfrac{35}{384}\lambda^2 + O(\lambda^3).$$

Note that the first two terms of the expansion (1.26) exponentiate in such a way that $\ln \mathcal{Z}$ has only *connected* contributions, that is, contributions that cannot be factorized into a product of sums:

$$\ln \mathcal{Z}(\lambda) - \ln \mathcal{Z}(0) = -\tfrac{1}{8}\lambda \sum_i \Delta_{ii}^2 + \lambda^2 \sum_{i,j} \left(\tfrac{1}{16}\Delta_{ii}\Delta_{jj}\Delta_{ij}^2 + \tfrac{1}{48}\Delta_{ij}^4\right) + O(\lambda^3).$$

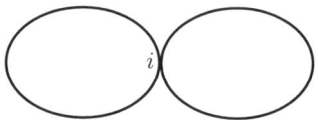

Fig. 1.1 Feynman diagram: the contribution $\langle x^4 \rangle_0$ at order λ in the example (1.25).

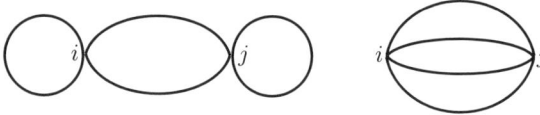

Fig. 1.2 Feynman diagrams: connected contributions from $\langle x_i^4 x_j^4 \rangle_0$ at order λ^2 in the example (1.25).

1.3.2 Feynman diagrams

To each different contribution generated by Wick's theorem can be associated a graph called a Feynman diagram. Each monomial contributing to $V(\mathbf{x})$ is represented by a point (a vertex) to which are attached a number of lines equal to the degree of the monomial. Each pairing is represented by a line joining the vertices to which the corresponding variables belong.

We have introduced the notion of connected contributions. To this notion corresponds a property of graphs. A connected contribution also corresponds to a connected diagram. Connected contributions to the normalization (1.22) in the example (1.25) are displayed up to order λ^2 in Figs 1.1 and 1.2, the indices i and j indicating the summations in (1.26).

1.3.3 Connected contributions

We now discuss, more generally, the notion of connected contribution that we have just introduced. We use, below, the subscript c to indicate the connected part of an expectation value. Then, for example,

$$\langle V(\mathbf{x}) \rangle = \langle V(\mathbf{x}) \rangle_c, \quad \langle V^2(\mathbf{x}) \rangle = \langle V^2(\mathbf{x}) \rangle_c + \langle V(\mathbf{x}) \rangle_c^2$$
$$\langle V^3(\mathbf{x}) \rangle = \langle V^3(\mathbf{x}) \rangle_c + 3 \langle V^2(\mathbf{x}) \rangle_c \langle V(\mathbf{x}) \rangle_c + \langle V(\mathbf{x}) \rangle_c^3, \cdots.$$

More generally, at order k, one finds

$$\tfrac{1}{k!} \langle V^k(\mathbf{x}) \rangle = \tfrac{1}{k!} \langle V^k(\mathbf{x}) \rangle_c + \text{non-connected terms.}$$

A non-connected term is a product of the form

$$\langle V^{k_1}(\mathbf{x}) \rangle_c \langle V^{k_2}(\mathbf{x}) \rangle_c \cdots \langle V^{k_p}(\mathbf{x}) \rangle_c, \quad k_1 + k_2 + \cdots + k_p = k,$$

with a weight $1/k!$ coming from the expansion of the exponential function and multiplied by a combinatorial factor corresponding to all possible different ways to

group k objects into subsets of $k_1 + k_2 + \cdots + k_p$ objects, if all k_i are distinct. One finds
$$\frac{1}{k!} \times \frac{k!}{k_1! k_2! \ldots k_p!} = \frac{1}{k_1! k_2! \ldots k_p!}.$$

If m values k_i are equal, it is necessary to divide by an additional combinatorial factor $1/m!$ because, otherwise, the same term is counted $m!$ times.

One then notices that the perturbative expansion can be written as
$$\mathcal{W}(\lambda) = \ln \mathcal{Z}(\lambda) = \ln \mathcal{Z}(0) + \sum_k \frac{(-\lambda)^k}{k!} \langle V^k(\mathbf{x}) \rangle_c. \tag{1.27}$$

1.4 Expectation values. Generating function. Cumulants

We now calculate moments of the distribution $\mathrm{e}^{-A(\mathbf{x},\lambda)}/\mathcal{Z}(\lambda)$, where $A(\mathbf{x}, \lambda)$ is a polynomial (1.21):
$$A(\mathbf{x}, \lambda) = \sum_{i,j=1}^n \tfrac{1}{2} x_i A_{ij} x_j + \lambda V(\mathbf{x}).$$

Expectation values of the form $\langle x_{i_1} x_{i_2} \ldots x_{i_\ell} \rangle_\lambda$, which we will call ℓ-point functions as it is customary in the context of path integrals, are given by the ratios
$$\langle x_{i_1} x_{i_2} \ldots x_{i_\ell} \rangle_\lambda = \mathcal{Z}^{-1}(\lambda) \mathcal{Z}_{i_1 i_2 \ldots i_\ell}(\lambda), \tag{1.28a}$$
$$\mathcal{Z}_{i_1 i_2 \ldots i_\ell}(\lambda) = \int \mathrm{d}^n x \, x_{i_1} x_{i_2} \ldots x_{i_\ell} \exp\left[-A(\mathbf{x}, \lambda)\right]. \tag{1.28b}$$

1.4.1 The two-point function

As an illustration, we give a few elements of the calculation of the two-point function $\langle x_{i_1} x_{i_2} \rangle_\lambda$ up to order λ^2. One first expands the integral
$$\mathcal{Z}_{i_1 i_2}(\lambda) = \int \mathrm{d}^n x \, x_{i_1} x_{i_2} \exp\left[-A(\mathbf{x}, \lambda)\right].$$

In the example (1.25), at order λ^2, one finds
$$\mathcal{Z}_{i_1 i_2}(\lambda)/\mathcal{Z}(0) = \Delta_{i_1 i_2} - \tfrac{1}{24} \lambda \Delta_{i_1 i_2} \sum_i \langle x_i^4 \rangle_0 - \tfrac{1}{2} \lambda \sum_i \Delta_{i i_1} \Delta_{ii} \Delta_{i i_2}$$
$$+ \frac{\lambda^2}{2!(4!)^2} \sum_{i,j} \Delta_{i_1 i_2} \langle x_i^4 x_j^4 \rangle_0 + \frac{\lambda^2}{2!4!} \sum_{i,j} \Delta_{i i_1} \Delta_{ii} \Delta_{i i_2} \langle x_j^4 \rangle_0$$
$$+ \lambda^2 \sum_{i,j} \left(\tfrac{1}{4} \Delta_{i i_1} \Delta_{i i_2} \Delta_{ij}^2 \Delta_{jj} + \tfrac{1}{6} \Delta_{i_1 i} \Delta_{j i_2} \Delta_{ij}^3 \right.$$
$$\left. + \tfrac{1}{4} \Delta_{i_1 i} \Delta_{j i_2} \Delta_{ij} \Delta_{ii} \Delta_{jj} \right) + O(\lambda^3).$$

One then calculates the ratio

$$\langle x_{i_1} x_{i_2}\rangle_\lambda = \mathcal{Z}_{i_1 i_2}(\lambda)/\mathcal{Z}(\lambda).$$

In the ratio of the two series, the non-connected terms cancel and one left with

$$\langle x_{i_1} x_{i_2}\rangle_\lambda = \Delta_{i_1 i_2} - \tfrac{1}{2}\lambda \sum_i \Delta_{i i_1}\Delta_{ii}\Delta_{i i_2} + \lambda^2 \sum_{i,j}\left(\tfrac{1}{4}\Delta_{i_1 i}\Delta_{j i_2}\Delta_{ij}\Delta_{ii}\Delta_{jj}\right.$$
$$\left.+\tfrac{1}{4}\Delta_{i i_1}\Delta_{i i_2}\Delta_{ij}^2\Delta_{jj} + \tfrac{1}{6}\Delta_{i_1 i}\Delta_{j i_2}\Delta_{ij}^3\right) + O(\lambda^3). \tag{1.29}$$

In terms of Feynman diagrams, the contributions of order 1, λ and λ^2 are displayed in Figs 1.3 and 1.4, respectively.

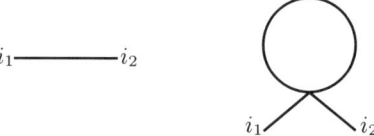

Fig. 1.3 The two-point function: contributions of order 1 and λ.

Fig. 1.4 The two-point function: contributions of order λ^2.

One could, similarly, calculate the four-point function, that is, the expectation value of the general monomial of degree 4. One would find many contributions. However, the results simplify if one directly calculates the cumulants of the distribution. For this purpose, it is convenient to first introduce a generating function of the expectation values $\langle x_{i_1} x_{i_2} \ldots x_{i_p}\rangle_\lambda$.

1.4.2 Generating functions. Cumulants

We now introduce the function

$$\mathcal{Z}(\mathbf{b}, \lambda) = \int d^n x \, \exp\left[-A(\mathbf{x}, \lambda) + \mathbf{b}\cdot\mathbf{x}\right], \tag{1.30}$$

which generalizes the function (1.8) of the gaussian example. It is proportional to the generating function of the expectation values (1.28a) (see Section 1.1)

$$\langle e^{\mathbf{b}\cdot\mathbf{x}}\rangle_\lambda = \mathcal{Z}(\mathbf{b}, \lambda)/\mathcal{Z}(\lambda)$$

1.4 Gaussian integrals

that generalizes the function (1.14). Then, by differentiating one obtains

$$\langle x_{i_1} x_{i_2} \ldots x_{i_\ell} \rangle_\lambda = \mathcal{Z}^{-1}(\lambda) \left[\frac{\partial}{\partial b_{i_1}} \frac{\partial}{\partial b_{i_2}} \cdots \frac{\partial}{\partial b_{i_\ell}} \mathcal{Z}(\mathbf{b}, \lambda) \right]_{\mathbf{b}=0}. \tag{1.31}$$

We now introduce the function

$$\mathcal{W}(\mathbf{b}, \lambda) = \ln \mathcal{Z}(\mathbf{b}, \lambda). \tag{1.32}$$

In a probabilistic interpretation, $\mathcal{W}(\mathbf{b}, \lambda)$ is a generating function of the cumulants of the distribution.

Note that, in the gaussian example, $\mathcal{W}(\mathbf{b})$ reduces to a quadratic form in $\mathbf{b}$. Moreover, it follows from equation (1.27) that the perturbative expansion of cumulants is much simpler because it contains only connected contributions. In particular, all diagrams corresponding to the normalization integral (1.26) can only appear in $\mathcal{W}(0, \lambda)$. Therefore, they cancel in the ratio $\mathcal{Z}(\mathbf{b}, \lambda)/\mathcal{Z}(\lambda)$, as we have already noticed in the calculation of the two-point function in Section 1.4.1.

Remark. In statistical physics, expectation values of products of the form $\langle x_{i_1} x_{i_2} \ldots x_{i_\ell} \rangle$ are called ℓ-point correlation functions and the cumulants

$$W^{(\ell)}_{i_1 i_2 \ldots i_\ell} = \left[\frac{\partial}{\partial b_{i_1}} \frac{\partial}{\partial b_{i_2}} \cdots \frac{\partial}{\partial b_{i_\ell}} \mathcal{W}(\mathbf{b}, \lambda) \right]_{\mathbf{b}=0},$$

are the *connected* correlation functions.

Examples. Expanding the relation (1.32) in powers of $\mathbf{b}$, one finds that the one-point functions are identical:

$$W^{(1)}_i = \langle x_i \rangle_\lambda.$$

For the two-point functions, one finds

$$W^{(2)}_{i_1 i_2} = \langle x_{i_1} x_{i_2} \rangle_\lambda - \langle x_{i_1} \rangle_\lambda \langle x_{i_2} \rangle_\lambda = \langle (x_{i_1} - \langle x_{i_1} \rangle_\lambda)(x_{i_2} - \langle x_{i_2} \rangle_\lambda) \rangle_\lambda.$$

Thus, the connected two-point function is the two-point function of the variables to which their expectation values have been subtracted.

In the case of an even perturbation $V(\mathbf{x}) = V(-\mathbf{x})$, as in the example (1.25),

$$W^{(2)}_{i_1 i_2} = \langle x_{i_1} x_{i_2} \rangle_\lambda,$$
$$W^{(4)}_{i_1 i_2 i_3 i_4} = \langle x_{i_1} x_{i_2} x_{i_3} x_{i_4} \rangle_\lambda - \langle x_{i_1} x_{i_2} \rangle_\lambda \langle x_{i_3} x_{i_4} \rangle_\lambda - \langle x_{i_1} x_{i_3} \rangle_\lambda \langle x_{i_2} x_{i_4} \rangle_\lambda$$
$$- \langle x_{i_1} x_{i_4} \rangle_\lambda \langle x_{i_3} x_{i_2} \rangle_\lambda. \tag{1.33}$$

The connected four-point function, which vanishes exactly for a gaussian measure, gives a first evaluation of the deviation from a gaussian measure (Fig. 1.5).

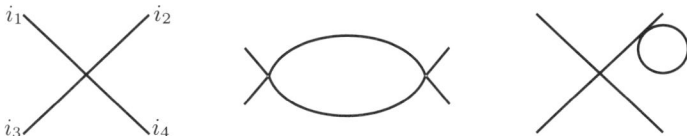

Fig. 1.5 Four-point function: connected contributions of order λ and λ^2 in the example (1.25).

In the example (1.25), at order λ^2, one then finds

$$W^{(4)}_{i_1 i_2 i_3 i_4} = -\lambda \sum_i \Delta_{i_1 i} \Delta_{i_2 i} \Delta_{i_3 i} \Delta_{i_4 i} + \tfrac{1}{2} \lambda^2 \sum_{i,j} \Delta_{i_1 i} \Delta_{i_2 i} \Delta_{i_3 j} \Delta_{i_4 j} \Delta^2_{ij}$$

$$+ \tfrac{1}{2} \lambda^2 \sum_{i,j} \Delta_{i_1 i} \Delta_{i_3 i} \Delta_{i_2 j} \Delta_{i_4 j} \Delta^2_{ij} + \tfrac{1}{2} \lambda^2 \sum_{i,j} \Delta_{i_1 i} \Delta_{i_4 i} \Delta_{i_3 j} \Delta_{i_2 j} \Delta^2_{ij}$$

$$+ \tfrac{1}{2} \lambda^2 \sum_{i,j} (\Delta_{ii} \Delta_{ij} \Delta_{i_1 i} \Delta_{i_2 j} \Delta_{i_3 j} \Delta_{i_4 j} + 3 \text{ terms}) + O(\lambda^3). \qquad (1.34)$$

1.5 Steepest descent method

Th steepest descent method is an approximation scheme to evaluate certain types of contour integrals in a complex domain. It involves approximating integrals by a sum of gaussian expectation values.

We first describe the method in the case of real integrals over one variable and then generalize to complex integrals. Finally, we generalize the method to an arbitrary number of variables.

1.5.1 Real integrals

We consider the integral

$$\mathcal{I}(\lambda) = \int_a^b \mathrm{d}x \, \mathrm{e}^{-A(x)/\lambda}, \qquad (1.35)$$

where the function $A(x)$ is a real function, analytic in a neighbourhood of the real interval (a, b), and λ a positive parameter. We want to evaluate the integral in the limit $\lambda \to 0_+$. In this limit, the integral is dominated by the maxima of the integrand and thus the minima of $A(x)$. Two situations can occur:

(i) The minimum of $A(x)$ corresponds to an end-point of the interval. One then expands $A(x)$ near the end-point and integrates. This is not the case we are interested in here.

(ii) The function $A(x)$ has one, or several, minima in the interval (a, b). The minima correspond to points x^c that are solutions of

$$A'(x^c) = 0,$$

where, generically, $A''(x^c) > 0$ (the case $A''(x^c) = 0$ requires a separate analysis). For reasons that will become clearer later, such points are called saddle points (see example (ii)). When several saddle points are found, the largest contribution comes from the absolute minimum of $A(x)$.

Moreover, if corrections of order $\exp[-\mathrm{const.}/\lambda]$ are neglected, the integration domain can be restricted to a neighbourhood $(x^c - \varepsilon, x^c + \varepsilon)$ of x^c, where ε is finite but otherwise arbitrarily small. Indeed, contributions coming from the exterior of the interval are bounded by

$$(b-a) \, \mathrm{e}^{-A''(x^c) \varepsilon^2 / 2\lambda},$$

where the property $\varepsilon \ll 1$ has been used, in such a way that
$$A(x) - A(x^c) \sim \tfrac{1}{2} A''(x^c)(x - x^c)^2 \, .$$

More precisely, the region that contributes is of order $\sqrt{\lambda}$. Thus, it is convenient to change variables, $x \mapsto y$:
$$y = (x - x^c)/\sqrt{\lambda}.$$

The expansion of the function A then reads
$$A/\lambda = A(x^c)/\lambda + \tfrac{1}{2} y^2 A''(x^c) + \tfrac{1}{6}\sqrt{\lambda} A'''(x^c) y^3 + \tfrac{1}{24}\lambda A^{(4)}(x^c) y^4 + O(\lambda^{3/2}).$$

One sees that, at leading order, it is sufficient to keep the quadratic term. This reduces the calculation to the restricted gaussian integral
$$\mathcal{I}(\lambda) \sim \sqrt{\lambda}\, \mathrm{e}^{-A(x^c)/\lambda} \int_{-\varepsilon/\sqrt{\lambda}}^{\varepsilon/\sqrt{\lambda}} \mathrm{d}y\, \mathrm{e}^{-A''(x^c) y^2/2} \, .$$

The integration range can be extended to the whole real axis $[-\infty, +\infty]$ because, for similar reasons, contributions from outside the integration domain are exponentially small in $1/\lambda$. The leading contribution is thus given by the gaussian integral, which yields
$$\mathcal{I}(\lambda) \sim \sqrt{2\pi\lambda/A''(x^c)}\, \mathrm{e}^{-A(x_c)/\lambda} \, . \tag{1.36}$$

To calculate higher order corrections, one expands the exponential in powers of λ and integrates term by term. Setting
$$\mathcal{I}(\lambda) = \sqrt{2\pi\lambda/A''(x^c)}\, \mathrm{e}^{-A(x_c)/\lambda}\, \mathcal{J}(\lambda),$$

one finds, for example, at next order,
$$\mathcal{J}(\lambda) = 1 - \frac{\lambda}{24} A^{(4)} \langle y^4 \rangle + \frac{\lambda}{2 \times 6^2} A'''^2 \langle y^6 \rangle + O(\lambda^2)$$
$$= 1 + \frac{\lambda}{24}\left(5\frac{A'''^2}{A''^3} - 3\frac{A^{(4)}}{A''^2}\right) + O(\lambda^2),$$

where $\langle \bullet \rangle$ means gaussian expectation value.

Remarks.
(i) The steepest descent method generates a formal expansion in powers of λ:
$$\mathcal{J}(\lambda) = 1 + \sum_{k=1}^{\infty} J_k \lambda^k,$$

which, in general, diverges for all values of the expansion parameter. The divergence can easily be understood: if one changes the sign of λ in the integral, the maximum of the integrand becomes a minimum, and the saddle point no longer gives the leading contribution to the integral.

Nevertheless, the series is useful because, for small enough λ, partial sums satisfy

$$\exists \lambda_0 > 0, \{M_K\}: \quad \forall K \text{ and } 0_+ \leq \lambda \leq \lambda_0 \quad \left| \mathcal{J}(\lambda) - \sum_{k=0}^{K} J_k \lambda^k \right| \leq M_K \lambda^{K+1},$$

where the coefficients M_k generically grow like $k!$. Such a series is called an asymptotic series. At fixed λ, if the index K is chosen such that the bound is minimum, the function is determined up to corrections of order $\exp[-\text{const.}/\lambda]$. Note that such a bound can be extended to a sector in the λ complex plane, $|\text{Arg}\,\lambda| < \theta$.

(ii) Often integrals have the more general form

$$\mathcal{I}(\lambda) = \int dx\, \rho(x)\, e^{-A(x)/\lambda}.$$

Then, provided $\ln \rho(x)$ is analytic at the saddle point, it is not necessary to take into account the factor $\rho(x)$ in the saddle point equation. Indeed, this would induce a shift $x - x_c$ of the position of the saddle point, solution of

$$A''(x_c)(x - x_c) \sim \lambda \rho'(x_c)/\rho(x_c),$$

and, thus, of order λ while the contribution to the integral comes from a much larger region of order $\sqrt{\lambda} \gg \lambda$.

One can, thus, still expand all expressions around the solution of $A'(x) = 0$. At leading order, one then finds

$$\mathcal{I}(\lambda) \sim \sqrt{2\pi\lambda/A''(x^c)}\, \rho(x_c)\, e^{-A(x_c)/\lambda}.$$

Let us now apply the method to two classical examples, the Γ function that generalizes $n!$ to complex arguments, and the modified Bessel function.

Examples.

(i) A classical example is the asymptotic evaluation of the function

$$\Gamma(s) = \int_0^\infty dx\, x^{s-1}\, e^{-x},$$

for $s \to +\infty$. The integral does not immediately have the canonical form (1.35), but it takes it after a linear change of variables: $x = (s-1)x'$. One identifies $s - 1 = 1/\lambda$. Then,

$$\Gamma(s) = (s-1)^{s-1} \int_0^\infty dx\, e^{-(x - \ln x)/\lambda}$$

and, thus, $A(x) = x - \ln x$. The position of the saddle point is given by

$$A'(x) = 1 - 1/x = 0 \;\Rightarrow\; x^c = 1.$$

The second derivative at the saddle point is $A''(x^c) = 1$. The result, at leading order, is
$$\Gamma(s) \underset{s \to \infty}{\sim} \sqrt{2\pi}(s-1)^{s-1/2} e^{1-s} \sim \sqrt{2\pi} s^{s-1/2} e^{-s}, \tag{1.37}$$
an expression also called Stirling's formula. Note that with the help of the complex generalization of the method, which we explain later, the result can be extended to all complex values of s such that $|\arg s| < \pi$.

(ii) We now evaluate the modified Bessel function
$$I_0(x) = \frac{1}{2\pi} \int_{-\pi}^{\pi} d\theta \, e^{x \cos \theta},$$
$(= J_0(ix))$ for $x \to +\infty$ (the function is even).

This integral has the canonical form for the application of the steepest descent method ($x = 1/\lambda$), and the integrand is an entire function.

The saddle points are given by
$$\sin \theta = 0 \Rightarrow \theta = 0 \pmod{\pi}.$$

For $x \to +\infty$, the leading saddle point is $\theta = 0$. One expands at the saddle point
$$x \cos \theta = x - \tfrac{1}{2} x \theta^2 + \tfrac{1}{24} x \theta^4 + O(\theta^6).$$

The region that contributes to the integral is of order $\theta = O(1/\sqrt{x})$. Thus,
$$I_0(x) = \frac{1}{2\pi} e^x \int_{-\infty}^{\infty} d\theta \, e^{-x\theta^2/2} \left(1 + \tfrac{1}{24} x \theta^4\right) + O(e^x / x^2)$$
$$= \frac{1}{\sqrt{2\pi x}} e^x \left(1 + \frac{1}{8x} + O\left(\frac{1}{x^2}\right)\right).$$

Let us use this example to justify the denomination *saddle point*. For this purpose, it is necessary to examine the function $\cos \theta$, which appears in the integrand, in the complex plane in the vicinity of the saddle point $\theta = 0$. The curves of constant modulus of the integrand are the curves $\text{Re} \cos \theta$ constant:
$$\text{Re} \cos \theta - 1 \sim -\tfrac{1}{2} \left[(\text{Re} \, \theta)^2 - (\text{Im} \, \theta)^2 \right] = \text{const.}$$

Locally, these curves are hyperbolae. They cross only at the saddle point. The hyperbola corresponding to a vanishing constant degenerates into two straight lines (see Fig. 1.6). The modulus of the integrand thus has a saddle point structure.

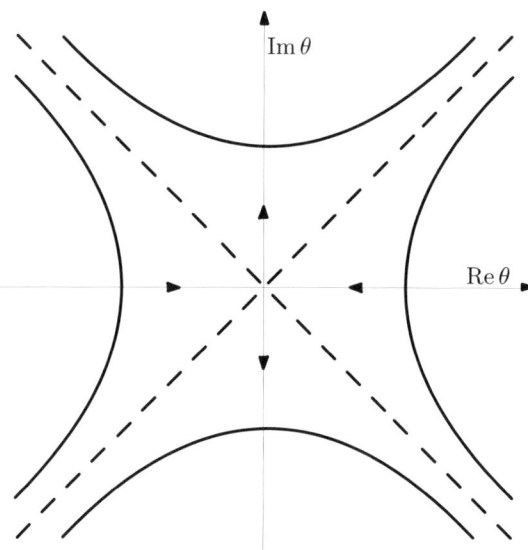

Fig. 1.6 Function I_0: constant modulus curves of the integrand near the saddle point $\theta = 0$.

1.5.2 Complex integrals

One wants now to evaluate the integral

$$\mathcal{I}(\lambda) = \oint_C \mathrm{d}x \, \mathrm{e}^{-A(x)/\lambda}, \tag{1.38}$$

where $A(x)$ is an analytic function of the complex variable x and λ a real positive parameter, for $\lambda \to 0_+$. The contour C goes from a point a to a point b in the complex plane, and is contained within the domain of analyticity of A. As a limiting case, one can consider the situation where the points a and b go to infinity in the complex plane.

At first sight, one could think that the integral is again dominated by the points where the modulus of the integrand is maximum and, thus, the real part of $A(x)$ is minimum. However, the contribution of the neighbourhood of such points in general cancels because the phase also varies rapidly (an argument that leads to the stationary phase method).

The steepest descent method then involves deforming the contour C, within the domain of analyticity of A (and without crossing a singularity), to minimize the maximum modulus of the integrand on the contour, that is, to maximize the minimum of $\mathrm{Re}\, A(x)$.

If it is possible to deform the contour C into an equivalent contour along which $\mathrm{Re}\, A(x)$ is monotonous, then the integral is dominated by an end-point. Otherwise, the real part of A has a minimum. On an optimal contour, the minimum corresponds either to a singularity of the function or to a regular point where the derivative of

A vanishes:
$$A'(x) = 0.$$
This is the case we want to study. A point x^c where $A'(x) = 0$ is again, generically, a *saddle point* with respect to the curves of constant $\mathrm{Re}\, A(x)$ (Fig. 1.6). Such a structure of the integrand can be better understood if one remembers that the curves with $\mathrm{Re}\, A$ and $\mathrm{Im}\, A$ constant form two sets of bi-orthogonal curves. The only possible double points of these curves are singularities and saddle points. Indeed, let us expand the function at x^c:
$$A(x) - A(x^c) = \tfrac{1}{2} A''(x^c)(x-x^c)^2 + O\!\left((x-x^c)^3\right).$$
Then, in terms of the real coordinates u, v defined by
$$u + iv = (x - x^c) e^{i \mathrm{Arg}\, A''(x^c)/2},$$
one finds
$$\mathrm{Re}[A(x) - A(x^c)] \sim \tfrac{1}{2} |A''(x^c)|(u^2 - v^2).$$
Near the saddle point, one can choose a contour that follows a curve with $\mathrm{Im}\, A$ constant and, thus, along which the phase of the integrand remains constant: no cancellation can occur. The leading contribution to the integral comes from the neighbourhood of the saddle point. Neglected contributions decay faster than any power of λ. The remaining part of the argument and the calculation are the same as in the real case.

Example. Consider the integral representation of the usual Bessel function
$$J_0(x) = \frac{1}{2i\pi} \oint_C \frac{dz}{z} e^{x(z - 1/z)/2},$$
where C is a simple closed contour, which encloses the origin. We evaluate the integral for x real, $x \to +\infty$, by the steepest descent method.

We set
$$A(z) = (1/z - z)/2.$$
Saddle points are solutions of
$$2A'(z) = -\frac{1}{z^2} - 1 = 0 \;\Rightarrow\; z = \pm i.$$
The two saddle points are relevant. For the saddle point $z = i$, we set $z = i + e^{3i\pi/4} s$. Then,
$$A(z) = -i + s^2/2 + O(s^3).$$
The contribution of the saddle point is
$$\frac{1}{2\pi} e^{ix - i\pi/4} \int_{-\infty}^{+\infty} ds\, e^{-xs^2/2} = \frac{1}{\sqrt{2\pi x}} e^{ix - i\pi/4}.$$
The second saddle point gives the complex conjugate contribution. One thus finds
$$J_0(x) \underset{x \to +\infty}{\sim} \sqrt{\frac{2}{\pi x}} \cos(x - \pi/4).$$

1.6 Steepest descent method: Several variables, generating functions

We now consider the general n-dimensional integral

$$\mathcal{I}(\lambda) = \int \mathrm{d}^n x \, \exp\left[-\frac{1}{\lambda} A(x_1, \ldots, x_n)\right], \tag{1.39}$$

where, for simplicity, we assume that A is an entire function and the integration domain is $\mathbb{R}^n$.

In the limit $\lambda \to 0_+$, the integral is dominated by saddle points, solutions of

$$\frac{\partial}{\partial x_i} A(x_1, x_2, \ldots, x_n) = 0, \; \forall i. \tag{1.40}$$

When several saddle points are found, one orders them according to the values of $\mathrm{Re}\, A$. Often the relevant saddle point corresponds to the minimum value of $\mathrm{Re}\, A$, but this is not necessarily the case since the saddle point may not reachable by a deformation of the initial domain of integration. Unfortunately, in the case of several complex variables, deforming contours is, generally, not a simple exercise.

To evaluate the leading contribution of a saddle point $\mathbf{x}^c$, it is convenient to change variables, setting

$$\mathbf{x} = \mathbf{x}^c + \mathbf{y}\sqrt{\lambda}.$$

One then expands $A(\mathbf{x})$ in powers of λ (and thus $\mathbf{y}$):

$$\frac{1}{\lambda} A(x_1, \ldots, x_n) = \frac{1}{\lambda} A(\mathbf{x}^c) + \frac{1}{2!} \sum_{i,j} \frac{\partial^2 A(\mathbf{x}^c)}{\partial x_i \partial x_j} y_i y_j + R(\mathbf{y}) \tag{1.41}$$

with

$$R(\mathbf{y}) = \sum_{k=3}^{\infty} \frac{\lambda^{k/2-1}}{k!} \sum_{i_1, i_2, \ldots, i_k} \frac{\partial^k A(\mathbf{x}^c)}{\partial x_{i_1} \ldots \partial x_{i_k}} y_{i_1} \ldots y_{i_k}. \tag{1.42}$$

After the change of variables, the term quadratic in $\mathbf{y}$ becomes independent of λ. The integral then reads

$$\mathcal{I}(\lambda) = \lambda^{n/2} \mathrm{e}^{-A(\mathbf{x}^c)/\lambda} \int \mathrm{d}^n y \, \exp\left[-\frac{1}{2!} \sum_{i,j} \frac{\partial^2 A(\mathbf{x}^c)}{\partial x_i \partial x_j} y_i y_j - R(\mathbf{y})\right]. \tag{1.43}$$

Once the integrand is expanded in powers of $\sqrt{\lambda}$, the calculation of each term is reduced to the gaussian expectation value of a polynomial. At leading order, one finds

$$\mathcal{I}(\lambda) \underset{\lambda \to 0}{\sim} (2\pi\lambda)^{n/2} \left[\det \mathbf{A}^{(2)}\right]^{-1/2} \mathrm{e}^{-A(\mathbf{x}^c)/\lambda}, \tag{1.44}$$

where $\mathbf{A}^{(2)}$ is the matrix of second partial derivatives:

$$[\mathbf{A}^{(2)}]_{ij} \equiv \frac{\partial^2 A(\mathbf{x}^c)}{\partial x_i \partial x_j}.$$

1.6.1 Generating function and steepest descent method

We now introduce the function

$$\mathcal{Z}(\mathbf{b}, \lambda) = \int d^n x \, \exp\left[-\frac{1}{\lambda}\left(A(\mathbf{x}) - \mathbf{b} \cdot \mathbf{x}\right)\right], \quad (1.45)$$

where $A(\mathbf{x})$ is now a regular function of the x_i. We also define

$$\mathcal{N} = 1/\mathcal{Z}(0, \lambda).$$

The function $\mathcal{Z}(\mathbf{b}, \lambda)$ has the general form (1.30), and is proportional to the generating function of the moments of a distribution $\mathcal{N} e^{-A(\mathbf{x})/\lambda}$.

Expectation values of polynomials with the weight $\mathcal{N} e^{-A(\mathbf{x})/\lambda}$,

$$\langle x_{k_1} x_{k_2} \ldots x_{k_\ell} \rangle \equiv \mathcal{N} \int d^n x \, x_{k_1} x_{k_2} \ldots x_{k_\ell} \, e^{-A(\mathbf{x})/\lambda}, \quad (1.46)$$

are obtained by differentiating $\mathcal{Z}(\mathbf{b}, \lambda)$ (see equation (1.31)):

$$\langle x_{k_1} x_{k_2} \ldots x_{k_\ell} \rangle = \lambda^\ell \mathcal{N} \left[\frac{\partial}{\partial b_{k_1}} \frac{\partial}{\partial b_{k_2}} \cdots \frac{\partial}{\partial b_{k_\ell}} \mathcal{Z}(\mathbf{b}, \lambda)\right]\bigg|_{\mathbf{b}=0}.$$

Steepest descent method. We now apply the steepest descent method to the integral (1.45). The saddle point equation is

$$b_i = \frac{\partial A}{\partial x_i}, \quad \forall\, i. \quad (1.47)$$

We expand $A(\mathbf{x})$ at a saddle point $\mathbf{x}^c$, as explained in Section 1.5, and use the leading order result (1.44):

$$\mathcal{Z}(\mathbf{b}, \lambda) \underset{\lambda \to 0}{\sim} (2\pi\lambda)^{n/2} \left[\det \mathbf{A}^{(2)}\right]^{-1/2} \exp\left[-\frac{1}{\lambda}\left(A(\mathbf{x}^c) - \mathbf{b} \cdot \mathbf{x}^c\right)\right]$$

with

$$[\mathbf{A}^{(2)}]_{ij} \equiv \frac{\partial^2 A(\mathbf{x}^c)}{\partial x_i \partial x_j}.$$

We now introduce $\mathcal{W}(\mathbf{b}, \lambda)$, the generating function of the cumulants of the distribution, which are also the connected correlation functions (equation (1.32) but with a different normalization)

$$\mathcal{W}(\mathbf{b}, \lambda) = \lambda \ln \mathcal{Z}(\mathbf{b}, \lambda).$$

Using identity (3.51): $\ln \det \mathbf{M} = \operatorname{tr} \ln \mathbf{M}$, valid for any matrix $\mathbf{M}$, we can write the first order terms of the expansion of $\mathcal{W}$ as

$$\mathcal{W}(\mathbf{b}, \lambda) = -A(\mathbf{x}_c) + \mathbf{b} \cdot \mathbf{x}^c + \tfrac{1}{2} n \ln(2\pi\lambda) - \tfrac{1}{2}\lambda \operatorname{tr} \ln \frac{\partial^2 A(\mathbf{x}^c)}{\partial x_i \partial x_j} + O(\lambda^2). \quad (1.48)$$

Since

$$\langle x_{k_1} x_{k_2} \ldots x_{k_\ell} \rangle_c = \lambda^{\ell-1} \left[\frac{\partial}{\partial b_{k_1}} \frac{\partial}{\partial b_{k_2}} \cdots \frac{\partial}{\partial b_{k_\ell}} \mathcal{W}(\mathbf{b}, \lambda)\right]\bigg|_{\mathbf{b}=0},$$

successive derivatives of the expansion (1.48) with respect to $\mathbf{b}$ (taking into account that $\mathbf{x}_c$ is a function of $\mathbf{b}$ through equation (1.47)), calculated at $\mathbf{b} = 0$, yield the corresponding expansions of the cumulants of the distribution.

1.7 Gaussian integrals: Complex matrices

In Section 1.2, we have proved the result (1.5) only for real matrices. Here, we extend the proof to complex matrices.

The proof based on diagonalization, used for real matrices, has a complex generalization. Indeed, any complex symmetric matrix $\mathbf{A}$ has a decomposition of the form

$$\mathbf{A} = \mathbf{U}\mathbf{D}\mathbf{U}^{\mathrm{T}}, \qquad (1.49)$$

where $\mathbf{U}$ is a unitary matrix and $\mathbf{D}$ a diagonal positive matrix. In the integral (1.4), one then changes variables $\mathbf{x} \mapsto \mathbf{y}$:

$$x_i = \sum_{i=j}^{n} U_{ij} y_j \,.$$

This change of variables is a direct complex generalization of the orthogonal transformation (1.7). The integral (1.4) then factorizes and the result is the product of the integral and the (here non-trivial) jacobian of the change of variables. Thus,

$$\mathcal{Z}(\mathbf{A}) = (2\pi)^{n/2} (\det \mathbf{D})^{-1/2} / \det \mathbf{U}\,.$$

Since

$$\det \mathbf{A} = \det \mathbf{D} (\det \mathbf{U})^2,$$

one recovers the result (1.5).

Complex matrix representations. Since the existence of the representation (1.49) may not be universally known, as an exercise, we give here a general proof.

(i) *Polar decomposition.* A complex matrix $\mathbf{M}$ has a 'polar' decomposition:

$$\mathbf{M} = \mathbf{U}\mathbf{H} \text{ with } \mathbf{U}^{\dagger}\mathbf{U} = \mathbf{1}\,,\ \mathbf{H} = \mathbf{H}^{\dagger}. \qquad (1.50)$$

If a matrix has no vanishing eigenvalue the proof is simple. The representation (1.50) implies a relation between hermitian positive matrices:

$$\mathbf{M}^{\dagger}\mathbf{M} = \mathbf{H}^2.$$

One chooses for $\mathbf{H}$ the matrix $(\mathbf{M}^{\dagger}\mathbf{M})^{1/2}$ with positive eigenvalues. One then verifies immediately that the matrix $\mathbf{U} = \mathbf{M}\mathbf{H}^{-1}$ is unitary:

$$\mathbf{U}^{\dagger}\mathbf{U} = \mathbf{1}\,.$$

This decomposition implies another, equivalent one. A hermitian matrix can be diagonalized by a unitary transformation and thus can be written as

$$\mathbf{H} = \mathbf{V}^{\dagger}\mathbf{D}\mathbf{V}\,,\ \mathbf{V}^{\dagger}\mathbf{V} = \mathbf{1}$$

1.7 Gaussian integrals

with $\mathbf{D}$ diagonal: $D_{ij} = h_i \delta_{ij}$, $h_i > 0$. It follows that

$$\mathbf{M} = \mathbf{U}\mathbf{V}^\dagger \mathbf{D}\mathbf{V},$$

or, in a simpler notation,

$$\mathbf{M} = \mathbf{U}_2^\dagger \mathbf{D}\mathbf{U}_1, \tag{1.51}$$

where $\mathbf{U}_1$ and $\mathbf{U}_2$ are two unitary matrices.

(ii) *Symmetric unitary matrices.* We now prove a decomposition of symmetric unitary matrices. We thus consider a matrix $\mathbf{U}$ satisfying

$$\mathbf{U}^\dagger \mathbf{U} = \mathbf{1}, \quad \mathbf{U} = \mathbf{U}^T.$$

We decompose $\mathbf{U}$ into real and imaginary parts:

$$\mathbf{U} = \mathbf{X} + i\mathbf{Y}.$$

The two matrices $\mathbf{X}$ and $\mathbf{Y}$ are real and symmetric and satisfy, as a consequence of the unitarity relation,

$$\mathbf{X}^2 + \mathbf{Y}^2 = \mathbf{1}, \quad \mathbf{XY} - \mathbf{YX} = 0.$$

Two commuting real symmetric matrices can be diagonalized simultaneously. The corresponding eigenvalues $\{x_i, y_i\}$ (which are real) satisfy

$$x_i^2 + y_i^2 = 1,$$

which we parametrize as

$$x_i = r_i \cos \theta_i, \quad y_i = r_i \sin \theta_i, \quad r_i > 0.$$

If $\mathbf{O}$ is the common orthogonal matrix that diagonalizes $\mathbf{X}$ and $\mathbf{Y}$ and thus $\mathbf{U}$, the matrix $\mathbf{U}$ can be written as

$$\mathbf{U} = \mathbf{O}^T \mathbf{R} \mathbf{W} \mathbf{O}$$

with

$$W_{ij} = e^{i\theta_i} \delta_{ij}, \quad R_{ij} = r_i \delta_{ij}.$$

One then sets

$$V_{ij} = e^{i\theta_i/2} O_{ij} \Leftrightarrow \mathbf{V} = \mathbf{W}^{1/2}\mathbf{O},$$

and obtains the expected representation (1.49) for unitary matrices:

$$\mathbf{U} = \mathbf{V}^T \mathbf{R} \mathbf{V}, \quad \mathbf{R} \text{ diagonal } > 0, \quad \mathbf{V}^T \mathbf{V} = \mathbf{1}. \tag{1.52}$$

(iii) *Complex symmetric matrices.* We now prove the representation for general complex symmetric matrices

$$\mathbf{M} = \mathbf{M}^T.$$

Representation (1.51) must then satisfy

$$\mathbf{U}_2^\dagger \mathbf{D} \mathbf{U}_1 = \mathbf{U}_1^{\mathrm{T}} \mathbf{D} \mathbf{U}_2^*,$$

where * denotes complex conjugation. Introducing the unitary matrix

$$\mathbf{W} = \mathbf{U}_2 \mathbf{U}_1^{\mathrm{T}},$$

one obtains the constraint

$$\mathbf{D} = \mathbf{W}\mathbf{D}\mathbf{W}^* \;\Leftrightarrow\; \mathbf{D} = \mathbf{W}^{\mathrm{T}}\mathbf{D}\mathbf{W}^\dagger.$$

Multiplying the r.h.s. of the second equation by the r.h.s. of the first equation (in this order), one finds

$$\mathbf{D}^2 = \mathbf{W}^{\mathrm{T}}\mathbf{D}^2\mathbf{W}^* \;\Leftrightarrow\; \mathbf{D}^2 = \mathbf{W}^\dagger \mathbf{D}^2 \mathbf{W} \;\Leftrightarrow\; [\mathbf{W}, \mathbf{D}^2] = 0.$$

The latter equation, in component form reads

$$\left(h_i^2 - h_j^2\right) W_{ij} = 0,$$

and thus

$$W_{ij} = 0, \text{ for } h_i \neq h_j.$$

If all eigenvalues of $\mathbf{D}$ are simple, then $\mathbf{W}$ is a diagonal unitary matrix:

$$W_{ij} = \mathrm{e}^{i\theta_i}\, \delta_{ij}.$$

Introducing this result into the representation (1.51), eliminating $\mathbf{U_2}$ in terms of $\mathbf{W}$, one finds

$$\mathbf{M} = \mathbf{U}_1^{\mathrm{T}} \mathbf{W}^* \mathbf{D} \mathbf{U}_1.$$

Finally, setting

$$\mathbf{U}_0 = [\mathbf{W}^{1/2}]^* \mathbf{U}_1,$$

one obtains the representation of a complex symmetric matrix in terms of a positive diagonal matrix $\mathbf{D}$ and a unitary matrix $\mathbf{U}_0$:

$$\mathbf{M} = \mathbf{U}_0^{\mathrm{T}} \mathbf{D} \mathbf{U}_0. \tag{1.53}$$

If $\mathbf{D}$ has degenerate eigenvalues, from (1.51) in the corresponding subspace the matrix $\mathbf{M}$ is proportional to a symmetric unitary matrix. One then uses the result (1.52) and this shows that the decomposition (1.53) still holds.

Exercises

Exercise 1.1

One considers two stochastic correlated variables x, y with gaussian probability distribution. One finds the five expectation values

$$\langle x \rangle = \langle y \rangle = 0, \ \langle x^2 \rangle = 5, \ \langle xy \rangle = 3, \ \langle y^2 \rangle = 2.$$

Calculate the expectation values $\langle x^4 \rangle$, $\langle x^3 y \rangle$, $\langle x^2 y^2 \rangle$, $\langle xy^5 \rangle$, $\langle y^6 \rangle$, $\langle x^3 y^3 \rangle$, using Wick's theorem.

Determine the corresponding gaussian distribution.

Solution.

$$75, 45, 28, 180, 120, 432.$$

The gaussian distribution is proportional to

$$e^{-(2x^2 - 6xy + 5y^2)/2}.$$

Exercise 1.2

One considers three stochastic correlated variables x, y, z with gaussian probability distribution. One finds the nine expectation values

$$\langle x \rangle = \langle y \rangle = \langle z \rangle = 0, \ \langle x^2 \rangle = \langle y^2 \rangle = \langle z^2 \rangle = a, \ \langle xy \rangle = b, \ \langle xz \rangle = \langle zy \rangle = c.$$

Calculate the expectation values $\langle x^4 \rangle$, $\langle x^6 \rangle$, $\langle x^3 y \rangle$, $\langle x^2 y^2 \rangle$, $\langle x^2 yz \rangle$ as functions of a, b, c.

Determine for $a = 2, b = 1, c = 0$ the corresponding gaussian distribution.

Solution.

$$\langle x^4 \rangle = 3a^2, \ \langle x^6 \rangle = 15a^3, \ \langle x^3 y \rangle = 3ab, \ \langle x^2 y^2 \rangle = a^2 + 2b^2, \ \langle x^2 yz \rangle = ac + 2bc.$$

For $a = 2, b = 1, c = 0$ the gaussian distribution is proportional to

$$\exp\left[-\tfrac{1}{12}\left(4x^2 + 4y^2 + 3z^2 - 4xy\right)\right].$$

Exercise 1.3

Inductive algebraic proof of result (1.5). The determinant of a general matrix $\mathbf{A}^{(n)}$ $n \times n$, of elements $A_{ij}^{(n)}$ can be calculated inductively by subtracting from all rows a multiple of the last row in order to cancel the last column (assuming that $A_{nn}^{(n)} \neq 0$, otherwise one interchanges rows or columns). This method leads to a relation between determinants:

$$\det \mathbf{A}^{(n)} = A_{nn}^{(n)} \det \mathbf{A}^{(n-1)},$$

where $\mathbf{A}^{(n-1)}$ is a matrix $(n-1) \times (n-1)$ with elements

$$A_{ij}^{(n-1)} = A_{ij}^{(n)} - A_{in}^{(n)} A_{nj}^{(n)} / A_{nn}^{(n)}, \quad i,j = 1, \ldots, n-1. \tag{1.54}$$

Show that the result (1.6), combined with this identity, leads to the general result (1.5).

Solution. One considers the integral (1.4) and integrates over one variable, which one can call x_n (assuming that Re $A_{nn} > 0$), using the result (1.6):

$$\int \mathrm{d}x_n \exp\left(-\tfrac{1}{2} A_{nn} x_n^2 - x_n \sum_{i=1}^{n-1} A_{ni} x_i\right) = \sqrt{\frac{2\pi}{A_{nn}}} \exp\left(\tfrac{1}{2} \sum_{i,j=1}^{n-1} \frac{A_{in} A_{nj}}{A_{nn}} x_i x_j\right).$$

The remaining integral is a gaussian integral over $n-1$ variables:

$$\mathcal{Z}(\mathbf{A}) = \sqrt{\frac{2\pi}{A_{nn}}} \int \left(\prod_{i=1}^{n-1} \mathrm{d}x_i\right) \exp\left(-\sum_{i,j=1}^{n-1} \tfrac{1}{2} x_i \left(A_{ij} - A_{in} A_{nn}^{-1} A_{nj}\right) x_j\right).$$

One then notes that, by iterating this partial integration, one obtains a form of $1/\sqrt{\det \mathbf{A}}$ as generated by identity (1.54). One concludes

$$\mathcal{Z}(\mathbf{A}) = (2\pi)^{n/2} (\det \mathbf{A})^{-1/2}. \tag{1.55}$$

Exercise 1.4

Use the steepest descent method to evaluate the integral

$$I_n(\alpha) = \int_0^1 \mathrm{d}x \, x^{\alpha n} (1-x)^{\beta n},$$

with $\beta = 1 - \alpha$, $\alpha > 0$, $\beta > 0$, in the limit $n \to \infty$.

Solution. The saddle point is $x_c = \alpha$ and thus

$$I_n(\alpha) \sim \sqrt{2\pi \alpha(1-\alpha)/n} \, \alpha^{n\alpha} (1-\alpha)^{n(1-\alpha)}.$$

Exercise 1.5

One considers the integral

$$Z(g) = \int \mathrm{d}^3 q \exp\left[\frac{1}{g}\left(\frac{\mathbf{q}^2}{2} - \frac{(\mathbf{q}^2)^2}{4}\right)\right],$$

where $\mathbf{q}$ is a two-component vector $\mathbf{q} = (q_1, q_2)$. Evaluate the integral for $g \to 0_+$ by the steepest descent method (this exercise involves a subtle point).

Solution. For some indications, see Section 8.3.1:

$$Z(g) \sim 4\pi^{3/2} g^{1/2} \mathrm{e}^{1/4g}.$$

Exercise 1.6

Hermite polynomials $\mathcal{H}_n$ appear in the eigenfunctions of the quantum harmonic oscillator. One integral representation is

$$\mathcal{H}_n(z) = \sqrt{\frac{n}{2\pi}} \int_{-\infty}^{+\infty} ds \, e^{-ns^2/2} (z - is)^n.$$

Evaluate the polynomials for $n \to \infty$ and z real by the steepest descent method.

Solution. The polynomials $\mathcal{H}_n(z)$ are alternatively even or odd:

$$\mathcal{H}_n(-z) = (-1)^n \mathcal{H}_n(z).$$

Thus, one can restrict the problem to the values $z \geq 0$.
  One sets

$$A(s) = \tfrac{1}{2}s^2 - \ln(z - is).$$

The saddle points are given by

$$A'(s) = s - \frac{1}{s + iz} = 0 \;\Rightarrow\; s_\pm = -\tfrac{1}{2}iz \pm \sqrt{1 - z^2/4}.$$

Moreover,

$$A''(s) = 1 + \frac{1}{(s+iz)^2} = s^2 + 1.$$

It is then necessary to distinguish between the two cases $0 \leq z < 2$ and $z > 2$ ($z = 2$ requires a special treatment).
  (ii) $z > 2$. It is convenient to set $z = 2\cosh\theta$ with $\theta > 0$. Then,

$$s_\pm = -i\,e^{\pm\theta} \;\Rightarrow\; e^{-nA} = \exp\left[\tfrac{1}{2}n\,e^{\pm 2\theta} \mp n\theta\right].$$

It is simple to verify by contour deformation that the relevant saddle point is s_- (s_+ is a saddle point between the hole at $s = -iz$ and the saddle point s_-) and thus

$$\mathcal{H}_n(z) \underset{n\to\infty}{\sim} \frac{1}{\sqrt{1 - e^{-2\theta}}} \exp\left[\tfrac{1}{2}n\,e^{-2\theta} + n\theta\right].$$

In contrast, for $|z| < 2$, the two saddle points are relevant. Setting $z = 2\cos\theta$, one finds

$$\mathcal{H}_n(z) \underset{n\to\infty}{\sim} \frac{1}{\sqrt{1 - e^{-2i\theta}}} e^{n\,e^{-2i\theta}/2 + ni\theta} + \text{complex conjugate}.$$

Exercise 1.7

Evaluate, using the steepest descent method, the integral

$$I_n(a) = \int_{-\infty}^{+\infty} dx \, e^{-nx^2/2 + nax} \cosh^n x$$

as a function of the real parameter a in the limit $n \to \infty$. Express the result in a parametric form as a function of the saddle point position.

Solution. One notices that the integral can be written as

$$I_n(a) = \int_{-\infty}^{+\infty} dx \, e^{nf(x)},$$
$$f(x) = -x^2/2 + ax + \ln \cosh x.$$

The saddle point position is given by

$$f'(x) = 0 = -x + a + \tanh x,$$

with

$$f''(x) = -\tanh^2(x).$$

The saddle point equation thus has a unique solution $x(a)$ for all a. To parametrize the result in terms of $x(a)$, one substitutes

$$a = x - \tanh x.$$

At the saddle point

$$f(x) = x^2/2 - x \tanh x + \ln \cosh x,$$

and thus

$$I_n(a) = \frac{(2\pi)^{1/2}}{|\tanh x|} e^{nf(x)}.$$

Note that the steepest descent method does not apply for $a = 0$ where $f''(x)$ vanishes. It is then necessary to expand $f(x)$ to order x^4 and integrate directly.

2 PATH INTEGRALS IN QUANTUM MECHANICS

The path integral formalism, which we explain here in the context of quantum mechanics, leads to a representation of physical quantities as averages with an appropriate weight (real or complex) over a set of paths or trajectories. Thus, it gives a rather concrete meaning to the concept of quantum fluctuations. In fact, to a large extent, we could define quantum mechanics directly in terms of path integrals, quite interesting a viewpoint, but which would require the introduction of a mathematical formalism more complicated than differential equations for the study even of simple quantum systems. Note, however, that when the number of degrees of freedom is large, as in statistical physics or in quantum field theory, the advantage of the path integral formalism becomes overwhelming.

Here instead, for pedagogical reasons, we assume some basic background in quantum mechanics, as the notion of Hilbert space (the notation of bras and will be used for vectors), on which act operators corresponding to the different physical observables, like position, momentum, the time evolution operator or the density matrix. Moreover, the evolution of wave functions is governed by the Schrödinger equation. (But we give, for completeness, a short summary of this background material in Appendix A.)

We first describe a general strategy that leads to a representation of the matrix elements (in a special basis) of some evolution operators by path integrals. The existence of such a representation is based on two fundamental properties:

(i) A markovian evolution, that is, without memory, a characteristic property of isolated systems or without influence on their environment.

(ii) The locality of the evolution for short time intervals, a property that we explain later.

These two properties are satisfied both by the unitary operator $e^{-itH/\hbar}$ (this form assumes that H is a time-independent hamiltonian) that describes quantum evolution and the statistical operator $e^{-\beta H}$, which is the density matrix at thermal equilibrium. But they are shared also by some stochastic processes of diffusion type, which are not related to quantum physics, as brownian motion, which historically has led to the introduction of path integrals (Wiener's integral).

We then construct a path integral representation of the matrix elements of the statistical operator in the simplest example of hamiltonians of the form $H = \hat{p}^2/2m + V(\hat{q})$, where $\hat{q}, \hat{p}$ are the position and momentum operators, respectively.

Integrands in path integrals can be considered as probability distributions for paths and allow defining expectation values. In particular, moments of the distribution are then called correlation functions. Therefore, as we have already done in Chapter 1, we introduce generating functionals of correlation functions. We also define functional derivatives, which then allow recovering correlation functions by differentiating generating functionals.

We then calculate explicitly gaussian path integrals, gaussian expectation values and formulate Wick's theorem in this context.

2.1 Local markovian processes

We now define, in the framework of quantum mechanics, markovian evolution and locality.

2.1.1 Markovian evolution

We consider a bounded operator in a Hilbert space, $U(t, t')$, $t \geq t'$, which describes the evolution from time t' to time t and which satisfies a Markov property in time:

$$U(t, t'')U(t'', t') = U(t, t'), \text{ for } t \geq t'' \geq t' \text{ and } U(t', t') = \mathbf{1}. \tag{2.1}$$

This property, which is also characteristic of some stochastic processes as we show in Section 5.5, means that the evolution associated with the operator U has no memory, that is, the evolution from time t'' to time t depends only on the state of the system at time t'', and not on the preceding evolution.

Moreover, we assume that $U(t, t')$ is differentiable with a continuous derivative. We set

$$\left.\frac{\partial U(t, t')}{\partial t}\right|_{t=t'} = -H(t)/\hbar,$$

where $\hbar$ is a real parameter that will be identified later with Planck's constant. We then differentiate equation (2.1) with respect to t and take the $t'' = t$ limit. We find

$$\hbar\frac{\partial U}{\partial t}(t, t') = -H(t)U(t, t'). \tag{2.2}$$

When the operator H is time-independent, it is the generator of time translations; the evolution operator then takes the special form $U(t, t') \equiv U(t - t') = \mathrm{e}^{-(t-t')H/\hbar}$ and property (2.1) becomes a semi-group property: $U(t)U(t') = U(t + t')$.

Quite generally, the Markov property (2.1) allows writing $U(t'', t')$ as a product of n operators corresponding to time intervals $\varepsilon = (t'' - t')/n$ that can be chosen arbitrarily small by increasing n:

$$U(t'', t') = \prod_{m=1}^{n} U[t' + m\varepsilon, t' + (m-1)\varepsilon], \qquad n\varepsilon = t'' - t'. \tag{2.3}$$

The product (2.3) is *time ordered* as in equation (2.1). When H is time-independent, $U(t + \varepsilon, t) = \mathrm{e}^{-\varepsilon H/\hbar}$ and one recognizes Trotter's formula.

Note that the operators $U(t + \varepsilon, t)$ play a role analogous to transfer matrices in classical statistical lattice models (see Chapter 4).

2.1.2 Matrix elements and locality

Position operator and matrix elements. We now introduce the basis in Hilbert space in which we assume local evolution for short time intervals. In the quantum context, it is the basis in which the position operator $\hat{q}$ is diagonal. (As in the case of plane waves, this is a generalized basis because the basis vectors do not belong to the Hilbert space.) Using the bra and ket notation, usual in quantum mechanics, we denote by $|q\rangle$ the eigenvector of $\hat{q}$ with eigenvalue q:

$$\hat{q}\,|q\rangle = q\,|q\rangle. \tag{2.4}$$

In quantum mechanics, to each physical observable is associated a hermitian operator. This applies to position and, thus, the eigenvalues of $\hat{q}$ are real and the eigenvectors are orthogonal:

$$\langle q'|q\rangle = \delta(q - q'),$$

where δ is Dirac's function. Moreover, the basis is complete, which implies

$$\int dq\,|q\rangle\,\langle q| = \mathbf{1}. \tag{2.5}$$

In terms of matrix elements in this basis, identity (2.1) for times $t_3 \geq t_2 \geq t_1$ takes the form

$$\langle q_3|\,U(t_3, t_1)\,|q_1\rangle = \int dq_2\,\langle q_3|\,U(t_3, t_2)\,|q_2\rangle\,\langle q_2|\,U(t_2, t_1)\,|q_1\rangle,$$

where we have used the decomposition (2.5) of the identity. Generalizing this identity, we rewrite equation (2.3) as

$$\langle q''\,|U(t'', t')|\,q'\rangle = \int \prod_{k=1}^{n-1} dq_k \prod_{k=1}^{n} \langle q_k\,|U(t_k, t_{k-1})|\,q_{k-1}\rangle \tag{2.6}$$

with the conventions

$$t_k = t' + k\varepsilon,\ q_0 = q',\ q_n = q''.$$

In this expression, we can take the limit $n \to \infty$, or $\varepsilon \to 0$, reducing the evaluation of expression (2.6) to the asymptotic evaluation (but with enough precision) of matrix elements $\langle q|U(t+\varepsilon, t)|q'\rangle$ for infinitesimal time intervals.

Locality of short time evolution. If *the operator H is local in the basis in which the position operator $\hat{q}$ is diagonal*, which means that the matrix elements $\langle q_1|H(t)|q_2\rangle$ have a support limited to $q_1 = q_2$, one can use identity (2.3) to construct a *path integral* representation for the matrix elements $\langle q''|U(t'', t')|q'\rangle$. Indeed, in the limit $\varepsilon \to 0$, as a consequence of the locality of H, only the matrix elements with $|q - q'| \ll 1$ contribute to expression (2.6). This property is satisfied when the operator $H(t)$ can be expressed in terms of the usual momentum and position operators $\hat{p}$ and $\hat{q}$ of quantum mechanics: $H(t) \equiv H(\hat{p}, \hat{q}; t)$, and is a polynomial in $\hat{p}$ (for details see also

Section 5.5). Equivalently, this means that H acts on wave functions $\langle q|\psi\rangle \equiv \psi(q)$ as a differential operator.

The operator H. In this work, we meet three types of operators:

If the operator U describes quantum evolution in time, it is unitary and the operator H is anti-hermitian, $H = i\tilde{H}$, where $\tilde{H}$ is the quantum hamiltonian. The locality of quantum evolution is also in direct correspondence with the locality of classical evolution. Quantum evolution will be discussed in Chapter 9.

In this chapter, the operator H itself is hermitian and the interpretation is different. For example, if H is a time-independent hamiltonian, the operator $U(\hbar\beta, 0)$ is the statistical quantum operator, that is, the density matrix of a statistical quantum system at thermal equilibrium at a temperature $T = 1/\beta$ (the equilibrium can be induced by a weak coupling to a thermal bath). Nevertheless, we call below the evolution variable t time (or euclidean time), even though, from the viewpoint of quantum evolution, it is an imaginary time. Indeed, if one changes the time t in it, one recovers the usual evolution operator of quantum mechanics. The same analytic continuation allows transposing the algebraic part of the calculations of this chapter to quantum evolution.

Finally, in Section 5.5, H will be the so-called Fokker–Planck hamiltonian associated with a diffusion equation: such a hamiltonian is real but, in general, not hermitian.

2.1.3 Example: free evolution or brownian motion

We first illustrate these remarks with the example of the statistical operator corresponding to a free particle of mass m, which is mathematically equivalent to the example of brownian motion (Section 5.5). We now assume that the operator $\hat{\mathbf{q}}$ is a d-component space vector ($\mathbf{q} \in \mathbb{R}^d$).

To study the problem, it is useful to also introduce the momentum operator $\hat{\mathbf{p}}$. The canonical commutation relations between the components of $\hat{\mathbf{q}}$ and the d components of the momentum operator $\hat{\mathbf{p}}$ are

$$[\hat{q}_\alpha, \hat{p}_\beta] = i\hbar \delta_{\alpha\beta}. \tag{2.7}$$

The hamiltonian of the free motion then can be written as

$$H_0 = \hat{\mathbf{p}}^2/2m.$$

To calculate the matrix elements of the operator $U(t, t') = e^{-(t-t')H_0/\hbar}$, it is convenient to introduce also the basis in which $\hat{\mathbf{p}}$ is diagonal, which is related to the position basis by Fourier transformation.

Fourier transformation: convention. Denoting by $\widetilde{|\mathbf{p}\rangle}$ the vectors of the basis in which the momentum operator $\hat{\mathbf{p}}$ is diagonal, we define the Fourier transformation by

$$\int d^d q \, e^{i\mathbf{q}\cdot\mathbf{p}/\hbar} |\mathbf{q}\rangle = \widetilde{|\mathbf{p}\rangle}. \tag{2.8}$$

The matrix elements of the identity operator in this basis are then

$$\widetilde{\langle\mathbf{p}''|}\mathbf{1}\widetilde{|\mathbf{p}'\rangle} = \widetilde{\langle\mathbf{p}''|}\widetilde{|\mathbf{p}'\rangle} = (2\pi\hbar)^d \delta^{(d)}(\mathbf{p}'' - \mathbf{p}').$$

Therefore, in a product of operators in momentum representation, the integration measure is $\mathrm{d}^d p/(2\pi\hbar)^d$:

$$\widetilde{\langle\mathbf{p}''|}U_2 U_1 \widetilde{|\mathbf{p}'\rangle} = \int \frac{\mathrm{d}^d p}{(2\pi\hbar)^d} \widetilde{\langle\mathbf{p}''|}U_2\widetilde{|\mathbf{p}\rangle}\widetilde{\langle\mathbf{p}|}U_1\widetilde{|\mathbf{p}'\rangle}.$$

Moreover, this normalization implies for the wave functions $\psi(\mathbf{q}) \equiv \langle\mathbf{q}|\psi\rangle$:

$$\widetilde{\langle\mathbf{p}|}\psi\rangle \equiv \tilde{\psi}(\mathbf{p}) = \int \mathrm{d}^d q\, e^{-i\mathbf{q}\cdot\mathbf{p}/\hbar}\, \psi(\mathbf{q}) \;\Rightarrow\; \psi(\mathbf{q}) = \frac{1}{(2\pi\hbar)^d}\int \mathrm{d}^d p\, e^{i\mathbf{q}\cdot\mathbf{p}/\hbar}\, \tilde{\psi}(\mathbf{p}).$$

Free hamiltonian. One can solve equation (2.2) in the mixed position-momentum representation:

$$-\hbar\frac{\partial}{\partial t}\widetilde{\langle\mathbf{p}|}U(t,t')|\mathbf{q}'\rangle = \frac{\mathbf{p}^2}{2m}\widetilde{\langle\mathbf{p}|}U(t,t')|\mathbf{q}'\rangle \quad \text{with} \quad \widetilde{\langle\mathbf{p}|}U(t',t')|\mathbf{q}'\rangle = e^{-i\mathbf{q}'\cdot\mathbf{p}/\hbar}.$$

The solution is

$$\widetilde{\langle\mathbf{p}|}U(t,t')|\mathbf{q}'\rangle = e^{-i\mathbf{q}'\cdot\mathbf{p}/\hbar}\, e^{-(t-t')\mathbf{p}^2/2m\hbar}.$$

Inverting the Fourier transformation, one concludes

$$\langle\mathbf{q}|U(t,t')|\mathbf{q}'\rangle = \int \frac{\mathrm{d}^d p}{(2\pi\hbar)^d}\exp\left[\frac{1}{\hbar}\left(i(\mathbf{q}-\mathbf{q}')\cdot\mathbf{p} - (t-t')\frac{\mathbf{p}^2}{2m}\right)\right]$$

$$= \left(\frac{m}{2\pi\hbar(t-t')}\right)^{d/2}\exp\left[-\frac{1}{\hbar}\frac{m(\mathbf{q}-\mathbf{q}')^2}{2(t-t')}\right]. \tag{2.9}$$

This expression clearly exhibits the locality property: when $t - t' \to 0$, the region in which $\langle\mathbf{q}|U(t,t')|\mathbf{q}'\rangle$ is not negligible reduces to $|\mathbf{q}-\mathbf{q}'| = O(\sqrt{t-t'})$.

2.2 Solution of the evolution equation for short times

In the rest of the chapter, the operator $H(t)$, which appears in equation (2.2), is identified with a quantum hamiltonian of the form

$$H = \hat{\mathbf{p}}^2/2m + V(\hat{\mathbf{q}}, t), \tag{2.10}$$

(where $\mathbf{p}, \mathbf{q}$ are d-component vectors). Such hamiltonians, like, more generally, all hamiltonians that are polynomials in $\hat{\mathbf{p}}$, are local. More general hamiltonians will be discussed in Chapters 5 and 10.

We then evaluate expression (2.6) in the limit $n \to \infty$, $\varepsilon \to 0$ at $n\varepsilon$ fixed.

In terms of the matrix elements of the operators H and U in the $|q\rangle$ basis, equation (2.2) becomes a partial differential equation, which formally is a Schrödinger equation for quantum evolution in imaginary time:

$$-\hbar\frac{\partial}{\partial t}\langle \mathbf{q}|U(t,t')|\mathbf{q}'\rangle = \left[-\frac{\hbar^2}{2m}\nabla_\mathbf{q}^2 + V(\mathbf{q},t)\right]\langle \mathbf{q}|U(t,t')|\mathbf{q}'\rangle \qquad (2.11)$$

with the boundary conditions

$$\langle \mathbf{q}|U(t',t')|\mathbf{q}'\rangle = \delta^{(d)}(\mathbf{q}-\mathbf{q}').$$

To solve equation (2.11) in the limit $t-t'\to 0$, it is convenient to set

$$\langle \mathbf{q}|U(t,t')|\mathbf{q}'\rangle = \exp\left[-\sigma(\mathbf{q},\mathbf{q}';t,t')/\hbar\right].$$

Equation (2.11) then becomes

$$\partial_t \sigma = -\frac{1}{2m}(\nabla_q\sigma)^2 + \frac{\hbar}{2m}\nabla_q^2\sigma + V(\mathbf{q},t). \qquad (2.12)$$

One knows from the solution (2.9) of the free case that the solution is singular for $|t-t'|\to 0$. If one assumes that the singular terms do not depend on the potential, one is led to set

$$\sigma(\mathbf{q},\mathbf{q}';t,t') = m\frac{(\mathbf{q}-\mathbf{q}')^2}{2(t-t')} + \frac{d}{2}\hbar\ln[2\pi\hbar(t-t')/m] + \sigma_1(\mathbf{q},\mathbf{q}';t,t'). \qquad (2.13)$$

The partial derivatives of σ that appear in equation (2.12) then become

$$\partial_t \sigma = -\frac{m}{2(t-t')^2}(\mathbf{q}-\mathbf{q}')^2 + \frac{d\hbar}{2(t-t')} + \partial_t\sigma_1(\mathbf{q},\mathbf{q}';t,t'),$$
$$\nabla_q \sigma = \frac{m}{t-t'}(\mathbf{q}-\mathbf{q}') + \nabla_q\sigma_1,$$
$$\nabla_q^2 \sigma = \frac{dm}{t-t'} + \nabla_q^2\sigma_1.$$

Introducing these expressions into equation (2.12), one verifies the consistency of the ansatz and one finds that the function σ_1 is of order $t-t'$. The leading terms in the equation, which are of order $|t-t'|^0$, reduce to

$$\partial_t \sigma_1 = -\frac{1}{t-t'}(\mathbf{q}-\mathbf{q}')\cdot\nabla_q\sigma_1 + V(\mathbf{q},t),$$

where the neglected terms are at least of order $t-t'$. It is more suggestive to write the equation as

$$[(t-t')\partial_t + (\mathbf{q}-\mathbf{q}')\cdot\nabla_q]\sigma_1 = (t-t')V(\mathbf{q},t). \qquad (2.14)$$

Under this form, one realizes that the equation can be solved by introducing a dilatation parameter λ and a function $\phi(\lambda)$ obtained by substituting $t \mapsto t' + \lambda(t-t')$, $\mathbf{q} \mapsto \mathbf{q}' + \lambda(\mathbf{q} - \mathbf{q}')$ in σ_1. The function satisfies the equation

$$\lambda \frac{\mathrm{d}\phi}{\mathrm{d}\lambda} = \lambda(t - t')V\big(\mathbf{q}' + \lambda(\mathbf{q} - \mathbf{q}'), t' + \lambda(t - t')\big).$$

Integrating, one infers

$$\sigma_1 = \phi(1) = (t - t') \int_0^1 \mathrm{d}\lambda\, V\big(\mathbf{q}' + \lambda(\mathbf{q} - \mathbf{q}'), t' + \lambda(t - t')\big).$$

The solution can be cast into a more suitable form by changing variables, $\lambda \mapsto \tau$, with

$$\tau = t' + \lambda(t - t'),$$

and introducing the function $\mathbf{q}(\tau)$ that corresponds to the trajectory relating the point $\mathbf{q}'$ to $\mathbf{q}$ with constant velocity:

$$\mathbf{q}(\tau) = \mathbf{q}' + \frac{\tau - t'}{t - t'}(\mathbf{q} - \mathbf{q}'). \tag{2.15}$$

The solution of equation (2.14) can then be written as

$$\sigma_1(\mathbf{q}, \mathbf{q}'; t, t') = \int_{t'}^{t} \mathrm{d}\tau\, V\big(\mathbf{q}(\tau), \tau\big). \tag{2.16}$$

The free contribution can also be expressed in terms of $\mathbf{q}(\tau)$ ($\dot{q} \equiv \mathrm{d}q/\mathrm{d}\tau$):

$$\frac{m(\mathbf{q} - \mathbf{q}')^2}{2(t - t')} = \frac{1}{2}\int_{t'}^{t} \mathrm{d}\tau\, m\, \dot{\mathbf{q}}^2(\tau).$$

Finally, one verifies that $\langle \mathbf{q}\,|U(t,t')|\,\mathbf{q}'\rangle = \mathrm{e}^{-\sigma/\hbar}$, where σ has the form (2.13), satisfies the initial time condition: its limit for $t - t' \to 0$ is indeed $\delta^{(d)}(\mathbf{q} - \mathbf{q}')$. One concludes

$$\langle \mathbf{q}\,|U(t,t')|\,\mathbf{q}'\rangle = \left(\frac{m}{2\pi\hbar(t-t')}\right)^{d/2} \exp\left[-\mathcal{S}(\mathbf{q})/\hbar\right] \tag{2.17}$$

with

$$\mathcal{S}(\mathbf{q}) = \int_{t'}^{t} \mathrm{d}\tau\, \left[\tfrac{1}{2}m\dot{\mathbf{q}}^2(\tau) + V(\mathbf{q}(\tau),\tau)\right] + O\left((t-t')^2\right). \tag{2.18}$$

Note that if the potential does not depend on time, one can also use the remark ($\varepsilon = t - t'$)

$$U(\hbar\varepsilon, 0) = \mathrm{e}^{-\varepsilon H} = \mathrm{e}^{-\varepsilon \hat{p}^2}\, \mathrm{e}^{-\varepsilon V(\hat{q}) + O(\varepsilon^2)},$$

where the term of order ε^2 is proportional to the commutator $[\hat{p}^2, V(\hat{q})]$ (Baker–Hausdorf formula). Taking the matrix elements of both sides, one obtains the result, provided the commutator is not too singular.

Fundamental observation. In the expansion of σ, the most singular term for $\varepsilon = t - t' \to 0$ is $(\mathbf{q} - \mathbf{q}')^2/\varepsilon$ (which is independent of the potential). This implies that the support of the matrix element $\langle \mathbf{q}|U(t' + \varepsilon, t')|\mathbf{q}'\rangle$ corresponds to values $|\mathbf{q}' - \mathbf{q}| = O(\sqrt{\varepsilon})$, a property typical of brownian motion. Moreover, for $|\mathbf{q}' - \mathbf{q}| = O(\sqrt{\varepsilon})$, one finds

$$\sigma_1(\mathbf{q}, \mathbf{q}'; t' + \varepsilon, t') = \varepsilon V\big((\mathbf{q} + \mathbf{q}')/2, t'\big) + O\left(\varepsilon^2\right)$$
$$= \tfrac{1}{2}\varepsilon\big(V(\mathbf{q}, t) + V(\mathbf{q}', t')\big) + O\left(\varepsilon^2\right)$$
$$= \varepsilon V(\mathbf{q}, t) + O\left(\varepsilon^{3/2}\right),$$

if the potential is differentiable. Therefore, replacing in σ_1 expression (2.13) by $(t - t')V(\mathbf{q}, t)$, for example, modifies σ only by terms of order $(t - t')^{3/2}$, which are negligible, as we show below. More generally, for the three expressions to be equivalent, the potential must be at least continuous, more singular potentials requiring a special analysis.

2.3 Path integral representation

We now combine expression (2.6), in the large n limit, with the short time evaluation of the matrix elements of the statistical operator, to construct a path integral representation of the statistical operator and then, as a consequence, of the quantum partition function.

2.3.1 The statistical operator

Combining the short time evaluation (2.17, 2.18) with equation (2.6), one obtains ($\varepsilon = (t'' - t')/n$)

$$\langle \mathbf{q}''|U(t'', t')|\mathbf{q}'\rangle = \lim_{n\to\infty}\left(\frac{m}{2\pi\hbar\varepsilon}\right)^{dn/2}\int\prod_{k=1}^{n-1}d^d q_k\,\exp\left[-\mathcal{S}(\mathbf{q}, \varepsilon)/\hbar\right] \quad (2.19)$$

with

$$\mathcal{S}(\mathbf{q}, \varepsilon) = \sum_{k=0}^{n-1}\int_{t_k}^{t_{k+1}}dt\left[\tfrac{1}{2}m\dot{\mathbf{q}}^2(t) + V\big(\mathbf{q}(t), t\big)\right] + O(\varepsilon^2).$$

Introducing the piecewise linear trajectory defined by (Fig. 2.1)

$$\mathbf{q}(t) = \mathbf{q}_k + \frac{t - t_k}{t_{k+1} - t_k}(\mathbf{q}_{k+1} - \mathbf{q}_k)\ \text{for}\ t_k \leq t \leq t_{k+1},$$

which interpolates in time the variables $\mathbf{q}_k \equiv \mathbf{q}(t_k)$, one can rewrite the expression as

$$\mathcal{S}(\mathbf{q}, \varepsilon) = \int_{t'}^{t''}dt\left[\tfrac{1}{2}m\dot{\mathbf{q}}^2(t) + V\big(\mathbf{q}(t), t\big)\right] + O(n\varepsilon^2). \quad (2.20)$$

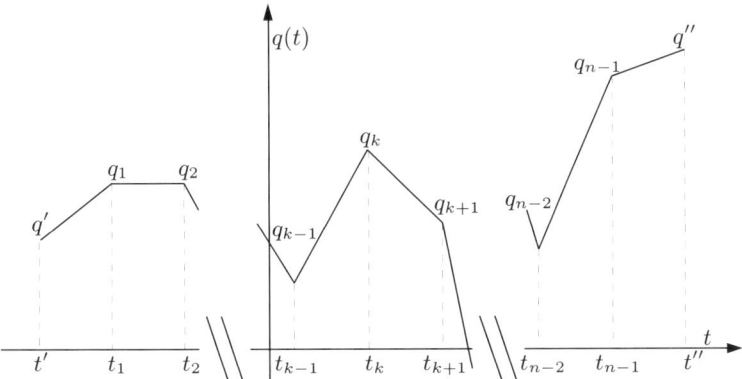

Fig. 2.1 A path contributing to the integral (2.19).

The integration over the variables $\mathbf{q}_k$ is thus equivalent to an integration over the points of a piecewise linear path of the kind displayed in Fig. 2.1.

The terms neglected in expression (2.20) are of order $n\varepsilon^2$. In the $n \to \infty$, $\varepsilon \to 0$ with $n\varepsilon = t'' - t'$ fixed, limit, they vanish. When $\varepsilon \to 0$, the limit of $\mathcal{S}(\mathbf{q}, \varepsilon)$ is the *euclidean* action

$$\mathcal{S}(\mathbf{q}) \equiv \int_{t'}^{t''} dt \left[\tfrac{1}{2} m \dot{\mathbf{q}}^2(t) + V(\mathbf{q}(t), t) \right], \tag{2.21}$$

which is the integral of the *euclidean* lagrangian associated with the hamiltonian. The euclidean action differs from the usual action of classical mechanics by the relative sign between the kinetic and potential terms. From a formal viewpoint, the euclidean action corresponds to a motion in *imaginary time* and this explains the sign difference.

We then take the formal limit of expression (2.19):

$$\langle \mathbf{q}'' | U(t'', t') | \mathbf{q}' \rangle = \int_{\mathbf{q}(t')=\mathbf{q}'}^{\mathbf{q}(t'')=\mathbf{q}''} [d\mathbf{q}(t)] \exp\left[-\mathcal{S}(\mathbf{q})/\hbar\right], \tag{2.22}$$

and call it *path integral* because the r.h.s. involves a sum over all paths satisfying the prescribed boundary conditions, with the weight $\exp[-\mathcal{S}/\hbar]$.

In what follows, we denote by $[d\mathbf{q}(t)]$ (with brackets) the integration measure to distinguish path integrals from ordinary integrals.

Remark. In the symbol $[d\mathbf{q}(t)]$ is hidden a normalization

$$\mathcal{N} = \left(\frac{m}{2\pi\hbar\varepsilon}\right)^{dn/2}$$

which diverges for $n \to \infty$ but does not depend on the potential. Therefore, in explicit calculations, path integrals are generally normalized by dividing them by reference path integrals whose value is already known (the free motion $V \equiv 0$, for example).

Generalization. The generalization of the preceding construction to a system with several particles corresponding to a hamiltonian

$$H = \sum_a \frac{\hat{\mathbf{p}}_a^2}{2m_a} + V(\hat{\mathbf{q}};t),$$

is immediate and leads to a path integral that involves the corresponding euclidean action

$$\mathcal{S}(\mathbf{q}) = \int dt \left[\sum_a \tfrac{1}{2} m_a (\dot{\mathbf{q}}_a)^2 + V(\mathbf{q}(t)) \right].$$

Discussion. In the action (2.21), the two terms play quite different roles. The potential determines the contribution of paths as a function of the value of $\mathbf{q}(t)$ at each time, and determines the physical properties of the quantum system.

The kinetic term $\int dt\, \dot{\mathbf{q}}^2$, in contrast, determines the space of paths that contribute to the integral and, thus, is essential for the very existence of the path integral. It selects paths regular enough, that is, as one sees in expression (2.20), those for which $[\mathbf{q}(t+\varepsilon) - \mathbf{q}(t)]^2 / \varepsilon$ remains finite when ε goes to zero. These paths are typical of brownian motion or random walk. The kinetic term really belongs to the integration measure. In fact, the explicit expression of the kinetic term, which involves $\dot{q}$, is somewhat formal since the typical paths that contribute to the path integral are continuous (they satisfy a Hölder condition of order $1/2$) but not differentiable.

However, a more mathematically correct notation would be less intuitive. In particular, it is clear in expression (2.22) that the maxima of the integrand correspond to paths that minimize the action (2.21), that is, that satisfy

$$\frac{\delta \mathcal{S}}{\delta q_i(t)} = 0 \text{ with } \frac{\delta^2 \mathcal{S}}{\delta q_i(t_1) \delta q_j(t_2)} \geq 0,$$

in the sense of operators. (δ indicates functional derivatives as defined in Section 2.5.3.) They are thus solutions of the equation of the classical motion, here euclidean (or in imaginary time), which are differentiable functions. This observation is specially relevant for $\hbar \to 0$ where classical solutions are saddle points in the sense of the steepest descent method applied to a path integral. Though only non-differentiable paths contribute to a path integral, in the semi-classical limit $\hbar \to 0$, the leading contributions come from paths in the neighbourhood of differentiable classical paths.

2.3.2 Partition function

In the following sections, we discuss one of the simplest quantities: the quantum partition function. The potential is now assumed to be *time-independent*.

The quantum partition function $\mathcal{Z}(\beta) = \operatorname{tr} e^{-\beta H}$ (β is the inverse temperature) has a path integral representation that follows directly from the representation of the matrix elements of the statistical operator $e^{-\beta H}$. Indeed,

$$\mathcal{Z}(\beta) = \operatorname{tr} e^{-\beta H} \equiv \operatorname{tr} U(\hbar\beta, 0) = \int dq''\, dq'\, \delta(\mathbf{q}'' - \mathbf{q}') \langle \mathbf{q}''|U(\hbar\beta, 0)|\mathbf{q}'\rangle.$$

We now express $\langle \mathbf{q}''|U(\hbar\beta,0)|\mathbf{q}'\rangle$ in terms of the path integral representation (2.22). The δ function and the two integrals over $\mathbf{q}' = \mathbf{q}(0)$ and $\mathbf{q}'' = \mathbf{q}(\hbar\beta)$ imply that the paths that contribute to the integral satisfy the periodic boundary conditions $\mathbf{q}(0) = \mathbf{q}(\hbar\beta)$, and one integrates over all values of $\mathbf{q}(0)$:

$$\mathcal{Z}(\beta) = \int_{\mathbf{q}(0)=\mathbf{q}(\hbar\beta)} [\mathrm{d}\mathbf{q}(t)] \exp\left[-\mathcal{S}(\mathbf{q})/\hbar\right]. \tag{2.23}$$

These conditions define closed paths, which also correspond to the set of periodic functions $\mathbf{q}(t)$ with period $\hbar\beta$.

The action (2.21) now depends implicitly on $\hbar$:

$$\mathcal{S}(\mathbf{q}) = \int_0^{\hbar\beta} \mathrm{d}t \left[\tfrac{1}{2}m\dot{\mathbf{q}}^2(t) + V(\mathbf{q}(t))\right].$$

Changing t to $t/\hbar$, one can rewrite it as

$$\mathcal{S}(q)/\hbar = \int_0^\beta \mathrm{d}t \left[\tfrac{1}{2}m\dot{\mathbf{q}}^2(t)/\hbar^2 + V(\mathbf{q}(t))\right]. \tag{2.24}$$

Trace of an operator. We have used, above, the expression of the trace of an operator O defined in terms of matrix elements in the position basis. The argument goes as follows:

$$\int \mathrm{d}q \, \langle q| O |q\rangle = \int \mathrm{d}q \, \mathrm{tr}\left(|q\rangle\langle q| O\right) = \mathrm{tr} \int \mathrm{d}q \, |q\rangle\langle q| O = \mathrm{tr}\, O,$$

where the completeness relation (2.5) has been used.

Low temperature limit. The ground state energy E_0 of the hamiltonian can be obtained from the partition function by studying the low temperature, $\beta \to \infty$ limit. Indeed,

$$\lim_{\beta \to \infty} \frac{1}{\beta} \ln \mathcal{Z}(\beta) = -E_0.$$

This remark will be used more systematically in Chapter 3.

Note that in the path integral one can impose boundary conditions at $t = 0$ and $t = \beta$ as well as at $t = \pm\beta/2$ because the action is time-translation invariant. However, in the limit $\beta \to \infty$, in the first case one is led to integrate over all paths with $t \geq 0$ and an initial condition $q(0) = 0$, while in the latter case the resulting formalism is explicitly time-translation invariant since one integrates from $-\infty$ to $+\infty$. This formalism, clearly, is simpler.

2.4 Explicit calculation: gaussian path integrals

We have defined path integrals as formal limits of integrals with discrete times. One could thus fear that any calculation of a path integral would require returning to its definition as a limit of discrete time integrals. This is fortunately not the case; most calculations make no reference to the limiting process, and this gives a practical justification to the introduction of this new mathematical entity. This will be illustrated in this section, first with the example of free motion, then with the harmonic oscillator, both corresponding to gaussian path integrals.

2.4.1 Free motion

In the example of free motion (in one dimension for simplicity), $V \equiv 0$, the euclidean action is a simple quadratic form in $q(t)$:

$$\mathcal{S}(q) = \int_{t'}^{t''} \tfrac{1}{2} m \dot{q}^2(t) \mathrm{d}t$$

and, thus, the path integral (2.22) is gaussian. To calculate it explicitly, it is sufficient to adapt the method explained in Section 1.2. The role of the matrix $\mathbf{A}$ in the integral (1.4) is played here by the differential operator $-m\mathrm{d}^2/(\mathrm{d}t)^2$, as an integration by parts of the term $(\dot{q})^2$ shows:

$$\int_{t'}^{t''} \mathrm{d}t\, \dot{q}^2(t) = q(t'')\dot{q}(t'') - q(t')\dot{q}(t') - \int_{t'}^{t''} \mathrm{d}t\, q(t) \frac{\mathrm{d}^2}{(\mathrm{d}t)^2} q(t) \,.$$

Since $\mathbf{A}$ is here a differential operator, the determinant, as well as its inverse, depend on boundary conditions, and this is the main difference with respect to the calculation with a finite number of variables.

To calculate the integral, we change variables, $q(t) \mapsto r(t)$, (a translation at each time t):

$$q(t) = q_c(t) + r(t), \tag{2.25}$$

where the function q_c satisfies the boundary conditions

$$q_c(t') = q', \ q_c(t'') = q'' \Rightarrow r(t') = r(t'') = 0\,. \tag{2.26}$$

The action becomes

$$\mathcal{S}(q_c + r) = \mathcal{S}(q_c) + \mathcal{S}(r) + m \int_{t'}^{t''} \mathrm{d}t\, \dot{q}_c(t)\dot{r}(t)\,.$$

The term linear in r can be written as

$$\int_{t'}^{t''} \dot{q}_c(t) \dot{r}(t) \mathrm{d}t = - \int_{t'}^{t''} \ddot{q}_c(t) r(t) \mathrm{d}t\,, \tag{2.27}$$

where, in the integration by parts, the boundary conditions (2.26) have been used. The linear term thus vanishes if the function $q_c(t)$ is a solution of the equation of the free classical motion:
$$m\ddot{q}_c(t) = 0.$$

The solution that satisfies the conditions (2.26) is
$$q_c(t) = q' + \frac{t - t'}{t'' - t'}(q'' - q').$$

One infers
$$\mathcal{S}(q_c) = \frac{m}{2}\frac{(q'' - q')^2}{t'' - t'}.$$

Since the change of variables is a translation, the jacobian of the transformation is equal to 1 and $[dq(t)] = [dr(t)]$. We thus obtain
$$\langle q''|e^{-(t''-t')\hat{p}^2/2m\hbar}|q'\rangle = \mathcal{N}\exp\left[-\frac{m}{2\hbar}\frac{(q''-q')^2}{t''-t'}\right],$$

where
$$\mathcal{N} = \int [dr(t)]\exp\left[-\frac{m}{2\hbar}\int_{t'}^{t''}dt\,\dot{r}^2(t)\right] \tag{2.28}$$

with $r(t') = r(t'') = 0$. Since the normalization $\mathcal{N}$ does not depend on q', q'', the expression is consistent with the exact result (2.9). The path integral $\mathcal{N}$ is the value of $\langle q''|e^{-(t''-t')\hat{p}^2/2m\hbar}|q'\rangle$ for $q' = q'' = 0$. To determine it, one must either return to discrete time intervals, or use some independent information.

When the potential does not vanish, an exact calculation is, in general, no longer feasible. However, at short times, the kinetic term dominates. In the potential, at leading order, one can replace $q(t)$ by the classical trajectory $q_c(t)$. One then recovers exactly the form (2.18) of the action.

2.4.2 The harmonic oscillator

The hamiltonian of the one-dimensional harmonic oscillator can be written as
$$H_0 = \frac{1}{2m}\hat{p}^2 + \frac{1}{2}m\omega^2\hat{q}^2, \tag{2.29}$$

where ω is a constant. The corresponding euclidean action is
$$\mathcal{S}_0(q) = \int_{t'}^{t''}\left[\tfrac{1}{2}m\dot{q}^2(t) + \tfrac{1}{2}m\omega^2 q^2(t)\right]dt, \tag{2.30}$$

which, again, is a quadratic form in $q(t)$. Therefore, the path integral (2.22),
$$\langle q''|U_0(t'',t')|q'\rangle = \int_{q(t')=q'}^{q(t'')=q''}[dq(t)]\,e^{-\mathcal{S}_0(q)/\hbar}, \tag{2.31}$$

remains gaussian.

In the path integral, the role of the matrix $\mathbf{A}$ in the integral (1.4) here is played by the differential operator $m(-\mathrm{d}^2/(\mathrm{d}t)^2 + \omega^2)$.

It is convenient to divide the calculation into several steps. The easiest part, and to some extent the most useful one, is the determination of the explicit dependence on the boundary conditions, that is, on q', q''.

One changes variables $q(t) \mapsto r(t)$ (a translation at each time t), setting

$$q(t) = q_c(t) + r(t), \tag{2.32}$$

where the functions q_c and $r(t)$ satisfy the boundary conditions (2.26) and q_c is determined below. The action becomes

$$\mathcal{S}_0(q_c + r) = \mathcal{S}_0(q_c) + \mathcal{S}_0(r) + m\int_{t'}^{t''} \mathrm{d}t \left(\dot{q}_c(t)\dot{r}(t) + \omega^2 q_c(t) r(t)\right).$$

In the term linear in r, one integrates by parts (equation (2.27)). One then chooses the function $q_c(t)$, solution of the equation

$$-m\ddot{q}_c + m\omega^2 q_c = 0, \tag{2.33}$$

in such a way that the term linear in $r(t)$ cancels. One notices that $q_c(t)$ satisfies the classical motion equation derived from the action $\mathcal{S}_0$, that is, a motion in an inverted harmonic potential $-\frac{1}{2}m\omega^2 q^2$ (the $-$ sign reflecting the euclidean time).

The action (2.30) then reduces to the sum of two terms:

$$\mathcal{S}_0(q) = \mathcal{S}_0(q_c) + \mathcal{S}_0(r), \tag{2.34}$$

where $\mathcal{S}_0(q_c)$ is the classical action evaluated on the trajectory q_c. The path integral becomes (the jacobian is 1)

$$\langle q''|U_0(t'',t')|q'\rangle = \mathcal{N}(\omega,\tau)\,\mathrm{e}^{-\mathcal{S}_0(q_c)/\hbar} \tag{2.35}$$

with

$$\mathcal{N}(\omega,\tau) = \int [\mathrm{d}r(t)]\exp\left[-\frac{1}{\hbar}\int_{t'}^{t''}\mathrm{d}t\left(\tfrac{1}{2}m\dot{r}^2(t) + \tfrac{1}{2}m\omega^2 r^2(t)\right)\right], \tag{2.36}$$

where the paths now satisfy the boundary conditions (2.26): $r(t') = r(t'') = 0$.

The classical action. Setting $\tau = t'' - t'$, one can write the solution of equation (2.33) as

$$q_c(t) = \frac{1}{\sinh(\omega\tau)}\left[q'\sinh\bigl(\omega(t''-t)\bigr) + q''\sinh\bigl(\omega(t-t')\bigr)\right]. \tag{2.37}$$

In expression (2.30), an integration by parts,

$$\int \dot{q}^2\, \mathrm{d}t = q\dot{q} - \int q\ddot{q}\, \mathrm{d}t,$$

combined with equation (2.33), then simplifies the calculation of the classical action $\mathcal{S}_0(q_c)$. One finds

$$\mathcal{S}_0(q_c) = \frac{m\omega}{2\sinh\omega\tau}\left[(q'^2 + q''^2)\cosh\omega\tau - 2q'q''\right]. \qquad (2.38)$$

Normalization. The remaining gaussian integral (2.36) does not depend on q', q'' anymore and gives a simple normalization. Since, as expected, a direct calculation involves an infinite factor, for essentially pedagogical reasons, we complete the calculation only in Section 2.7.

The complete expression is

$$\langle q''|\, U_0(t'',t')\, |q'\rangle$$
$$= \left(\frac{m\omega}{2\pi\hbar\sinh\omega\tau}\right)^{1/2} \exp\left\{-\frac{m\omega}{2\hbar\sinh\omega\tau}\left[(q'^2 + q''^2)\cosh\omega\tau - 2q'q''\right]\right\}. \qquad (2.39)$$

2.5 Correlation functions: generating functional

In the path integral (2.23), the integrand $\mathrm{e}^{-\mathcal{S}(q)/\hbar}$ defines a positive measure for paths. To this measure correspond expectation values, defined by

$$\langle \mathcal{F}(q) \rangle \equiv \mathcal{N} \int [\mathrm{d}q(t)]\mathcal{F}(q)\exp[-\mathcal{S}(q)/\hbar].$$

where the normalization $\mathcal{N}$ is chosen in such a way that $\langle 1 \rangle = 1$.

2.5.1 Correlation functions

Moments of the measure of the form

$$\langle q(t_1)q(t_2)\ldots q(t_n)\rangle = \mathcal{N}\int[\mathrm{d}q(t)]q(t_1)q(t_2)\ldots q(t_n)\exp[-\mathcal{S}(q)/\hbar], \qquad (2.40)$$

are called *correlation functions*: a correlation function that depends on n different times is called an *n-point function*.

On a finite time interval β, the path integral definition requires some boundary conditions. For simplicity, we choose in most of this work periodic boundary conditions: $q(\beta/2) = q(-\beta/2)$. Then, the normalization is related to the quantum partition function:

$$\mathcal{N} = \mathcal{Z}^{-1}(\beta). \qquad (2.41)$$

Of course, it is possible to define correlation functions corresponding to other boundary conditions in the path integral but we will not discuss them here, all explicit expressions in the case of gaussian integrals being more complicated.

A more physical interpretation of these correlation functions will be given in Chapter 4. In this chapter, they appear only as useful mathematical quantities.

In the discussion of algebraic properties of correlation functions, generating functionals, generalization of the generating function introduced in Section 1.1, and functional derivatives provide very useful tools. We define them now.

2.5.2 Generating functional

We consider an infinite set $\{F^{(n)}(t_1,\ldots,t_n)\}$, $n = 0, 1, \ldots$, of *symmetric* functions of their arguments. We introduce a new function of one variable $f(t)$ and the following formal series in f:

$$\mathcal{F}(f) = \sum_{n=0}^{\infty} \frac{1}{n!} \int dt_1 \ldots dt_n \, F^{(n)}(t_1,\ldots,t_n) f(t_1) \ldots f(t_n). \tag{2.42}$$

The sum $\mathcal{F}(f)$ is a generating functional of the set of functions $F^{(n)}$.

The function $f(t)$ plays the role of the vector b_i of Section 1.1 and t the role of the index i.

In fact, more generally, we will admit for $F^{(n)}$'s also mathematical distributions (like the Dirac δ function and its derivatives). Then, the function $f(t)$ must belong to the corresponding set of test functions and, thus, must be enough differentiable.

The correlation functions that we have just introduced,

$$Z^{(n)}(t_1,\ldots,t_n) \equiv \langle q(t_1) \ldots q(t_n) \rangle,$$

are, by definition, symmetric functions of their arguments. A generating functional of correlation functions is then

$$\begin{aligned}\mathcal{Z}(f) &= \sum_{n=0}^{\infty} \frac{1}{n!} \int dt_1 \ldots dt_n \, Z^{(n)}(t_1,\ldots,t_n) f(t_1) \ldots f(t_n) \\ &= \sum_{n=0}^{\infty} \frac{1}{n!} \int dt_1 \ldots dt_n \, \langle q(t_1) \ldots q(t_n) \rangle f(t_1) \ldots f(t_n).\end{aligned}$$

The latter series can be formally summed, by exchanging integrals and expectation value (because it is a linear operation). One finds the simple formula

$$\mathcal{Z}(f) = \left\langle \exp\left[\int dt \, q(t) f(t)\right] \right\rangle. \tag{2.43}$$

2.5.3 Functional derivative

The functions $F^{(n)}$ can be recovered from the generating functional $\mathcal{F}(f)$ by *functional differentiation*, an operation that we now define.

Functional differentiation with respect to a function $f(t)$, which we denote by $\delta/\delta f(t)$ to distinguish it from normal derivatives, is defined by the usual algebraic properties, linearity and Leibnitz rule:

$$\begin{aligned}\frac{\delta}{\delta f(t)}[\mathcal{F}_1(f) + \mathcal{F}_2(f)] &= \frac{\delta}{\delta f(t)}\mathcal{F}_1(f) + \frac{\delta}{\delta f(t)}\mathcal{F}_2(f), \\ \frac{\delta}{\delta f(t)}[\mathcal{F}_1(f)\mathcal{F}_2(f)] &= \mathcal{F}_1(f)\frac{\delta}{\delta f(t)}\mathcal{F}_2(f) + \mathcal{F}_2(f)\frac{\delta}{\delta f(t)}\mathcal{F}_1(f)\end{aligned} \tag{2.44}$$

together with
$$\frac{\delta}{\delta f(u)} f(t) = \delta(t-u), \tag{2.45}$$

where $\delta(t)$ is Dirac's function (more accurately distribution).

For example, the first derivative of $\mathcal{F}(f)$ is

$$\frac{\delta}{\delta f(u)}\mathcal{F}(f) = \sum_{n=0}^{\infty} \frac{1}{n!} \int dt_1 \ldots dt_n\, F^{(n+1)}(u, t_1, \ldots, t_n) f(t_1) \ldots f(t_n). \tag{2.46}$$

Differentiating now p times and taking the limit $f \equiv 0$, one obtains

$$F^{(p)}(t_1, \ldots, t_p) = \left\{ \left(\prod_{i=1}^{p} \frac{\delta}{\delta f(t_i)} \right) \mathcal{F} \right\} \bigg|_{f \equiv 0}.$$

In example (2.43), these identities become

$$\frac{\delta \mathcal{Z}(f)}{\delta f(t_1)} = \left\langle q(t_1) \exp\left[\int dt\, q(t) f(t) \right] \right\rangle,$$

and, more generally,

$$\left(\prod_{i=1}^{p} \frac{\delta}{\delta f(t_i)} \right) \mathcal{Z}(f) = \left\langle \prod_{i=1}^{p} q(t_i) \exp\left[\int dt\, q(t) f(t) \right] \right\rangle.$$

In the limit $f \equiv 0$, one thus obtains

$$\left\langle \prod_{i=1}^{p} q(t_i) \right\rangle = \left(\prod_{i=1}^{p} \frac{\delta}{\delta f(t_i)} \right) \mathcal{Z}(f) \bigg|_{f \equiv 0}.$$

Remark. This formalism also applies when the $F^{(n)}$ are not functions in the strict sense but also distributions. For example,

$$\frac{\delta}{\delta f(u)} \frac{df(t)}{dt} = \frac{\delta}{\delta f(u)} \frac{d}{dt} \int dv\, \delta(t-v) f(v)$$
$$= \frac{d}{dt} \delta(t-u),$$

where $\delta(t)$ is Dirac's δ function. With this extension, the classical equations of motion in the lagrangian formalism can be obtained by functional differentiation of the action.

For example, the functional derivative of the action

$$S(q) = \int dt \left[\tfrac{1}{2} (\dot{q}(t))^2 - V(q(t)) \right]$$

is

$$\frac{\delta S}{\delta q(\tau)} = \int dt \left[\dot{q}(t) \frac{d}{dt} \delta(t-\tau) - V'(q) \delta(t-\tau) \right] = -\ddot{q}(\tau) - V'(q(\tau)).$$

The equation of the classical motion can thus be written as $\delta S/\delta q(\tau) = 0$.

2.6 General gaussian path integral and correlation functions

In Section 1.2, we have shown how to derive all expectation values with a gaussian weight from a gaussian integral where a linear term is added to the quadratic form (expression (1.8)). We now generalize the method to correlation functions.

2.6.1 General gaussian path integral

We consider the path integral

$$\operatorname{tr} U_G(\tau/2, -\tau/2; b) = \int [dq(t)] \exp[-\mathcal{S}_G(q,b)/\hbar] \qquad (2.47)$$

with periodic boundary conditions: $q(\tau/2) = q(-\tau/2)$ and

$$\mathcal{S}_G(q,b) = \int_{-\tau/2}^{\tau/2} dt \left[\tfrac{1}{2} m \dot{q}^2(t) + \tfrac{1}{2} m \omega^2 q^2(t) - b(t) q(t) \right]. \qquad (2.48)$$

The integral, which is a functional of $b(t)$, can be calculated by first eliminating the linear term in $\mathcal{S}_G(q,b)$. Again, the calculation is based on a change variables, a translation of the path $q(t) \mapsto r(t)$ with

$$q(t) = q_c(t) + r(t), \quad q_c(\tau/2) = q_c(-\tau/2) \Rightarrow r(\tau/2) = r(-\tau/2), \qquad (2.49)$$

where the function $q_c(t)$ is determined below. Then,

$$\mathcal{S}_G(q,b) = \mathcal{S}_0(r) + \mathcal{S}_G(q_c, b) + \mathcal{S}_{\text{lin.}}(r,b),$$

where $\mathcal{S}_{\text{lin.}}(r,b)$, the linear term in r, is

$$\mathcal{S}_{\text{lin.}}(r,b) = \int_{-\tau/2}^{\tau/2} dt \left[m \dot{q}_c(t) \dot{r}(t) + m \omega^2 q_c(t) r(t) - b(t) r(t) \right].$$

In $\mathcal{S}_{\text{lin.}}(r,b)$, one integrates by parts:

$$\int_{-\tau/2}^{\tau/2} dt\, \dot{q}_c(t) \dot{r}(t) = \dot{q}_c(\tau/2) r(\tau/2) - \dot{q}_c(-\tau/2) r(-\tau/2) - \int_{-\tau/2}^{\tau/2} dt\, \ddot{q}_c(t) r(t).$$

Taking into account the boundary conditions (2.49) $r(\tau/2) = r(-\tau/2)$, one obtains

$$\mathcal{S}_{\text{lin.}}(r,b) = \int_{-\tau/2}^{\tau/2} dt \left[-m \ddot{q}_c(t) + m \omega^2 q_c(t) - b(t) \right] r(t)$$
$$+ m r(\tau/2) \left[\dot{q}_c(\tau/2) - \dot{q}_c(-\tau/2) \right].$$

Thus, the coefficient of the term linear in r vanishes if the function $q_c(t)$ is a solution of the classical equation of motion

$$-\ddot{q}_c(t) + \omega^2 q_c(t) = b(t)/m$$

and, in addition, if $\dot{q}_c(\tau/2) = \dot{q}_c(-\tau/2)$. The solution that satisfies, in addition, the boundary conditions (2.49) can be written as

$$q_c(t) = \frac{1}{m} \int_{-\tau/2}^{\tau/2} \Delta(t-u) b(u) du, \qquad (2.50)$$

where the function $\Delta(t)$ is the solution of the equation

$$-\ddot{\Delta} + \omega^2 \Delta = \delta(t) \qquad (2.51)$$

with periodic boundary conditions:

$$\Delta(\tau/2) = \Delta(-\tau/2), \quad \dot{\Delta}(\tau/2) = \dot{\Delta}(-\tau/2).$$

One finds

$$\Delta(t) = \frac{1}{2\omega \sinh(\omega\tau/2)} \cosh\big(\omega(\tau/2 - |t|)\big). \qquad (2.52)$$

This result can be verified directly. We recall here that, in the sense of distributions,

$$\frac{\mathrm{d}}{\mathrm{d}t}|t| = \mathrm{sgn}(t), \quad \frac{\mathrm{d}}{\mathrm{d}t}\mathrm{sgn}(t) = 2\delta(t),$$

where $\mathrm{sgn}(t)$ is the sign function, $\mathrm{sgn}(t) = 1$ for $t > 0$, $\mathrm{sgn}(t) = -1$ for $t < 0$, and $\delta(t)$ is Dirac's δ function.

The function $\Delta(t)$, which plays an essential role in perturbative expansions around the harmonic oscillator (see Section 2.8), is also called *propagator*. It has a simple limit when $\tau \to \infty$ (assuming $\omega > 0$):

$$\Delta(t) = \frac{1}{2\omega} \mathrm{e}^{-\omega|t|} = \frac{1}{2\pi} \int_{-\infty}^{+\infty} \mathrm{d}\kappa \, \frac{\mathrm{e}^{i\kappa t}}{\kappa^2 + \omega^2}, \qquad (2.53)$$

which is the solution of equation (2.51) that decreases at infinity.

Equation (2.50) is the continuum analogue of equation (1.9). The kernel $\Delta(t_1 - t_2)$ is the inverse of the differential operator $-\mathrm{d}_t^2 + \omega^2$ with periodic boundary conditions. It depends only on the difference $t_2 - t_1$ because periodic boundary conditions are time-translation invariant; the interval $[0, \tau]$ can be identified with a circle and time with an angular variable.

A general remark then simplifies the calculation of the classical action. One calculates the action corresponding to $\lambda q_c(t)$:

$$\mathcal{S}_G(\lambda q_c, b) = \int_{-\tau/2}^{\tau/2} \mathrm{d}t \left\{ \lambda^2 \left[\tfrac{1}{2} m \dot{q}_c^2(t) + \tfrac{1}{2} m\omega^2 q_c^2(t) \right] - \lambda b(t) q_c(t) \right\}.$$

Since q_c is a solution of the classical equation, the action $\mathcal{S}_G(\lambda q_c, b)$ is stationary at the solution, that is for $\lambda = 1$:

$$\left. \frac{\mathrm{d}}{\mathrm{d}\lambda} \mathcal{S}_G(\lambda q_c, b) \right|_{\lambda=1} = 0.$$

One infers
$$\int_{-\tau/2}^{\tau/2} dt \left[\tfrac{1}{2}m\dot{q}_c^2(t) + \tfrac{1}{2}m\omega^2 q_c^2(t)\right] = \frac{1}{2}\int_{-\tau/2}^{\tau/2} dt\, b(t)q_c(t)$$

and, thus,
$$\mathcal{S}_G(q_c,b) = -\frac{1}{2}\int_{-\tau/2}^{\tau/2} dt\, q_c(t)b(t) = -\frac{1}{2m}\int_{-\tau/2}^{\tau/2} dt\,du\, b(t)\Delta(t-u)b(u).$$

The remaining integration over the path $r(t)$ yields a normalization that is identical to $\operatorname{tr} e^{-\tau H_0/\hbar}$ (H_0 is the hamiltonian (2.29)):
$$\operatorname{tr} U_G(\tau/2, -\tau/2; b) = \operatorname{tr} e^{-\tau H_0/\hbar}\, e^{-\mathcal{S}_G(q_c,b)/\hbar}.$$

Substituting $\tau = \hbar\beta$, one finds
$$\begin{aligned}\mathcal{Z}_G(\beta,b) &= \operatorname{tr} U_G(\hbar\beta/2, -\hbar\beta/2; b)\\ &= \mathcal{Z}_0(\beta)\exp\left[\frac{1}{2m\hbar}\int_{-\hbar\beta/2}^{\hbar\beta/2} du\,dv\, b(v)\Delta(v-u)b(u)\right],\end{aligned}\quad(2.54)$$

where $\mathcal{Z}_0(\beta)$ is the partition function of the harmonic oscillator (2.29), which can be derived from expression (2.39), but which we calculate more directly later (equation (2.61)).

2.6.2 Gaussian correlation functions. Wick's theorem

We now introduce gaussian expectation values, which we denote by $\langle\bullet\rangle_0$, corresponding to the normalized distribution $e^{-\mathcal{S}_0/\hbar}/\mathcal{Z}_0(\beta)$ with periodic boundary conditions. The functional (2.54), which can then be written as
$$\mathcal{Z}_G(\beta,b) = \mathcal{Z}_0(\beta)\left\langle \exp\left[\frac{1}{\hbar}\int_{-\hbar\beta/2}^{\hbar\beta/2} dt\, b(t)q(t)\right]\right\rangle_0,$$

generates the corresponding correlation functions (equation (2.43)). If one acts by functional differentiation (functional derivatives have been defined in Section 2.5.3) on both sides, in the limit $b \equiv 0$ one obtains
$$\hbar^\ell \left[\prod_{j=1}^\ell \frac{\delta}{\delta b(t_j)}\right] \mathcal{Z}_G(\beta,b)\bigg|_{b\equiv 0} = \mathcal{Z}_0(\beta)\langle q(t_1)q(t_2)\ldots q(t_\ell)\rangle_0.\quad(2.55)$$

More generally, a differential operator $\mathcal{F}(\hbar\delta/\delta b)$ generates on the r.h.s. the expectation value of the functional $\mathcal{F}(q)$ since
$$\mathcal{F}(\hbar\delta/\delta b)\mathcal{Z}_G(\beta,b)|_{b\equiv 0} = \int [dq(t)]\mathcal{F}(q)\exp[-\mathcal{S}_0(q)/\hbar].\quad(2.56)$$

Replacing $\mathcal{Z}_G(b,\beta)$ by the explicit result (2.54) on the l.h.s. of equation (2.55), one then finds

$$\langle q(t_1)\ldots q(t_\ell)\rangle_0 = \prod_{j=1}^{\ell} \hbar \frac{\delta}{\delta b(t_j)} \exp\left[\frac{1}{2m\hbar}\int du\,dv\, b(v)\Delta(v-u)b(u)\right]\bigg|_{b\equiv 0}.$$

In particular, the second derivative yields the two-point correlation function

$$\langle q(t)q(u)\rangle_0 = \mathcal{Z}_0^{-1}(\beta)\hbar^2 \frac{\delta^2}{\delta b(t)\delta b(u)} \mathcal{Z}_G(\beta,b)\bigg|_{b\equiv 0} = \frac{\hbar}{m}\Delta(t-u). \qquad (2.57)$$

More generally, a functional differentiation of the exponential of a quadratic form generates a factor b:

$$\hbar\frac{\delta}{\delta b(t_1)}\exp\left[\frac{1}{2m\hbar}\int du\,dv\, b(v)\Delta(v-u)b(u)\right]$$
$$= \frac{1}{m}\int du_1\, \Delta(t_1-u_1)b(u_1)\exp\left[\frac{1}{2m\hbar}\int du\,dv\, b(v)\Delta(v-u)b(u)\right].$$

The arguments of Section 1.2 then apply again here. In the limit $b\equiv 0$, the only surviving terms correspond to pairings of functional differentiations. We recover a characteristic property of the centred gaussian measure: as we have already explained, all correlation functions can be expressed in terms of the two-point function in a form expressed by Wick's theorem:

$$\langle q(t_1)q(t_2)\ldots q(t_\ell)\rangle_0 = \sum_{P\{1,2,\ldots,\ell\}} \left(\frac{\hbar}{m}\right)^\ell \Delta(t_{P_1}-t_{P_2})\ldots\Delta(t_{P_{\ell-1}}-t_{P_\ell})$$
$$= \sum_{P\{1,2,\ldots,\ell\}} \langle q(t_{P_1})q(t_{P_2})\rangle_0 \cdots \langle q(t_{P_{\ell-1}})q(t_{P_\ell})\rangle_0, \qquad (2.58)$$

where $P\{1,2,\ldots,\ell\}$ are all possible pairings of $\{1,2,\ldots,\ell\}$.

Regularity properties of generic paths. The preceding results allow calculating the expectation value of the quantity

$$\left\langle \big(q(t+\varepsilon)-q(t)\big)^2\right\rangle_0 = \frac{2\hbar}{m}\left[\Delta(0)-\Delta(\varepsilon)\right].$$

For $\varepsilon \to 0$, one finds

$$\left\langle \big(q(t+\varepsilon)-q(t)\big)^2\right\rangle_0 \sim |\varepsilon|\frac{\hbar}{m}. \qquad (2.59)$$

This result confirms that generic trajectories are not differentiable (though continuous), and that the behaviour of $q(t+\varepsilon)-q(t)$ for $\varepsilon \to 0$ does not depend on the potential but only on the kinetic term in the action. It also implies that the average action $\langle S(q)\rangle_0$ is infinite, because the kinetic term diverges. A proof, valid for an arbitrary potential, is given in Section 4.5.2.

2.7 Harmonic oscillator: the partition function

We can now complete the calculation of the partition function $\mathcal{Z}_0(\beta)$ of the harmonic oscillator, which appears in the normalization of correlation functions in equations (2.40).

We first determine the dependence of the partition function on the parameter ω. Differentiating the path integral (2.31) with boundary conditions (2.23), one obtains

$$\frac{\partial}{\partial \omega}\mathcal{Z}_0(\beta) = -\frac{1}{\hbar}\int [dq]\, e^{-\mathcal{S}_0(q)/\hbar}\,\frac{\partial \mathcal{S}_0}{\partial \omega}$$

$$= -\frac{m\omega}{\hbar}\int [dq]\, e^{-\mathcal{S}_0(q)/\hbar}\int_{-\hbar\beta/2}^{\hbar\beta/2} dt\, q^2(t)$$

with $q(\hbar\beta/2) = q(-\hbar\beta/2)$. Dividing both sides by $\mathcal{Z}_0(\beta)$, one finds, by definition, a gaussian expectation value with the weight $e^{-\mathcal{S}_0(q)/\hbar}/\mathcal{Z}_0$:

$$\frac{\partial}{\partial \omega}\ln \mathcal{Z}_0(\beta) = -\frac{m\omega}{\hbar}\int_{-\hbar\beta/2}^{\hbar\beta/2} dt\, \langle q^2(t)\rangle_0 = -\hbar\beta\omega\Delta(0) = -\frac{\hbar\beta}{2}\frac{\cosh(\omega\hbar\beta/2)}{\sinh(\omega\hbar\beta/2)}, \tag{2.60}$$

where the explicit form (2.52) has been used. Thus,

$$\mathcal{Z}_0(\beta) = \mathcal{N}'\frac{1}{\sinh(\beta\hbar\omega/2)}.$$

For dimensional reasons, the constant $\mathcal{N}'$ is a pure number. It can be determined by taking the limit $\beta \to \infty$, where one has to find $e^{-\beta E_0}$. The complete result thus is

$$\mathcal{Z}_0(\beta) = \frac{1}{2\sinh(\beta\hbar\omega/2)} = \frac{e^{-\beta\hbar\omega/2}}{1 - e^{-\beta\hbar\omega}}, \tag{2.61}$$

where, indeed, one recognizes the partition function of the harmonic oscillator.

Normalization of the gaussian path integral. Equation (2.35) allows relating the normalization of the path integral (2.31) to the partition function $\mathcal{Z}_0(\beta)$ of the harmonic oscillator. Taking the trace of $U_0(\hbar\beta, 0)$ in equation (2.35), one finds

$$\mathcal{Z}_0(\beta) = \text{tr}\, U_0(\hbar\beta, 0) \equiv \int dq\, \langle q|\, U_0(\hbar\beta, 0)\, |q\rangle$$

$$= \mathcal{N}(\omega, \beta)\left(\frac{\pi\hbar}{m\omega \tanh(\beta\hbar\omega/2)}\right)^{1/2}. \tag{2.62}$$

This completes the calculation of the normalization and confirms expression (2.39).

2.7.1 Direct calculation of the gaussian partition function

It is useful, mainly in view of semi-classical calculations (solitons, instantons), to introduce a more direct method for calculating the gaussian partition function. However, since we expect problems due to infinite normalizations, we perform the calculation first with discrete time intervals. We set here $\hbar = m = 1$. Then,

$$\mathcal{Z}_0(\beta,\varepsilon) = (2\pi\varepsilon)^{-n/2} \int dq_0 dq_n \delta(q_n - q_0) \int \prod_{k=1}^{n-1} dq_k \, e^{-\mathcal{S}_0(q,\varepsilon)}, \tag{2.63}$$

where a possible choice for the discretized action is

$$\mathcal{S}_0(q,\varepsilon) = \sum_{k=1}^{n} \left[\frac{(q_k - q_{k-1})^2}{2\varepsilon} + \frac{1}{2}\varepsilon\omega^2 q_k^2 \right].$$

The form (2.18) would have given, instead,

$$\mathcal{S}_0(q,\varepsilon) = \sum_{k=1}^{n} \left[\frac{(q_k - q_{k-1})^2}{2\varepsilon} + \frac{1}{6}\varepsilon\omega^2 (q_{k-1}^2 + q_k^2 + q_{k-1}q_k) \right], \tag{2.64}$$

which is equivalent in the limit $\varepsilon \to 0$ and $|q_k - q_{k-1}| = O(\sqrt{\varepsilon})$.

The function $\mathcal{S}_0$ is a quadratic form in the variables q_k, which can easily be diagonalized. Indeed, periodic boundary conditions imply translation invariance. We thus introduce a discrete Fourier series expansion, setting

$$q_k = \frac{1}{\sqrt{n}} \sum_{\ell=0}^{n-1} e^{2i\pi k\ell/n} c_\ell \tag{2.65}$$

with the reality conditions

$$c_0 = \bar{c}_0, \quad \bar{c}_{n-\ell} = c_\ell. \tag{2.66}$$

Then,

$$\mathcal{S}_0(q,\varepsilon) = \sum_{\ell=0}^{n-1} \bar{c}_\ell \left[(1 - \cos(2\pi\ell/n))/\varepsilon + \tfrac{1}{2}\omega^2\varepsilon \right] c_\ell, \tag{2.67}$$

where the orthogonality relations

$$\frac{1}{n} \sum_{k=0}^{n-1} e^{2i\pi k\ell/n} = \begin{cases} 1 & \text{for } \ell = 0 \pmod{n}, \\ 0 & \text{otherwise}, \end{cases}$$

have been used. These relations also show that the transformation is unitary and the jacobian of the change of variables is a pure phase.

The integral now has the form (6.5), but relation (2.66) implies that only half the complex variables are independent. One finds

$$\mathcal{Z}_0(\beta,\varepsilon) = (2\varepsilon)^{-n/2}\left[\prod_{\ell=0}^{n-1}\left(1 - \cos(2\pi\ell/n)\right)/\varepsilon + \tfrac{1}{2}\omega^2\varepsilon\right]^{-1/2}. \qquad (2.68)$$

The product can be evaluated explicitly. Setting

$$\cosh\theta = 1 + \omega^2\varepsilon^2/2,$$

one uses the identity

$$\prod_{\ell=0}^{n-1} 2\left[1 - \cos(2\pi\ell/n) + \tfrac{1}{2}\omega^2\varepsilon^2\right] = \prod_{\ell=0}^{n-1} 2\left[\cosh\theta - \cos(2\pi\ell/n)\right]$$
$$= 2\left(\cosh n\theta - 1\right),$$

which can be verified by comparing the roots and asymptotic behaviour of both polynomials in the variable $\cosh\theta$.

In the limit $\varepsilon \to 0$ with $n\varepsilon = \beta$ fixed, $n\theta \to \beta\omega$ and, thus,

$$\mathcal{Z}_0(\beta) = \frac{\mathrm{e}^{-\beta\omega/2}}{1 - \mathrm{e}^{-\beta\omega}},$$

which is the partition function of the harmonic oscillator.

2.7.2 Calculation with continuous time

The calculation with discrete time intervals suggests how to proceed directly in the continuum. Actually, we follow the method of Section 1.2, and calculate the determinant of the differential operator $\mathbf{A} = -\mathrm{d}_t^2 + \omega^2$ with appropriate boundary conditions. In the continuum limit, the change of variables (2.65) becomes an expansion of the path on an orthonormal basis of *square integrable* functions, periodic on $(-\beta/2,\beta/2)$:

$$q(t) = \frac{1}{\sqrt{\beta}}\sum_\ell c_\ell\, \mathrm{e}^{2i\pi\ell t/\beta}, \quad c_{-\ell} = \bar{c}_\ell.$$

The jacobian is then trivial:

$$[\mathrm{d}q(t)] \mapsto \mathrm{d}c_0 \prod_{\ell>0} \mathrm{d}c_\ell \mathrm{d}\bar{c}_\ell. \qquad (2.69)$$

The action becomes

$$\mathcal{S}_0 = \tfrac{1}{2}\omega^2 c_0^2 + \sum_{\ell\geq 1}\bar{c}_\ell(\omega^2 + 4\pi^2\ell^2/\beta^2)c_\ell.$$

The integration is immediate and yields

$$Z_0(\beta) \propto \frac{1}{\omega} \prod_{\ell \geq 1} [(\omega^2 + 4\pi^2 \ell^2/\beta^2)]^{-1}. \tag{2.70}$$

Then, a problem arises: the infinite product on the r.h.s. of equation (2.70) is divergent. This divergence is related to the infinite factor contained in the integration measure: as we have already several times emphasized, it is necessary to divide a path integral by a reference integral in order to obtain a finite result. We cannot choose here the integral corresponding to free motion because the partition function does not exist in this case. Nevertheless, we can compare path integrals corresponding to different values of ω or calculate a derivative:

$$\frac{\partial}{\partial \omega} \ln Z_0(\beta) = -\frac{1}{\omega} - \sum_{\ell > 0} \frac{2\omega}{\omega^2 + 4\pi^2 \ell^2/\beta^2} = -\frac{\beta}{2\tanh(\omega\beta/2)},$$

which is the result (2.60).

Other method. To show that this form of the calculation in the continuum is robust, we repeat the calculation using a slightly different method, for $\omega = 1$. We first fix the value of the path at $t = \pm\beta/2$. Calling this value λ, we impose the boundary conditions

$$q(-\beta/2) = q(\beta/2) = \lambda. \tag{2.71}$$

The solution q_c of equation

$$\mathbf{A}f \equiv -\ddot{f}(t) + f(t) = 0$$

with the conditions (2.71) is

$$q_c(t) = \lambda \frac{\cosh t}{\cosh(\beta/2)}.$$

Then, changing variables,

$$q(t) \mapsto r(t) = q(t) - q_c(t) \Rightarrow r(-\beta/2) = r(\beta/2) = 0,$$

one obtains

$$\mathcal{S}_0(q) = \tanh(\beta/2)\lambda^2 + \mathcal{S}_0(r).$$

One thus needs the eigenvalues of the differential operator $\mathbf{A}$ when acting on functions vanishing at $\pm\beta/2$. The eigenfunctions f are linear combinations of two exponentials since

$$\mathbf{A}f \equiv -\ddot{f}(t) + f(t) = af(t).$$

The solutions that satisfy the conditions $f(\beta/2) = f(-\beta/2) = 0$ are

$$f_\ell(t) = \sqrt{2/\beta} \sin(\ell\pi(t - \beta/2)/\beta) \Rightarrow a_\ell = 1 + \ell^2\pi^2/\beta^2,$$

for all integers $\ell \geq 1$.

Up to a normalization, the path integral is the product of the integral over λ by a gaussian integrals that involve the product of the eigenvalues of $\mathbf{A}$:

$$\mathcal{Z}_0(\beta) \propto 1/\sqrt{\tanh(\beta/2) \det \mathbf{A}}.$$

From the discrete calculation, we know that an infinite normalization must be factorized and that the non-trivial contribution is

$$\prod_{\ell \geq 1}(1 + \beta^2/\pi^2\ell^2) = \sinh\beta/\beta.$$

Then,

$$\sinh(\beta)\tanh(\beta/2) = 2\sinh^2(\beta/2).$$

The result is thus proportional to $1/\sinh(\beta/2)$, which is the partition function of the harmonic oscillator.

2.8 Perturbed harmonic oscillator

We now consider a potential that is the sum of a harmonic potential and a perturbation. The hamiltonian can be written as

$$H = \frac{1}{2m}p^2 + \frac{m}{2}\omega^2 q^2 + V_{\mathrm{I}}(q), \qquad (2.72)$$

where we assume that the perturbation

$$V_{\mathrm{I}}(q) = \sum_{n=1} v_n q^n$$

is a polynomial in the variable q, even though some results generalize to all functions expandable in powers of q.

The corresponding partition function is given by (in this section we set $\hbar = 1$)

$$\mathcal{Z}(\beta) = \int [\mathrm{d}q] \exp\left\{-\int_{-\beta/2}^{\beta/2} \left[\tfrac{1}{2}m\dot{q}^2(t) + \tfrac{1}{2}m\omega^2 q^2(t) + V_{\mathrm{I}}(q(t))\right] \mathrm{d}t\right\} \qquad (2.73)$$

with $q(-\beta/2) = q(\beta/2)$.

The integrand (2.73) can be expanded in powers of $V_{\mathrm{I}}(q)$ and this leads to

$$\frac{\mathcal{Z}(\beta)}{\mathcal{Z}_0(\beta)} = \sum_{k=0} \frac{(-1)^k}{k!} \left\langle \left[\int \mathrm{d}t\, V_{\mathrm{I}}(q(t))\right]^k \right\rangle_0$$

$$= \sum_{k=0} \frac{(-1)^k}{k!} \int \mathrm{d}t_1 \mathrm{d}t_2 \ldots \mathrm{d}t_k \left\langle V_{\mathrm{I}}(q(t_1)) \ldots V_{\mathrm{I}}(q(t_k)) \right\rangle_0,$$

where $\langle \bullet \rangle_0$ means expectation value with respect to the gaussian measure $e^{-S_0/\hbar}/\mathcal{Z}_0$ (equation (2.55)) with periodic boundary conditions. The arguments given in Section 1.2 immediately apply here also. If $V_I(q)$ is a polynomial, successive terms in the expansion can be calculated using Wick's theorem (1.18) in the form (2.58). This is the basis of perturbation theory.

Formal expression. A generalization of identities (1.22, 1.24) leads to a formal representation of the perturbative expansion. Identity (2.56) with

$$\mathcal{F}(q) = \exp\left[-\int_{-\beta/2}^{\beta/2} V_I(q(t))\,dt\right],$$

leads to the representation

$$\mathcal{Z}(\beta) = \left\{\exp\left[-\int_{-\beta/2}^{\beta/2} dt\, V_I\left(\frac{\delta}{\delta b(t)}\right)\right]\mathcal{Z}_G(b,\beta)\right\}\bigg|_{b=0}, \tag{2.74}$$

where the functional $\mathcal{Z}_G(b,\beta)$, which corresponds to the action (2.48), has been calculated in Section 2.6. In terms of the explicit expression (2.54), one finds

$$\frac{\mathcal{Z}(\beta)}{\mathcal{Z}_0(\beta)} = \exp\left[-\int_{-\beta/2}^{\beta/2} dt\, V_I\left(\frac{\delta}{\delta b(t)}\right)\right]$$
$$\times \exp\left[\frac{1}{2m}\int_{-\beta/2}^{\beta/2} du\,dv\, b(u)\Delta(u-v)b(v)\right]\bigg|_{b=0}. \tag{2.75}$$

Perturbation theory and the minimum of the potential. To each decomposition of the potential into a sum of a quadratic term and a remainder $V_I(q)$ is associated a perturbative expansion. However, the integrand is the largest in the vicinity of the paths that minimize the action. Clearly, periodic functions that minimize the action are constant functions $q(t) \equiv q_0$, to minimize the kinetic term, where the value q_0 must minimize the potential $V(q)$ and thus

$$V'(q_0) = 0, \quad V''(q_0) > 0.$$

Therefore, the optimal decomposition is

$$V(q) = V(q_0) + \tfrac{1}{2}V''(q_0)(q-q_0)^2 + V_I(q).$$

Specific problems associated with a degeneracy of the minimum of the potential will be discussed in Chapter 8.

2.9 Perturbative expansion in powers of $\hbar$

In this section, we reinstate Planck's constant $\hbar$. We consider a general hamiltonian of the form
$$H = \frac{1}{2m}p^2 + V(q), \tag{2.76}$$
where the potential $V(q)$ is a polynomial with a unique minimum at $q = 0$ where
$$V(q) = \tfrac{1}{2}m\omega^2 q^2 + V_{\rm I}(q), \quad V_{\rm I}(q) = O(q^3).$$

We have discussed, in Section 2.8, perturbative calculations of the partition function in the form of expansions in powers of some interaction potential $V_{\rm I}(q)$. We have pointed out that an optimal harmonic approximation is provided by a quadratic expansion of the potential around its minimum.

A perturbative expansion around this harmonic approximation can be reorganized in a simple way as an expansion in powers of $\hbar$. The latter expansion differs from the simple perturbative expansion as soon as the perturbation $V_{\rm I}$ is no longer a monomial.

Any expansion in powers of $\hbar$ can be called a semi-classical expansion. But, in fact, it is the expansion that we discuss later in Section 2.10 that corresponds better to the idea of a semi-classical approach.

We consider the partition function $\mathcal{Z}(\beta)$ and set $\beta = \tau/\hbar$. Then,
$$\mathcal{Z}(\tau/\hbar) = \int_{q(-\tau/2)=q(\tau/2)} [\mathrm{d}q(t)]\exp\left[-\mathcal{S}(q)/\hbar\right], \tag{2.77}$$
where $\mathcal{S}(q)$ is the euclidean action:
$$\mathcal{S}(q) = \int_{-\tau/2}^{\tau/2}\left[\tfrac{1}{2}m\dot{q}^2(t) + V(q(t))\right]\mathrm{d}t. \tag{2.78}$$

We now evaluate the path integral in the formal limit $\hbar \to 0$, with τ fixed. From a semi-classical viewpoint, this limit corresponds indeed to $\hbar \to 0$, but simultaneously $\beta = \tau/\hbar \to \infty$, that is, in a correlated way the temperature goes to zero, a limit that enhances small oscillations around the classical minimum and excitations close to the quantum ground state.

In this limit, the path integral has exactly a form that justifies using the steepest descent method (see Section 1.5). Since $q = 0$ corresponds to an absolute and non-degenerate minimum of the potential V, the path integral is dominated for $\hbar \to 0$ by the trivial saddle point $q(t) \equiv 0$, which minimizes both the potential and the kinetic term. The change of variables $q(t) \mapsto \sqrt{\hbar}q(t)$ in the path integral, allows a direct identification of the leading terms in the action. After the change,
$$\mathcal{S}(q)/\hbar = \frac{m}{2}\int_{-\tau/2}^{\tau/2}\left[\dot{q}^2(t) + \omega^2 q^2(t)\right] + \frac{1}{\hbar}\int_{-\tau/2}^{\tau/2}\mathrm{d}t\, V_{\rm I}\bigl(q(t)\sqrt{\hbar}\bigr).$$

Since $V_I(q\sqrt{\hbar})/\hbar$ is at least of order $\sqrt{\hbar}$, at leading order one recovers the partition function of a harmonic oscillator. At higher orders, the integral can be calculated by first expanding the integrand in powers of V_I and then collecting terms with the same powers of $\hbar$.

For any potential expandable in powers of q and at any finite order in $\hbar$, the integrand remains the product of a gaussian factor by a polynomial in q since the terms of degree q^n in the potential start contributing only at order $\hbar^{n/2-1}$. Therefore, for any expandable potential, the expansion of $\mathcal{Z}(\tau/\hbar)$ in powers of $\hbar$ relies on gaussian expectation values of polynomials, which can be calculated with the help of Wick's theorem.

2.10 Semi-classical expansion

We now consider another $\hbar \to 0$ limit, which, at leading order, corresponds better to the idea of a semi-classical approximation. We take the limit $\hbar \to 0$ at β or temperature fixed. One then expects the quantum partition function to converge toward the classical partition function. This is what we now verify, by calculating the leading term and the first correction in the semi-classical expansion of the partition function (2.23). In particular, the calculation will help us identify the expansion parameter, since $\hbar$ has a dimension and, thus, must be divided by another quantity that has the dimension of an action.

The partition function is given by

$$\mathcal{Z}(\beta) = \int [\mathrm{d}q(t)] \exp\left[-\mathcal{S}(q)/\hbar\right], \tag{2.79}$$

where the paths satisfy periodic boundary conditions: $q(0) = q(\beta)$ and (equation (2.24))

$$\mathcal{S}(q)/\hbar = \int_0^\beta \mathrm{d}t \left[\tfrac{1}{2}m\dot{q}^2(t)/\hbar^2 + V(q(t))\right]. \tag{2.80}$$

For $\hbar \to 0$, the leading term in the action is now the kinetic term. The leading trajectories are those that satisfy $\dot{q} = 0$, and thus correspond to all constant functions. From the viewpoint of the steepest descent method, one finds a one-parameter family of degenerate saddle points. In Sections 8.2 and 8.3.1, we discuss this problem and argue that a collective coordinate that parametrizes all saddle points has then to be introduced. In the present example, the implementation of this idea is simple: we first evaluate the diagonal matrix element

$$\langle q_0|\mathrm{e}^{-\beta H}|q_0\rangle = \int_{q(0)=q(\beta)=q_0} [\mathrm{d}q(t)] \exp\left[-\mathcal{S}(q)/\hbar\right], \tag{2.81}$$

to which only one saddle point, $q(t) \equiv q_0$, contributes. After the translation $q(t) \mapsto q(t) + q_0$, the path integral becomes

$$\langle q_0|\mathrm{e}^{-\beta H}|q_0\rangle = \int_{q(0)=q(\beta)=0} [\mathrm{d}q(t)] \exp\left[-\Sigma(q)\right] \tag{2.82}$$

with
$$\Sigma(q) = \int_0^\beta dt \left[\tfrac{1}{2} m \dot q^2(t)/\hbar^2 + V(q_0 + q(t))\right]. \tag{2.83}$$

The form of the action shows that $\dot q$ is formally of order $\hbar$. With the boundary conditions in (2.82), the saddle point is $q(t) \equiv 0$ and thus $q(t)$ itself is of order $\hbar$. The potential can then be expanded in powers of $q(t)$:

$$V(q_0 + q(t)) = V(q_0) + V'(q_0) q(t) + \tfrac{1}{2} V''(q_0) q^2(t) + O(\hbar^3),$$

as well as the integrand in the integral (2.82). Each term has the form of a gaussian expectation value:

$$\langle q_0 | \mathrm{e}^{-\beta H} | q_0 \rangle = \mathcal{N}(\beta) \, \mathrm{e}^{-\beta V(q_0)} \left[1 - V'(q_0) \int_0^\beta dt \, \langle q(t) \rangle_0 \right.$$
$$\left. + \tfrac{1}{2}(V'(q_0))^2 \int_0^\beta dt\, du \, \langle q(t) q(u) \rangle_0 - \tfrac{1}{2} V''(q_0) \int_0^\beta dt \, \langle q^2(t) \rangle_0 + O(\hbar^3) \right],$$

where $\langle \bullet \rangle_0$ refers to the expectation value with respect to the measure corresponding to the free action $\mathcal{S}_0(q) = \tfrac{1}{2} m \int dt \, \dot q^2(t)/\hbar$.

Gaussian expectation values involve the two-point function $\Delta(t, u)$ corresponding to the free action with the boundary conditions $q(0) = q(\beta) = 0$:

$$\langle q_0 | \mathrm{e}^{-\beta H} | q_0 \rangle = \mathcal{N}(\beta) \, \mathrm{e}^{-\beta V(q_0)} \left[1 + \frac{\hbar^2}{2m}(V'(q_0))^2 \int_0^\beta \Delta(t, u) dt\, du \right.$$
$$\left. - \frac{\hbar^2}{2m} V''(q_0) \int_0^\beta \Delta(t, t) dt + O(\hbar^3) \right].$$

The normalization is given by

$$\mathcal{N}(\beta) = \langle q = 0 | \mathrm{e}^{-\beta p^2/2m} | q = 0 \rangle = \int \frac{dp}{2\pi \hbar} \mathrm{e}^{-\beta p^2/2m}$$
$$= \frac{1}{\hbar} \sqrt{\frac{m}{2\pi\beta}}, \tag{2.84}$$

that is, the free expression (2.9) in which one replaces $t - t'$ by $\hbar\beta$ (and here $d = 1$).

The two-point function. One still has to determine the two-point $\Delta(t, u)$. It can be obtained by calculating the path integral

$$\mathcal{Z}_{\mathrm{G}}(b, \beta) = \int [dq(t)] \exp\left[-\frac{m}{2\hbar^2} \int_0^\beta \dot q^2(t) dt + \int_0^\beta b(t) q(t) dt\right] \tag{2.85}$$

with $q(0) = q(\beta) = 0$. A calculation quite analogous to the one presented in Section 2.6 yields

$$\mathcal{Z}_G(b, \beta)/\mathcal{Z}_G(0, \beta) = \exp\left[\frac{\hbar^2}{2m}\int_0^\beta dt\, du\, b(t)\Delta(t,u)b(u)\right], \quad (2.86)$$

where $\Delta(t, u) = \Delta(u, t)$ is the solution of the equation

$$-\ddot{\Delta}(t, u) = \delta(t - u) \text{ with } \Delta(0, u) = \Delta(\beta, u) = 0.$$

One finds

$$\Delta(t, u) = -\tfrac{1}{2}|t - u| + \tfrac{1}{2}(t + u - 2ut/\beta). \quad (2.87)$$

Partition function. Using the two-point function, one can evaluate the expectation values explicitly:

$$\langle q_0| e^{-\beta H} |q_0\rangle = \mathcal{N}(\beta) e^{-\beta V(q_0)}\left[1 + \frac{\hbar^2 \beta^3}{24m}(V'(q_0))^2 - \frac{\hbar^2 \beta^2}{12m}V''(q_0) + O(\hbar^3)\right]. \quad (2.88)$$

This result can be verified by expanding the function (2.39) of the harmonic oscillator with $V(q) = \tfrac{1}{2}m\omega^2 q^2$ for $q' = q'' = q, \tau = \hbar\beta$:

$$\langle q_0| e^{-\beta H} |q_0\rangle = \frac{1}{\hbar}\sqrt{\frac{m}{2\pi\beta}}\, e^{-\beta m\omega^2 q^2/2}\left[1 - \tfrac{1}{12}\omega^2\beta^2\hbar^2 + \tfrac{1}{24}m\omega^4\beta^3\hbar^2 q^2 + O(\hbar^4)\right].$$

Finally, the partition function takes the form of a simple integral corresponding to the integration over q_0:

$$\mathcal{Z}(\beta) = \int dq\, \langle q| e^{-\beta H} |q\rangle$$

$$= \frac{1}{\hbar}\sqrt{\frac{m}{2\pi\beta}}\int dq\, e^{-\beta V(q)}\left[1 + \frac{\hbar^2\beta^3}{24m}(V'(q))^2 - \frac{\hbar^2\beta^2}{12m}V''(q) + O(\hbar^3)\right]$$

$$= \frac{1}{\hbar}\sqrt{\frac{m}{2\pi\beta}}\int dq\, e^{-\beta V(q)}\left[1 - \frac{\hbar^2\beta^2}{24m}V''(q) + O(\hbar^3)\right]$$

$$= \frac{1}{\hbar}\sqrt{\frac{m}{2\pi\beta}}\int dq\, \exp\left[-\beta V(q) - \beta^2\hbar^2 V''(q)/24m + O(\hbar^3)\right], \quad (2.89)$$

where the second expression is obtained after an integration by parts of the term proportional to V'^2.

Alternative calculation. An alternative calculation is based on expanding $q(t)$ as a Fourier series on an orthonormal basis of periodic functions on the interval $[0, \beta]$:

$$q(t) = q_0 + \delta q(t), \quad \delta q(t) = \sqrt{2/\beta}\sum_{n>0}\left[a_n \cos(2\pi nt/\beta) + b_n \sin(2\pi nt/\beta)\right].$$

We have separated the mode q_0 because it gives no contribution to the time derivative and, thus, is not constrained by the limit $\hbar \to 0$. In contrast, the coefficients a_n and b_n, and thus $\delta q(t)$, are of order $\hbar$. It is thus justified to expand the potential in powers of $\delta q(t)$:

$$V(q(t)) = V(q_0) + \delta q(t) V'(q_0) + \tfrac{1}{2}(\delta q(t))^2 V''(q_0) + O\left((\delta q)^3\right).$$

We then integrate over t and use the orthogonality of the basis. As a consequence, the term of order δq vanishes. The action then becomes

$$\mathcal{S}(q)/\hbar = \beta V(q_0) + \sum_{n>0}(a_n^2 + b_n^2)\left[\frac{2mn^2\pi^2}{\beta^2\hbar^2} + \frac{1}{2}V''(q_0)\right] + O(a_n^3, a_n^2 b_n, \ldots).$$

As we have explained, using the example of the gaussian integral in Section 2.7 (equation (2.69)), one can then replace the integration over paths $q(t)$ by an integration over the coefficients of an expansion of $q(t)$ on an orthonormal basis, here q_0 (this mode is not normalized, but the jacobian is a constant), and on $\{a_n, b_n\}$. The gaussian integration over the coefficients a_n and b_n is straightforward. We normalize all integrals by dividing them by the integral for $V''(q) \equiv 0$. Then,

$$\int \mathrm{d}a_n \exp\left[-\left(\frac{2mn^2\pi^2}{\beta^2\hbar^2} + \frac{1}{2}V''(q_0)\right)a_n^2\right] \propto \left[1 + \frac{\beta^2\hbar^2}{4m\pi^2 n^2}V''(q_0)\right]^{-1/2}$$

$$\sim \exp\left[-\frac{\beta^2\hbar^2}{8m\pi^2 n^2}V''(q_0)\right].$$

Summing over n (and using $\sum_n 1/n^2 = \pi^2/6$), one finds the partition function, up to a normalization. The normalization $\mathcal{N}(\beta)$, which is independent of the potential, can be determined by noting that before integration over q_0 one has obtained a contribution to a diagonal element of the density matrix of the form $\langle q|\,\mathrm{e}^{-\beta H}\,|q\rangle$. For $V \equiv 0$, it is given by expression (2.84). One then recovers the result (2.89).

Discussion.

(i) The leading contribution for $\hbar \to 0$ is the classical partition function corresponding to the Boltzmann weight obtained by integrating over the momentum p the Boltzmann weight in phase space: $\mathrm{e}^{-\beta H(p,q)}$, where H is the classical hamiltonian. Indeed, for a hamiltonian $H = p^2/2m + V(q)$, the two forms of $\mathcal{N}(\beta)$ in equations (2.84), lead to the identity

$$\mathcal{Z}_{\mathrm{cl.}}(\beta) = \int \frac{\mathrm{d}p\,\mathrm{d}q}{2\pi\hbar}\,\mathrm{e}^{-\beta H(p,q)} = \frac{1}{\hbar}\sqrt{\frac{m}{2\pi\beta}}\int \mathrm{d}q\,\mathrm{e}^{-\beta V(q)}. \qquad (2.90)$$

(ii) Introducing the thermal wavelength

$$\lambda_{\mathrm{th.}} = \hbar\sqrt{\beta/m},$$

and a length scale typical of the space variations of the potential (which we assume to be characterized by only one length scale)

$$l_{\text{pot.}} = \sqrt{|\langle V(q) \rangle / \langle V''(q) \rangle|},$$

we see that the ratio between the classical term and the first quantum correction can be characterized by the ratio $\lambda_{\text{th.}}/l_{\text{pot.}}$. At high temperature $T = 1/\beta$, and thus small β, the thermal wavelength is small and the statistical behaviour classical. On the contrary, at low temperature quantum effects become dominant.

In this analysis, we have implicitly assumed that the potential is regular enough, at least twice differentiable. Idealized potentials, like the square-well, which one uses frequently in quantum mechanics, require a special analysis.

(iii) We note that formally the classical limit corresponds to a kind of dimensional reduction. The quantum partition function involves an integration over paths, that is, one-dimensional objects. The classical function, in contrast, involves only an integration over the zero mode (in the sense of the Fourier series), which corresponds to a point, and has zero dimension.

Exercises

Exercise 2.1

Partition function. Reproduce the calculation of Section 2.7 with the discrete form (2.64) of the action.

Exercise 2.2

Locality. One considers the operator $U(t)$ defined by its matrix elements:

$$\langle q| U(t) |q' \rangle = \frac{t/\pi}{t^2 + (q - q')^2}.$$

Verify the semi-group property $U(t_1)U(t_2) = U(t_1 + t_2)$.

Show that the corresponding hamiltonian (as defined by equation (2.2) with $\hbar = 1$) is non-local, that is has a support for $q \neq q'$.

Solution. The matrix elements of the hamiltonian are

$$\langle q| H |q' \rangle = -\frac{1}{\pi(q - q')^2}, \quad \text{for } q \neq q'.$$

Exercise 2.3

The square-well potential. Use the expression obtained for the statistical operator $e^{-\beta H}$ of the harmonic oscillator to derive the spectrum of the attractive square-well potential $H = p^2/2 + \mathcal{V}(x)$ with

$$\mathcal{V}(x) = 0, \text{ for } |x| > a/2, \quad \mathcal{V}(x) = V < 0, \text{ for } |x| < a/2,$$

(exercise requiring some ingenuity).

Solution. The basic idea is to calculate the Fredholm determinant $D(V, E)$ of the operator $H - E$, whose zeros determine the spectrum. The determinant can be expressed in terms of a gaussian path integral:

$$D^{-1/2}(V, E) \propto \int [dq(x)] \exp[-\mathcal{S}(q)],$$

$$\mathcal{S}(q) = \int dx \left[\tfrac{1}{2}(q'(x))^2 + (V(x) - E)q^2(x) \right].$$

This gaussian integral can be calculated explicitly in various ways. One method is as follows. First, it is convenient to consider the integral as a limit for $\beta \to \infty$ of the partition function in euclidean time β (that is at temperature $1/\beta$), which is obtained by imposing periodic boundary conditions $q(-\beta/2) = q(\beta/2)$:

$$D^{-1/2}(V, E) \propto \lim_{\beta \to \infty} \operatorname{tr} U(\beta)$$

with

$$\langle q'' | U(\beta) | q' \rangle = \int_{q(-\beta/2)=q'}^{q(\beta/2)=q''} [dq(x)] \exp\left[-\mathcal{S}(q)\right].$$

We then split the interval $[-\beta/2, \beta/2]$ into three sub-intervals in which the potential is constant: $[-\beta/2, -a/2]$, $[-a/2, a/2]$, $[a/2, \beta/2]$. We write $U(\beta)$ as a product of statistical operators corresponding to these various intervals. In each interval, the path integral corresponds to the statistical operator of a harmonic oscillator. For what follows, we introduce the notation

$$-E = \tfrac{1}{2}\omega_1^2, \quad V - E = \tfrac{1}{2}\omega_2^2.$$

We denote by U_1 and U_2 the operators corresponding to ω_1 and ω_2, respectively. Then,

$$\operatorname{tr} U(\beta) = \operatorname{tr} U_1(\beta/2 - a/2) U_2(a) U_1(\beta/2 - a/2) = \operatorname{tr} U_1(\beta - a) U_2(a),$$

where the cyclic property of the trace has been used. In the limit $\beta \to \infty$, the operator U_1 becomes the projector on to the ground state of the corresponding oscillator:

$$\operatorname{tr} U(\beta) \underset{\beta \to \infty}{\sim} \sqrt{\frac{\omega_1}{\pi}} e^{-\omega_1(\beta-a)/2} \int dq' dq'' \, e^{-\omega_1 q'^2/2} e^{-\omega_1 q''^2/2} \langle q'' | U_2(a) | q' \rangle.$$

One then uses the explicit result (2.35) and performs the two gaussian integrations over q' and q''. One finds

$$\operatorname{tr} U(\beta) \underset{\beta \to \infty}{\sim} \left(\frac{2\omega_1 \omega_2}{\sinh(a\omega_2)} \right)^{1/2} \left(\omega_1^2 + \omega_2^2 + 2\omega_1 \omega_2 \coth(a\omega_2) \right)^{-1/2} e^{-\omega_1(\beta-a)/2}.$$

To obtain a result that has a finite limit, it is necessary to normalize the determinant. Dividing it by its value for vanishing potential, one finds

$$\mathrm{e}^{a\omega_1} D(V,E)/D(0,E) = \cosh(a\omega_2) + \frac{\omega_1^2 + \omega_2^2}{2\omega_1\omega_2} \sinh(a\omega_2).$$

We have calculated for $V - E > 0$. The determinant can only vanish for $E > V$. We thus set $\omega_2 = i\kappa_2$. The equation that gives the energies of the bound states can then be written as

$$\tan(a\kappa_2) = \frac{2\omega_1\kappa_2}{\omega_1^2 - \kappa_2^2},$$

an equation that can easily be verified by solving the Schrödinger equation for the square-well directly and by combining the two spectral equations that correspond to even and odd eigenfunctions.

3 PARTITION FUNCTION AND SPECTRUM

A direct application of the calculation of the partition function is provided by the determination of the spectrum of a hamiltonian (which we assume to be discrete, for simplicity). Indeed,

$$\mathcal{Z}(\beta) = \operatorname{tr} e^{-\beta H} = \sum_{k=0}^{\infty} e^{-\beta E_k}.$$

In particular,

$$E_0 = \lim_{\beta \to \infty} -\frac{1}{\beta} \ln \mathcal{Z}(\beta).$$

The different methods for evaluating a path integral proposed in Chapter 2, thus also yield methods for calculating energy eigenvalues.

Note, however, that we will consider only situations where the hamiltonian eigenvalues are not degenerate. Degenerate cases, which in general reflect symmetries of the hamiltonian, require a more detailed analysis.

We then illustrate the power of the path integral formalism, beyond simple perturbation theory:

We use it to derive a variational principle.

We evaluate it, applying the steepest descent method, in the case of $O(N)$ symmetric hamiltonians, in the large N limit.

Steepest descent calculations of path integrals involve determinants of differential operators: we give in this chapter a perturbative definition of such determinants.

Finally, we prove that the ground state of simple quantum systems is not degenerate.

3.1 Perturbative calculation

For any potential that can be written as a sum of a harmonic well and a polynomial perturbation, perturbation theory, as introduced in Section 2.8, leads to a perturbative expansion of the ground state energy and, more generally, of all energy levels with fixed quantum number. Indeed, let us consider a potential of the form (2.72) (here we set $m = 1$)

$$V(q) = \tfrac{1}{2}\omega^2 q^2 + V_{\mathrm{I}}(q).$$

For small V_{I}, the energy eigenvalues E_k are close to those of the harmonic oscillator:

$$E_k = (k + \tfrac{1}{2})\omega\hbar + \delta E_k.$$

The expansion of the partition function thus takes the form

$$\mathcal{Z}(\beta) = \sum_{k=0}^{\infty} e^{-(k+1/2)\beta\omega\hbar} \left[1 - \beta\delta E_k + O\left(\beta^2 (\delta E_k)^2\right)\right]. \tag{3.1}$$

The successive perturbative corrections to energy eigenvalues can then be obtained by the following method: one expands the partition function in powers of the perturbation V_{I}. One then expands each term for large β, and one extracts the coefficient of $-\beta e^{-(k+1/2)\beta\omega\hbar}$.

Example. We consider the potential

$$V(q) = \frac{1}{2}q^2 + \frac{\lambda}{4!}q^4,$$

which has the same perturbative structure as the example discussed in Section 1.3.

The partition function at order λ is given by (in this calculation we set $\hbar = 1$)

$$\begin{aligned}
\mathcal{Z}(\beta) &= \mathcal{Z}_0(\beta) \left[1 - (\lambda/24) \int_0^\beta dt \, \langle q^4(t) \rangle_0 + O(\lambda^2) \right] \\
&= \mathcal{Z}_0(\beta) \left[1 - (\lambda/24) \times 3 \times \beta \Delta^2(0) \right] + O(\lambda^2) \\
&= \frac{1}{2\sinh(\beta/2)} \left[1 - \beta(\lambda/32) \left(\coth^2(\beta/2) \right) \right] + O(\lambda^2),
\end{aligned} \tag{3.2}$$

where the propagator (2.52) has been used.

One verifies the two algebraic identities:

$$\sum_{k=0}^{\infty} \exp\left[-\left(k + \tfrac{1}{2}\right)\beta\right] = 1/\left[2\sinh(\beta/2)\right]$$

and, differentiating twice with respect to β,

$$\sum_{k=0}^{\infty} \left(k + \tfrac{1}{2}\right)^2 \exp\left[-\left(k + \tfrac{1}{2}\right)\beta\right] = \frac{2\coth^2(\beta/2) - 1}{8\sinh(\beta/2)}.$$

Setting

$$E_k = \left(k + \tfrac{1}{2}\right) + \lambda \delta E_k + O(\lambda^2),$$

one obtains for the partition function

$$\begin{aligned}
\mathcal{Z}(\beta) &= \sum_{k=0}^{\infty} \exp\left[-\left(k + \tfrac{1}{2} + \lambda \delta E_k\right)\beta\right] \\
&= \frac{1}{2\sinh(\beta/2)} - \beta\lambda \sum_{k=0}^{\infty} \delta E_k \, e^{-(k+1/2)\beta}.
\end{aligned}$$

Identifying the coefficient of λ, one infers
$$E_k = k + \tfrac{1}{2} + \tfrac{1}{16}\lambda\left(k^2 + k + \tfrac{1}{2}\right) + O\left(\lambda^2\right).$$

Let us point out that if one is interested only in the ground state energy, one can neglect all terms that decrease exponentially for $\beta \to \infty$, and the expressions drastically simplify:
$$\mathcal{Z}(\beta) \sim e^{-\beta E_0} = e^{-\beta/2}\left(1 - \tfrac{1}{32}\beta\lambda\right) + O(\lambda^2).$$

At next order, one finds (see equation (1.26))
$$\mathcal{Z}(\beta)/\mathcal{Z}_0(\beta) = 1 - \tfrac{1}{8}\lambda\beta\Delta^2(0) + \tfrac{1}{128}\lambda^2\beta^2\Delta^4(0) + \tfrac{1}{16}\beta\lambda^2\Delta^2(0)\int_{-\beta/2}^{\beta/2} dt\,\Delta^2(t)$$
$$+ \tfrac{1}{48}\lambda^2\beta\int_{-\beta/2}^{\beta/2} dt\,\Delta^4(t) + O\left(\lambda^3\right), \tag{3.3}$$

where the periodicity of $\Delta(t)$ has been used. Calculating the logarithm of the partition function, which is proportional to the free energy, one verifies that the non-connected contribution cancels (Section 1.3):

$$-\frac{1}{\beta}\ln\mathcal{Z}(\beta) + \frac{1}{\beta}\ln\mathcal{Z}_0(\beta) = \tfrac{1}{8}\lambda\Delta^2(0) - \tfrac{1}{16}\lambda^2\Delta^2(0)\int_{-\beta/2}^{\beta/2} dt\,\Delta^2(t)$$
$$- \tfrac{1}{48}\lambda^2\int_{-\beta/2}^{\beta/2} dt\,\Delta^4(t) + O\left(\lambda^3\right), \tag{3.4}$$

For $\beta \to \infty$, one can replace the function Δ by its asymptotic form (2.53) because the corrections are exponentially small. Moreover, one can integrate over $t \in (-\infty, +\infty)$ with, again, exponentially small errors because $\Delta(t)$ decreases exponentially for $|t| \to \infty$. The initial expression for $\mathcal{Z}(\beta)$ contains terms of orders β and β^2. The term proportional to β^2, which corresponds to a non-connected contribution, cancels in the logarithm. Therefore, the expression has a limit for $\beta \to \infty$, which yields the ground state energy:
$$E_0 = \tfrac{1}{2} + \tfrac{1}{32}\lambda - \tfrac{7}{1536}\lambda^2 + O(\lambda^3).$$

Let us, however, point out that for simple quantum examples such as a particle in a potential, the most efficient method to generate perturbative expansions of energy eigenvalues does not rely on path integrals. It is more efficient to start from the Schrödinger equation and to transform it into a Riccati equation for the logarithmic derivative of the wave function (see Section 3.2.4).

Expansion in powers of $\hbar$. We have pointed out in Section 2.9 that when the potential V_I is not a simple monomial, it is possible to define two types of expansions, in powers of V_I or in powers of $\hbar$. Let us examine here the form of the latter expansion for energy eigenvalues.

The starting point now is an expansion in powers of $\hbar$, based on the steepest descent method, of

$$\mathcal{Z}(\tau/\hbar) = \int_{q(-\tau/2)=q(\tau/2)} [dq(t)] \exp\left[-\mathcal{S}(q)/\hbar\right] \tag{3.5}$$

with

$$\mathcal{S}(q) = \int_{-\tau/2}^{\tau/2} \left[\tfrac{1}{2}m\dot{q}^2(t) + V(q(t))\right] dt. \tag{3.6}$$

We assume that the potential V has an absolute and non-degenerate minimum at $q = 0$ where it vanishes, in such a way that, for $\hbar \to 0$, the path integral is dominated by the saddle point $q(t) \equiv 0$. It can thus be calculated by expanding the integrand in powers of the terms of degree larger than 2 in the potential and integrating them. This leads to an expansion of $\mathcal{Z}(\tau/\hbar)$ in powers of $\hbar$ and then, for $\tau \to \infty$, to an expansion of the energy eigenvalues.

Since the $(N+1)$th eigenvalue E_N of H, satisfies

$$E_N = (N + \tfrac{1}{2})\omega\hbar + \mathcal{O}\left(\hbar^2\right),$$

the perturbative expansion takes the form

$$\mathcal{Z}(\tau/\hbar) = \sum_{N=0} e^{-\tau E_N/\hbar}$$
$$= \sum_{N=0} e^{-(N+1/2)\omega\tau} \sum_{k=0} \frac{1}{k!} (-\tau)^k \left(E_N/\hbar - (\tfrac{1}{2} + N)\omega\right)^k. \tag{3.7}$$

We conclude that the coefficient of $\hbar^k$ is a polynomial of degree k in τ, that E_N can be derived from the coefficient of $e^{-(N+1/2)\omega\tau}$.

Remark. Note the difference between the perturbative and the semi-classical or WKB expansions, which we describe in Section 3.2.1. The perturbative expansion applies to energy eigenvalues which tend towards the minimum of the potential when $\hbar \to 0$. In contrast, the WKB expansion (see Section 2.10) is valid for energy eigenvalues that remain at a finite distance of the minimum of the potential when $\hbar \to 0$, that is, have a finite classical energy. As a consequence, the order N of such eigenvalues diverges. The WKB approximation corresponds to large quantum numbers.

3.2 Semi-classical or WKB expansion

In this section, we show how to calculate the spectrum of a hamiltonian in the semi-classical limit or WKB approximation, starting from the semi-classical expansion of the partition function obtained in Section 2.10.

3.2.1 Spectrum and poles of the resolvent

Again, we assume that the hamiltonian has a discrete spectrum and, thus, the partition function can be written as

$$\mathcal{Z}(\beta) = \operatorname{tr} e^{-\beta H} = \sum_{\ell=0} e^{-\beta E_\ell}$$

with $E_0 < E_1 < E_2 < \ldots$.

The Laplace transform of $\mathcal{Z}(\beta)$, which is also the trace of the resolvent of H, is then

$$G(E) = \int_0^\infty \mathrm{d}\beta \, e^{\beta E} \, \mathcal{Z}(\beta) = \operatorname{tr} \frac{1}{H-E} = \sum_\ell \frac{1}{E_\ell - E}, \qquad (3.8)$$

where the result on the r.h.s. can be derived from the integral by an analytic continuation to the whole complex energy plane, starting from the domain $\operatorname{Re} E < E_0$ where the integral converges.

The spectrum of the hamiltonian is then given by the poles of $G(E)$.

Note that the sum over all E_ℓ does not always converge. This difficulty is directly related to a divergence of the integral (3.8) for $\beta \to 0$. The problem can be solved by the following method: one differentiates the integrand with respect to E until the integral converges. One obtains

$$G^{(m)}(E) = \int_0^\infty \mathrm{d}\beta \, e^{\beta E} \, \beta^m \, \mathcal{Z}(\beta) = m! \sum_\ell \frac{1}{(E_\ell - E)^m}.$$

One then defines $G(E)$ by integrating m times over E. Various definitions of $G(E)$ differ by a polynomial of degree $m-1$ in E, but all have the same poles with the same residues.

In what follows, it is useful to also introduce the function

$$L(E) = \int^E G(E')\mathrm{d}E' = -\sum_{\ell=0} \ln(E_\ell - E) + \text{polynomial}(E). \qquad (3.9)$$

In the expression of $L(E)$, we assume that the function $\ln z$ is cut along the real negative axis.

Note that the function

$$\mathcal{D}(E) = e^{-L(E)}$$

vanishes on the spectrum of H and is, therefore, a regularized form of the Fredholm determinant $\det(H-E)$.

We now calculate the quantity (ε real positive)

$$\Delta L(E_k) = \frac{1}{2i\pi} \lim_{\varepsilon \to 0_+} L(E_k + i\varepsilon) - L(E_k - i\varepsilon)$$

$$= \frac{1}{2i\pi} \lim_{\varepsilon \to 0_+} \sum_{\ell=0} \left(\ln(E_\ell - E_k + i\varepsilon) - \ln(E_\ell - E_k - i\varepsilon)\right).$$

All singularities corresponding to E_ℓ, $0 \le \ell < k$, give a contribution $2i\pi$, which is the discontinuity of the logarithmic function. For the singularity $E = E_k$:
$$\ln(E_k - E + i\varepsilon) - \ln(E_k - E - i\varepsilon)|_{E=E_k} = \ln(i\varepsilon) - \ln(-i\varepsilon) = i\pi\,.$$
The singularities with $\ell > k$ give no contribution, and one concludes
$$\Delta L(E_k) = k + \tfrac{1}{2}\,. \tag{3.10}$$
This relation can now be used to determine the spectrum in the semi-classical approximation.

3.2.2 Semi-classical approximation

In this section, we consider one-dimensional systems with a hamiltonian of the form
$$H = \frac{1}{2m}p^2 + V(q).$$
In contrast with the perturbative expansion, the restriction to one dimension is important here.

The leading order contribution to the resolvent is obtained, in the semi-classical expansion, by replacing the partition function
$$\mathcal{Z}(\beta) = \operatorname{tr} e^{-\beta H}$$
in the Laplace transformation by its properly normalized classical limit $\mathcal{Z}_{\text{cl.}}(\beta)$, which is the leading term in (2.89):
$$\mathcal{Z}_{\text{cl.}}(\beta) = \frac{1}{\hbar}\sqrt{\frac{m}{2\pi\beta}} \int dq\, e^{-\beta V(q)}\,.$$
One then finds
$$G_{\text{cl.}}(E) = \frac{1}{\hbar}\sqrt{m/2} \int dq\,[V(q) - E]^{-1/2},$$
where the potential convergence problems for $\beta \to 0$ are transformed into problems at $|q| \to \infty$.

We have chosen to cut the function $\sqrt{z}$ along the real negative axis ($\sqrt{z \pm i\varepsilon} = \pm i\sqrt{-z}$ for z real negative).

We note that the function $G_{\text{cl.}}(E)$ has a cut on the real axis at the right of the minimum of $V(q)$, instead of the expected poles. This can easily be understood. In this semi-classical limit, $\hbar$ goes to zero but energies remain finite. The kth energy eigenvalue E_k can only remain finite if the quantum number k diverges. In contrast, the differences $E_{k+1} - E_k$ go to zero, which explains that the limiting spectrum is continuous.

An integration yields the function (3.9):
$$L_{\text{cl.}}(E) = \int^E dE'\, G_{\text{cl.}}(E') = -\frac{1}{\hbar}\int dq\,\sqrt{2m[V(q) - E]}.$$

3.2 Partition function and spectrum

The discontinuity follows

$$\lim_{\varepsilon \to 0_+} L_{\text{cl.}}(E + i\varepsilon) - L_{\text{cl.}}(E - i\varepsilon) = \frac{2i}{\hbar} \int dq\, \theta[E - V(q)] \sqrt{2m[E - V(q)]},$$

where $\theta(x)$ is the step (or Heaviside) function ($\theta(x > 0) = 1$, $\theta(x < 0) = 0$). Equation (3.10) thus reduces to the Bohr–Sommerfeld quantization condition:

$$\int dq\, \theta(E_k - V(q)) \sqrt{2m[E_k - V(q)]} = \hbar\pi(k + \tfrac{1}{2}). \tag{3.11}$$

Remarks.
(i) An alternative calculation is based on expression (2.90). One then finds

$$G_{\text{cl.}}(E) = \int \frac{dp\, dq}{2\pi\hbar} \frac{1}{H(p,q) - E},$$

where H is the classical hamiltonian: $H(p, q) = p^2/2m + V(q)$. The contour integral over E gives 0 or 1 depending whether the values of H are inside or outside the contour. An equivalent form of equation (3.11) follows:

$$\int \frac{dp\, dq}{2\pi\hbar}\, \theta(E - H(p, q)) = k + \tfrac{1}{2}.$$

(ii) Since the l.h.s. in equation (3.11) remains finite for $\hbar \to 0$, one verifies that this approximation is valid for large quantum numbers $E_k = O(1)$, $k\hbar = O(1)$, while the difference between eigenvalues goes to zero with $\hbar$. This is consistent with the domain of validity of the approximation (2.89): we have seen that the approximation is valid at high temperature where physical quantities are mainly sensitive to large quantum numbers.

First correction. Successive terms in the semi-classical expansion of the partition function generate corrections to the Bohr–Sommerfeld approximation. For example, the expansion to order $\hbar^2$ is given by equation (2.89):

$$\mathcal{Z}(\beta) = \frac{1}{\hbar} \sqrt{\frac{m}{2\pi\beta}} \int dq\, e^{-\beta V(q)} \left[1 - \beta^2 \hbar^2 V''(q)/24m + O(\hbar^4)\right].$$

The identity

$$\beta^2 e^{\beta E} = \left(\frac{\partial}{\partial E}\right)^2 e^{\beta E},$$

allows writing the first correction to $G_{\text{cl.}}(E)$ as

$$G^{(1)}(E) = -\frac{\hbar^2}{24m} \frac{\sqrt{m/2}}{\hbar} \left(\frac{\partial}{\partial E}\right)^2 \int dq\, V''(q) \left[V(q) - E\right]^{-1/2}.$$

Integrating $G^{(1)}(E)$, one finds

$$L^{(1)}(E) = \int^E dE'\, G^{(1)}(E') = -\frac{\hbar^2}{24m}\frac{\sqrt{m/2}}{\hbar}\frac{\partial}{\partial E}\int dq\,\frac{V''(q)}{\sqrt{V(q)-E}}.$$

The discontinuity of $L^{(1)}(E)$ along the real axis follows:

$$L^{(1)}(E+i0) - L^{(1)}(E-i0) = \frac{i\hbar}{12\sqrt{2m}}\frac{\partial}{\partial E}\int dq\,\theta(E-V(q))\frac{V''(q)}{\sqrt{E-V(q)}}$$

and, thus, the first correction to the Bohr–Sommerfeld approximation (3.11).

Notice that one cannot differentiate directly the integrand with respect to E because $E-V(q)$ vanishes at least linearly at the boundary and, thus, the singularity of $[E-V(q)]^{-3/2}$ is not integrable.

3.2.3 Examples

The harmonic oscillator. In the example of the harmonic potential $V(q) = \tfrac{1}{2}m\omega^2 q^2$, the function $G(E)$ defined by equation (3.8) does not exist since the sum over n does not converge. In contrast, the function $G'(E)$ has a convergent expansion:

$$G'(E) = \sum_{n=0}^{\infty}\frac{1}{\bigl(\hbar\omega(n+1/2)-E\bigr)^2} = \frac{1}{\hbar^2\omega^2}\psi'(1/2 - E/\hbar\omega),$$

where $\psi(z)$ is the logarithmic derivative of the Γ function. Therefore, one can choose

$$G(E) = -\frac{1}{\hbar\omega}\psi(1/2 - E/\hbar\omega) \Rightarrow L(E) = \ln\Gamma(1/2 - E/\hbar\omega) \tag{3.12}$$

and, thus $\mathcal{D}(E) = 1/\Gamma(1/2 - E/\hbar\omega)$.

The preceding results can be applied to obtain the first terms of the corresponding semi-classical expansion. After a simple change of variables, the function $G_{\mathrm{cl.}}(E)$ becomes, formally,

$$G_{\mathrm{cl.}}(E) = \frac{1}{\hbar\omega}\int\frac{dx}{(x^2-E)^{1/2}}.$$

The integral does not converge for $|x|\to\infty$. Differentiating once with respect to E, one obtains

$$G'_{\mathrm{cl.}}(E) = \frac{1}{2\hbar\omega}\int\frac{dx}{(x^2-E)^{3/2}} = -\frac{1}{\hbar\omega E}.$$

Possible primitives are

$$G_{\mathrm{cl.}}(E) = -\frac{1}{\hbar\omega}\ln(-E/\hbar\omega), \quad L_{\mathrm{cl.}}(E) = -\frac{E}{\hbar\omega}[\ln(-E/\hbar\omega) - 1].$$

The discontinuity of $\ln(-E)$ across its cut is $2i\pi$ and, thus, at leading order in the semi-classical limit

$$E_k = \hbar\omega(k + \tfrac{1}{2}),$$

which coincides with the exact result. This is a characteristic property of the quantum harmonic oscillator.

The first correction to $G(E)$ can easily be inferred from the preceding results and one finds
$$G^{(1)}(E) = -\hbar\omega \frac{1}{24E^2} \Rightarrow L^{(1)}(E) = \hbar\omega \frac{1}{24E}$$

For all homogeneous potentials, the semi-classical expansion is also a large E expansion, as the exact form (3.12) shows (see also the example below). Therefore, these expressions can be verified by expanding the exact result (3.12), that is by replacing the function $\psi(z)$ by an expansion that is derived from the calculation of the Γ function by the steepest descent method as explained in Section 1.5 (equation (1.37)):
$$\psi(z+1/2) = \ln z + 1/24z^2 + O(1/z^4), \text{ for } |\text{Arg}z| < \pi.$$

The function $L^{(1)}(E)$, in this example, has no discontinuity and, thus, does not contribute to the spectrum. This result generalizes to all orders and this explains why the Bohr–Sommerfeld approximation is exact for the harmonic oscillator.

Another example. We now consider the homogeneous potential $V(q) = q^{2N}$ and set $\hbar = m = 1$. The leading order result is
$$E_k = [C_N(k+\tfrac{1}{2})]^{2N/(N+1)}, \quad C_N = \frac{\pi N \Gamma(3/2 + 1/2N)}{\sqrt{2}\Gamma(3/2)\Gamma(1/2N)}.$$

This result reflects the property that the leading term in equation (3.10) is proportional to $E^{1/2+1/2N}$. The first correction then is proportional to $E^{-1/2-1/2N}$. In this homogeneous example, the semi-classical expansion of the r.h.s. of equation (3.10) is also an expansion for $E \to \infty$.

3.2.4 WKB approximation and Schrödinger equation

We now explain how semi-classical results obtained from path integrals, can also be derived from the Schrödinger equation. We assume that the potential $V(x)$ is analytic in a strip around the real axis $|\text{Im } x| < \delta$, because this class of potentials allows a rather complete analysis. Then, eigenfunctions are also analytic functions.

We start from the time-independent Schrödinger equation
$$-\frac{\hbar^2}{2m}\varphi''(x) + V(x)\varphi(x) = E\varphi(x). \tag{3.13}$$

The equation can be expressed in terms of the logarithmic derivative of $\varphi(x)$. We set
$$\frac{\varphi'(x)}{\varphi(x)} = -\frac{S(x)}{\hbar}, \tag{3.14}$$

which leads to
$$\frac{\varphi''(x)}{\varphi(x)} = \frac{S^2(x)}{\hbar^2} - \frac{S'(x)}{\hbar}. \tag{3.15}$$

The Schrödinger equation then leads to a Riccati equation:
$$\hbar S'(x) - S^2(x) + U(x) = 0 \text{ with } U(x) = 2m[V(x) - E]. \tag{3.16}$$

This equation can be expanded systematically in powers of $\hbar$, at a fixed E, starting at leading order from $S(x) = U^{1/2}(x)$. It is convenient to decompose $S(x)$ into parts that are odd and even in $\hbar$, setting
$$S(x, \hbar) = S_+(x, \hbar) + S_-(x, \hbar), \qquad S_\pm(x, -\hbar) = \pm S_\pm(x, \hbar). \tag{3.17}$$

The Riccati equation then yields
$$\hbar S'_- - S^2_+ - S^2_- + U = 0, \tag{3.18}$$
$$\hbar S'_+ - 2S_+ S_- = 0, \tag{3.19}$$

where at leading order $S_+ = U^{1/2}$ and $S_- = 0$. The second equation allows expressing an eigenfunction φ in terms of S_+ only:
$$\varphi(x) = \frac{1}{\sqrt{S_+(x)}} \exp\left[-\frac{1}{\hbar} \int_{x_0}^{x} dx'\, S_+(x')\right]. \tag{3.20}$$

One can prove that the eigenfunction corresponding to the kth excited state vanishes exactly k times on the real axis. The spectrum can thus be determined by the condition
$$\frac{1}{2i\pi} \oint_C dz \frac{\varphi'(z)}{\varphi(z)} = k, \tag{3.21}$$

where k is the number of nodes of the eigenfunction and C a complex contour that encloses them.

The equation can be written as
$$-\frac{1}{2i\pi\hbar} \oint_C dz\, S(z) = k.$$

In the semi-classical limit, C encloses the cut of $U^{1/2}(x)$, which joins the two turning points solutions of $U(x) = 0$. The contribution of S_- can be evaluated explicitly, at least to all orders in $\hbar$. Indeed, one verifies that only the leading order contributes and, therefore,
$$-\frac{1}{2i\pi\hbar} \oint_C dz\, S_-(z) = -\frac{1}{4i\pi} \oint_C dz \frac{S'_+(z)}{S_+(z)} = -\frac{1}{8i\pi} \oint_C dz \frac{U'(z)}{U(z)} = -\frac{1}{2}.$$

In terms of S_+, equation (3.21) becomes
$$-\frac{1}{2i\pi\hbar} \oint_C dz\, S_+(z) = k + \tfrac{1}{2}. \tag{3.22}$$

If one substitutes for S_+ the formal solution of the Riccati equation as an expansion in powers of $\hbar$, one obtains the WKB or semi-classical expansion of the energy eigenvalues.

3.3 Path integral and variational principle

Convexity inequality. Many variational principles in physics are based on the convexity property of the exponential function, which in its simplest form reads

$$e^{(x_1+x_2)/2} \leq \tfrac{1}{2}\left(e^{x_1}+e^{x_2}\right).$$

More generally, let us denote by $\langle \bullet \rangle$ the expectation value of $\bullet$ with respect to a positive normalized measure. Then, for any real function F, one derives the convexity inequality

$$e^{\langle F \rangle} \leq \langle e^F \rangle \Leftrightarrow \langle F \rangle \leq \ln \langle e^F \rangle.$$

We now apply this inequality in a situation where the measure is gaussian, and F a polynomial whose expectation value can thus be calculated using Wick's theorem.

Variational principle. We want to evaluate approximately the integral over $\mathbb{R}^n$:

$$\mathcal{Z} = \int \mathrm{d}^n x \, e^{-V(\mathbf{x})},$$

where $V(\mathbf{x})$ is a polynomial in the variables x_i. We introduce the normalized gaussian measure $e^{-A(\mathbf{x})}/\mathcal{Z}_0$, where the function $A(\mathbf{x})$ is a general second degree polynomial:

$$A(\mathbf{x}) = \frac{1}{2}\sum_{i,j=1}^{n} x_i A_{ij} x_j - \sum_{i=1}^{n} b_i x_i. \tag{3.23}$$

We choose

$$F(\mathbf{x}) = A(\mathbf{x}) - V(\mathbf{x}).$$

We then find

$$\frac{1}{\mathcal{Z}_0} \int \mathrm{d}^n x \left(A(\mathbf{x}) - V(\mathbf{x})\right) e^{-A(\mathbf{x})} \leq \ln \int \mathrm{d}^n x \, e^{-V(\mathbf{x})} - \ln \mathcal{Z}_0 \,,$$

an inequality that can be rewritten as

$$\ln \mathcal{Z} \geq \ln \mathcal{Z}_0 + \frac{1}{\mathcal{Z}_0} \int \mathrm{d}^n x \left(A(\mathbf{x}) - V(\mathbf{x})\right) e^{-A(\mathbf{x})}$$
$$\geq \ln \mathcal{Z}_0 + \frac{n}{2} - \frac{1}{\mathcal{Z}_0} \int \mathrm{d}^n x \, V(\mathbf{x}) e^{-A(\mathbf{x})}. \tag{3.24}$$

This inequality is the source of a variational principle: since the matrix $\mathbf{A}$ and the vector $\mathbf{b}$ are arbitrary, the best approximation to the integral $\mathcal{Z}$ is obtained by looking for a maximum of the r.h.s. over all $\mathbf{A}$ and $\mathbf{b}$. Since the r.h.s. can be calculated explicitly, using Wick's theorem, optimization becomes a purely algebraic problem.

Example. A simple example is provided for $n=1$ by $V(x)=x^4$. Then the exact value is $\tfrac{1}{2}\Gamma(1/4) = 1.812...$ and the variational result $(\pi^2 \, \mathrm{e}/3)^{1/4} = 1.729...$.

Path integral. We now consider the partition function ($\hbar = 1$)
$$\mathcal{Z}(\beta) = \int [\mathrm{d}q(t)] \exp\left[-\mathcal{S}(q)\right], \tag{3.25}$$
where the paths satisfy periodic boundary conditions: $q(0) = q(\beta)$ and (equation (2.24))
$$\mathcal{S}(q) = \int_0^\beta \mathrm{d}t \left[\tfrac{1}{2}m\dot{q}^2(t) + V(q(t))\right]. \tag{3.26}$$
We choose the gaussian measure corresponding to
$$\mathcal{S}_0(q) = \frac{m}{2} \int_0^\beta \mathrm{d}t \left[\dot{q}^2(t) + \omega^2 (q(t) - q_0)^2\right],$$
where ω and q_0 are two variational parameters. The quantity F is here
$$F(q) = \int_0^\beta \mathrm{d}t \left[\tfrac{1}{2}m\omega^2 (q(t) - q_0)^2 - V(q(t))\right].$$

It is convenient to change $q(t) - q_0$ into $q(t)$. In terms of the partition function $\mathcal{Z}_0$ of the harmonic oscillator (equation (2.61)), the convexity inequality can then be written as
$$\ln \mathcal{Z}(\beta) \geq \ln \mathcal{Z}_{\mathrm{var.}}(\beta)$$
with
$$\ln \mathcal{Z}_{\mathrm{var.}}(\beta) = \ln \mathcal{Z}_0(\beta) + \int_0^\beta \mathrm{d}t \left\langle \left[\tfrac{1}{2}m\omega^2 q^2(t) - V(q(t) + q_0)\right]\right\rangle_0$$
$$= \ln \mathcal{Z}_0(\beta) + \tfrac{1}{4}\omega\beta \coth(\omega\beta/2) - \int_0^\beta \mathrm{d}t \left\langle V(q(t) + q_0)\right\rangle_0, \tag{3.27}$$
where $\langle \bullet \rangle_0$ means expectation value with respect to the gaussian measure associated to the harmonic oscillator, and the two-point function is given by the expressions (2.57, 2.52).

One notices that the inequality has a natural expression in terms of the free energy, which is proportional to $\ln \mathcal{Z}$.

The special example
$$V(q) = \tfrac{1}{2}m\nu^2 q^2,$$
allows a simple verification of the method. For symmetry reasons, one can set $q_0 = 0$. Then, the variational partition function in the r.h.s. becomes
$$\ln \mathcal{Z}_{\mathrm{var.}} = -\ln \sinh s + \frac{\cosh s}{2 \sinh s}\left(s - \frac{\beta^2 \nu^2}{4s}\right),$$
where we have set $s = \omega\beta/2$. Differentiating, one verifies that the maximum is reached for $s = \beta\nu/2$, and one then recovers the exact result.

A more interesting example is provided by the potential

$$V(q) = \frac{1}{4!}gm^2q^4.$$

One still finds $q_0 = 0$ for symmetry reasons. Since the solution of the equations for β finite is somewhat complicated, we consider the limit $\beta \to \infty$, which yields the ground state energy of the corresponding hamiltonian. Then,

$$E_{\text{var.}} = -\lim_{\beta \to \infty} \frac{1}{\beta} \ln \mathcal{Z}_{\text{var.}} = \frac{\omega}{4} + \frac{g}{32\omega^2}.$$

The variational estimate now yields an upper bound to the ground state energy E. The minimum is given by $\omega = (g/4)^{1/3}$ and one finds

$$E \leq \frac{3}{8}\left(\frac{g}{4}\right)^{1/3} \approx 0.236235\, g^{1/3},$$

while the exact coefficient of $g^{1/3}$ is $0.23157\ldots$. This estimate and upper bound cannot derived from perturbation theory since the potential is quartic near its minimum.

Note, however, that, as usual, variational methods can easily lead to very good estimates but which are then difficult to improve further.

3.4 $O(N)$ symmetric quartic potential for $N \to \infty$

As another application of the path integral formalism, we study the properties of an $O(N)$ symmetric quantum system in the large N limit.

We consider a quantum particle in N space dimensions. The position $\mathbf{q}$ and momentum $\mathbf{p}$ are thus N-component vectors. We choose a hamiltonian of the special form

$$H = \tfrac{1}{2}\mathbf{p}^2 + NU(\mathbf{q}^2/N), \qquad (3.28)$$

where $\mathbf{p}^2, \mathbf{q}^2$ are the squared lengths of the vectors $\mathbf{p}, \mathbf{q}$, respectively. The potential U is radial, and a polynomial in the examples we discuss below. The hamiltonian, thus, has an $O(N)$ orthogonal symmetry (rotations and reflections in N-dimensional space).

One method to study the properties of the quantum system is to separate U into a quartic term and a remainder and to use perturbation theory, as proposed in Section 3.1. However, in the case of an $O(N)$ symmetric system of the form (3.28) another, non-perturbative, expansion scheme can be found. It is based on studying the system in the large N limit.

We know that the quantum hamiltonian (3.28) commutes with the $N(N-1)/2$ generators of the $SO(N)$ orthogonal group, and this allows reducing the initial Schrödinger equation to an one-dimensional radial Schrödinger equation. The number N of dimensions then appears simply as a parameter in the equation and it is simple to verify that the Schrödinger equation can be solved in the large N limit.

However, this method is peculiar to this one-particle hamiltonian. The path integral formulation, in contrast, suggests a simple and intuitive strategy with many interesting generalizations to other $O(N)$ symmetric models. The method that we describe below is rather general, even though explicit calculations are presented only for the example of a quartic perturbation of the harmonic oscillator

$$U(\mathbf{q}^2) = \tfrac{1}{2}\omega^2 \mathbf{q}^2 + \tfrac{1}{24} g (\mathbf{q}^2)^2. \tag{3.29}$$

Another physical interpretation. The hamiltonian (3.28) with the potential (3.29) has another possible physical interpretation: one can consider the variables q_i as being attached to the N sites of a lattice. Then, the potential (3.29) corresponds to the sum of a one-particle potential at each site and a pair interaction potential coupling all N sites of the lattice:

$$U(\mathbf{q}^2) = \sum_i \left(\tfrac{1}{2}\omega^2 q_i^2 + \tfrac{1}{24} g q_i^4 \right) + \tfrac{1}{12} g \sum_{i<j} q_i^2 q_j^2 . \tag{3.30}$$

Such a system has, in general, no thermodynamic (i.e. $N \to \infty$) limit because the potential energy grows as the number of pairs and, thus, the square of the 'volume' N. The dependence of the potential on the number of sites N, as in expression (3.28), ensures a thermodynamic limit.

Partition function. The partition function is given by the path integral

$$\mathcal{Z}(\beta) = \int [\mathrm{d}\mathbf{q}]\, \mathrm{e}^{-\mathcal{S}(\mathbf{q})}, \tag{3.31}$$

$\mathbf{q}(0) = \mathbf{q}(\beta)$, with the action

$$\mathcal{S}(\mathbf{q}) = \int_0^\beta \mathrm{d}t \left[\tfrac{1}{2}\dot{\mathbf{q}}^2(t) + N U(\mathbf{q}^2(t))/N \right]. \tag{3.32}$$

The action is manifestly $O(N)$ symmetric since it depends only on scalar products.

The large N limit. The method that we present here allows calculating the partition function and, thus, also determining the spectrum of the hamiltonian (3.28) in the limit $N \to \infty$ with $U(\mathbf{q}^2)$ fixed and, more generally, as an expansion in powers of $1/N$. In the example of the quartic potential (3.29), the method yields an information complementary to perturbation theory that leads to an expansion in powers of g.

The basic idea behind large N calculations is inspired from the central limit theorem of probabilities. One expects that for $N \to \infty$, the quantities invariant under the $O(N)$ group as

$$\mathbf{q}^2(t) = \sum_{i=1}^N q_i^2(t),$$

3.4 Partition function and spectrum

self-average and, therefore, do not fluctuate (this assumes, of course that the variables q_i are weakly correlated, a property that has to be verified). For example, one expects that

$$\langle \mathbf{q}^2(t)\mathbf{q}^2(t')\rangle \underset{N\to\infty}{\sim} \langle \mathbf{q}^2(t)\rangle \langle \mathbf{q}^2(t')\rangle.$$

Therefore, $\mathbf{q}^2(t)$ is a simpler dynamic variable than $\mathbf{q}(t)$. The idea then is to integrate over $\mathbf{q}(t)$ at a fixed $\mathbf{q}^2(t)$. Technically, this idea can be implemented by a set of transformations, which we first explain in the example of a simple integral.

3.4.1 Simple integrals for $N \to \infty$

We first consider the simple integral

$$\mathcal{I}_N = \int d^N\mathbf{q}\, e^{-NU(\mathbf{q}^2/N)}, \tag{3.33}$$

where $U(\rho)$ is a polynomial, which, for simplicity, we assume to be increasing with an increasing derivative for $\rho > 0$:

$$U'(\rho) > 0, \quad U''(\rho) \geq 0.$$

The integrand and the integration measure are invariant under $SO(N)$ transformations acting on the vector $\mathbf{q}$, as in the example of the path integral.

To evaluate the integral for $N \to \infty$, one faces a difficulty: the dependence of the integral on the parameter N is partially implicit through the number N of integration variables. The problem can be solved here in a simple way by going from cartesian to polar coordinates. Then,

$$\mathcal{I}_N = \Sigma_N \int_0^\infty \frac{dq}{q} q^N e^{-NU(q^2/N)}, \tag{3.34}$$

where Σ_N is the surface of the sphere S_{N-1}:

$$\Sigma_N = \frac{2\pi^{N/2}}{\Gamma(N/2)}.$$

It is also convenient to set $q^2/N = \rho$, which leads to

$$\mathcal{I}_N = N^{(N-1)/2}\Sigma_N \int_0^\infty \frac{d\rho}{\rho} e^{-N\sigma(\rho)} \tag{3.35}$$

with

$$\sigma(\rho) = U(\rho) - \tfrac{1}{2}\ln\rho. \tag{3.36}$$

The evaluation of the integral for $N \to \infty$ now relies on the steepest descent method (see Section 1.5). The saddle points are given by

$$\sigma'(\rho) = 0 \iff 2\rho U'(\rho) = 1.$$

With our assumptions concerning the function $U(\rho)$, the equation has a unique solution. The calculation then follows the general method explained in Section 1.5.

Although this method is rather straightforward in the case of the integral (3.33), it does not generalize to path integrals and we introduce, therefore, an alternative method, which may seem more complicated, but is easily generalizable.

General method. We first rewrite the integral as

$$\mathcal{I}_N = N \int d^N\mathbf{q}\, d\rho\, \delta(\mathbf{q}^2 - N\rho)\, e^{-NU(\rho)}.$$

We then substitute for Dirac's δ function its Fourier representation:

$$\delta(\mathbf{q}^2 - N\rho) = \frac{1}{2\pi} \int_{-\infty}^{+\infty} d\mu\, e^{i\mu(\mathbf{q}^2 - N\rho)}.$$

It is convenient, for what follows, to set $\mu = i\lambda/2$ so that the representation becomes

$$\delta(\mathbf{q}^2 - N\rho) = \frac{i}{4\pi} \int_{-i\infty}^{+i\infty} d\lambda\, e^{-\lambda(\mathbf{q}^2 - N\rho)/2},$$

where initially λ is purely imaginary. After this substitution, the integral (3.33) becomes

$$\mathcal{I}_N = \int d^N\mathbf{q}\, e^{-NU(\mathbf{q}^2/N)} = \frac{N}{4i\pi} \int d^N\mathbf{q}\, d\rho\, d\lambda\, e^{-\lambda(\mathbf{q}^2 - N\rho)/2 - NU(\rho)}. \qquad (3.37)$$

The **q**-integral now is gaussian and can be calculated provided the λ integration contour satisfies the condition $\operatorname{Re}\lambda \geq 0$:

$$\mathcal{I}_N = \pi^{N/2} \frac{N}{4i\pi} \int d\rho\, d\lambda\, e^{N\lambda\rho/2 - N\ln(\lambda)/2 - NU(\rho)}.$$

In the limit $N \to \infty$, the resulting integral can be evaluated by the standard steepest descent method with two complex variables. The saddle point equations, obtained by differentiating successively with respect to λ and ρ, are

$$\rho = 1/\lambda, \quad \lambda = 2U'(\rho).$$

Notice that the saddle point in λ is real but the integration contour in λ is parallel to the imaginary axis. We leave, as an exercise, the completion of the calculation. One verifies that the result is the same as the one obtained by using directly polar coordinates.

3.4.2 Path integral

Identity (3.37) can be generalized to an arbitrary number of variables and, thus, to a path integral. In the example of the integral (3.31), one introduces two paths $\rho(t), \lambda(t)$, which are periodic because one calculates a partition function: $\rho(\beta) = \rho(0)$, $\lambda(\beta) = \lambda(0)$. The transformed path integral is obtained by multiplying the integrand by

$$\mathcal{N}(\beta) \int [\mathrm{d}\rho][\mathrm{d}\lambda] \exp\left[\int_0^\beta \mathrm{d}t\, \lambda(t)(\mathbf{q}^2(t) - N\rho(t))/2\right] = 1, \qquad (3.38)$$

where the normalization $\mathcal{N}(\beta)$ depends on the discretization but not on the path $\mathbf{q}(t)$, and by replacing $\mathbf{q}^2/N$ by ρ in the potential U.

The partition function is then given by

$$\mathcal{Z}(\beta) = \int [\mathrm{d}\mathbf{q}\,\mathrm{d}\rho\,\mathrm{d}\lambda]\, e^{-\mathcal{S}(\mathbf{q},\rho,\lambda)}$$

(where the normalization factor is now implicit) with

$$\mathcal{S}(\mathbf{q}, \rho, \lambda) = \int_0^\beta \mathrm{d}t \left[\tfrac{1}{2}\dot{\mathbf{q}}^2 + NU(\rho) + \tfrac{1}{2}\lambda(\mathbf{q}^2 - N\rho)\right].$$

We notice that the integral over $\mathbf{q}(t)$ is again gaussian and can be performed. Since

$$\int_0^\beta \mathrm{d}t \left[\dot{\mathbf{q}}^2(t) + \lambda(t)\mathbf{q}^2(t)\right] = \sum_i \int_0^\beta \mathrm{d}t \left[\dot{q}_i^2(t) + \lambda(t)q_i^2(t)\right],$$

one obtains the product of N identical integrals. Each integral yields

$$\int [\mathrm{d}q] \exp\left\{-\tfrac{1}{2}\int_0^\beta \mathrm{d}t \left[\dot{q}^2(t) + \lambda(t)q^2(t)\right]\right\} \propto \left[\det\left(-\mathrm{d}_t^2 + \lambda(\bullet)\right)\right]^{-1/2}. \qquad (3.39)$$

The operator $-\mathrm{d}_t^2 + \lambda$, which appears in the r.h.s., has the form of a one-dimensional quantum hamiltonian, $-\mathrm{d}_t^2$ being the kinetic term and $\lambda(t)$ the potential. Being a differential operator, its determinant depends on the boundary conditions. Finally, the path integral has to be normalized. We choose, below, to divide it by its value for $\lambda = 0$.

Thus, one finds

$$\mathcal{Z}(\beta) = \int [\mathrm{d}\rho\,\mathrm{d}\lambda]\, e^{-\mathcal{S}_N(\rho,\lambda)} \qquad (3.40)$$

with ($\operatorname{tr}\ln = \ln\det$, equation (3.51))

$$\mathcal{S}_N(\rho, \lambda) = \int_0^\beta \mathrm{d}t \left[NU(\rho) - \tfrac{1}{2}N\lambda\rho\right] + \tfrac{1}{2}N\operatorname{tr}\ln\left(-\mathrm{d}_t^2 + \lambda(\bullet)\right) - \tfrac{1}{2}N\operatorname{tr}\ln\left(-\mathrm{d}_t^2\right). \qquad (3.41)$$

The integration has rendered the dependence of the partition function on N explicit. One notices that in the limit $N \to \infty$ at U fixed, and assuming that the integral is dominated by paths $\rho, \lambda = O(1)$, the action becomes proportional to N. In the limit $N \to \infty$, the path integral can thus be evaluated by the *complex* steepest descent method.

Due to time-translation invariance on the circle $[0, \beta]$ (a circle as a consequence of periodic boundary conditions), a saddle point $\{\lambda(t), \rho(t)\}$ is either degenerate if $\lambda(t), \rho(t)$ depend explicitly on time (a solution and all those obtained by translation), or unique if the functions are constant. One can show that the simplest situation is realized and that the saddle point is not degenerate.

The calculations can then be carried out for a general potential $U(\rho)$, but we consider here only the example (3.29). One then notices that the integral over ρ is gaussian. The minimum of the quadratic form in ρ is obtained for

$$U'(\rho) - \tfrac{1}{2}\lambda = 0 \;\Rightarrow\; \rho(t) = 6\bigl(\lambda(t) - \omega^2\bigr)/g\,.$$

After translation of ρ, the integration over ρ yields a determinant, which is a constant and can, thus, be absorbed into the normalization. After integration, one finds

$$\mathcal{Z}(\beta) = \int [\mathrm{d}\lambda]\,\mathrm{e}^{-\mathcal{S}_N(\lambda)} \tag{3.42}$$

with

$$\mathcal{S}_N(\lambda) = -\frac{3N}{2g}\int_0^\beta \mathrm{d}t\bigl(\lambda(t) - \omega^2\bigr)^2 + \tfrac{1}{2}N\,\mathrm{tr}\ln\bigl(-\mathrm{d}_t^2 + \lambda(\bullet)\bigr) - \tfrac{1}{2}N\,\mathrm{tr}\ln\bigl(-\mathrm{d}_t^2\bigr). \tag{3.43}$$

Since we look for a saddle point with $\lambda(t)$ constant, we set

$$\lambda(t) = r^2 + \mu(t)$$

and expand at first order in $\mu(t)$ to obtain the saddle point equation. First,

$$-\frac{3}{g}\int_0^\beta \mathrm{d}t\bigl(r^2 + \mu(t) - \omega^2\bigr)^2 = -\frac{3\beta}{g}(r^2 - \omega^2)^2 - \frac{6}{g}(r^2 - \omega^2)\int_0^\beta \mathrm{d}t\,\mu(t) + O(\mu^2).$$

Moreover (see equation (3.52)),

$$\begin{aligned}
\mathrm{tr}\ln(-\mathrm{d}_t^2 + \lambda) &= \mathrm{tr}\ln(-\mathrm{d}_t^2 + r^2 + \mu) \\
&= \mathrm{tr}\ln(-\mathrm{d}_t^2 + r^2) + \mathrm{tr}\ln\bigl(1 + \mu(-\mathrm{d}_t^2 + r^2)^{-1}\bigr) \\
&= \mathrm{tr}\ln(-\mathrm{d}_t^2 + r^2) + \mathrm{tr}\,\mu(-\mathrm{d}_t^2 + r^2)^{-1} + O(\mu^2) \\
&= \mathrm{tr}\ln(-\mathrm{d}_t^2 + r^2) + \int \mathrm{d}t\,\mu(t)\Delta(0) + O(\mu^2).
\end{aligned}$$

In this expression we have taken into account that $\mu(t)$ is a diagonal operator and that the diagonal matrix elements of Δ, the inverse of $-\mathrm{d}_t^2 + r^2$ with periodic boundary conditions, are constant (see (2.52)):

$$\mathrm{tr}\,\mu(-\mathrm{d}_t^2 + r^2)^{-1} = \int \mathrm{d}t\,\mathrm{d}u\,\langle t|\mu|u\rangle\,\Delta(t - u) = \int \mathrm{d}t\,\mu(t)\Delta(0).$$

The saddle point equation is then obtained by expressing that the coefficient of $\int \mu(t)dt$ in the action vanishes:

$$-\frac{6}{g}(r^2 - w^2) + \Delta(0) = -\frac{6}{g}(r^2 - w^2) + \frac{1}{2r\tanh(\beta r/2)} = 0. \quad (3.44)$$

The action involves the determinant of $-d_t^2 + r^2$ with periodic boundary conditions, which is directly related to the partition function of the harmonic oscillator, as the equation (3.39) shows. Thus,

$$\mathcal{S}_N(r^2) = -\frac{3\beta N}{2g}(r^2 - w^2)^2 - N \ln \mathcal{Z}_0(\beta),$$

$$= -\frac{3\beta N}{2g}(r^2 - w^2)^2 + N \ln \ln 2\sinh(\beta r/2) \quad (3.45)$$

with the choice $r \geq 0$.

One verifies that by differentiating with respect to r, one recovers the saddle point equation. After division by $6N\beta/g$, the equation reads

$$r(w^2 - r^2) + \frac{g}{12\tanh(\beta r/2)} = 0. \quad (3.46)$$

Using the solution, one finds, at leading order for $N \to \infty$,

$$\ln \mathcal{Z}(\beta) = -\mathcal{S}_N(r^2) = \frac{3N\beta}{2g}(r^2 - w^2)^2 - N \ln 2\sinh(\beta r/2). \quad (3.47)$$

The path integral is determined only up to a normalization. In the limit $g = 0$, since $r = w + O(g)$, one recovers the partition function of the harmonic oscillator and, thus, the result is correctly normalized.

In the interpretation (3.30), the result (3.47) represents the partition function in the thermodynamic limit. Dividing by the volume N and multiplying by the temperature $T = 1/\beta$, one obtains the free energy density (with our sign convention)

$$\mathcal{F} = \frac{1}{\beta N}\ln \mathcal{Z} = \frac{3}{2g}(r^2 - w^2)^2 - \frac{1}{\beta}\ln 2\sinh(\beta r/2).$$

3.4.3 Ground state energy

The ground state energy of the hamiltonian (3.28) with the potential (3.29), is obtained in the limit of vanishing temperature $\beta \to \infty$. In the limit, r is solution of the third degree algebraic equation

$$r^3 - w^2 r - \tfrac{1}{12}g = 0, \quad (3.48)$$

which has a unique solution. In terms of the solution, the ground state energy for $N \to \infty$ is

$$E_0 \underset{\beta \to \infty}{\sim} -\frac{1}{\beta}\ln \mathcal{Z}(\beta) = -\frac{3N}{2g}(r^2 - w^2)^2 + \tfrac{1}{2}Nr,$$

or, eliminating g between E_0 and the saddle point equation,

$$E_0 = \frac{N}{8r}(3r^2 + \omega^2).$$

For $g \to 0$, one finds $r = \omega + g/(24\omega^2) + O(g^2)$ and, thus,

$$E_0 = \tfrac{1}{2}N\omega + \tfrac{1}{96}Ng/\omega^2 + O(g^2),$$

a result consistent with the perturbative expansion

$$E_0(g) = \tfrac{1}{2}N\omega + \tfrac{1}{96}(N+2)g/\omega^2 + O(g^2).$$

For $g \to \infty$, one finds $r \sim (g/12)^{1/3}$ and $E_0(g)$ thus exhibits the $g^{1/3}$ behaviour that can be proved for finite N:

$$E_0 \sim \tfrac{3}{8}(12)^{-1/3}Ng^{1/3}. \qquad (3.49)$$

This result is also the limit of E_0 for $\omega = 0$. Therefore, it cannot be derived from a perturbative calculation.

Finally, the calculation for $N \to \infty$ is possible even when the coefficient of the harmonic term is negative. Changing $\omega \mapsto i\omega$, one finds

$$r^3 + \omega^2 r - \tfrac{1}{12}g = 0, \quad E_0 = \frac{N}{8r}(3r^2 - \omega^2).$$

For example, for $g \to 0$, r is of order g and one obtains

$$\frac{1}{N}E_0 = -\frac{3\omega^4}{2g} + \frac{g}{48\omega^2} + O(g^3).$$

This result can also be derived by perturbative methods, but in a more sophisticated version than the one introduced so far because the potential is minimum on a whole sphere.

By expanding the expressions (3.46, 3.47) for $\beta \to \infty$, it is also possible to calculate the difference between the ground state energy and the energies of excited states.

Remark. The saddle point equation (3.48) has a solution up to a value g_c of g, which is negative:

$$g \geq g_c, \quad g_c = -\tfrac{8}{3}\sqrt{3}\omega^3.$$

This result is somewhat surprising since, for $g < 0$, the potential is not bounded from below. In classical physics, it is, of course, possible for a particle to sit in a relative minimum of a potential, but in quantum mechanics the corresponding state is unstable and tunnels out of the well. One shows, however, that the decay rate through barrier penetration vanishes in the limit $N \to \infty$, as it is implemented here, for all negative values of g such that $g > g_c$.

3.4.4 Beyond leading order

The successive terms of the expansion in powers of $1/N$ are then given by the successive corrections in the steepest descent method. It is convenient to change variables, setting
$$\lambda(t) - r^2 = \mu(t), \tag{3.50}$$
where one integrates over imaginary values of μ, and one expands the integrand in powers of $\mu(t)$, keeping the quadratic term in the exponential.

The first correction is generated by the gaussian integration. In particular, the calculation of the quadratic term in μ allows verifying that the solution of the saddle point equations is a local maximum of the modulus of the integrand on the contour. The calculation of the quadratic term involves the expansion of (see equation (3.52))

$$\begin{aligned}
&\operatorname{tr}\ln\left[(-\mathrm{d}_t^2 + r^2 + \mu(\bullet))(-\mathrm{d}_t^2 + r^2)^{-1}\right] \\
&= \int \mathrm{d}t\, \mu(t)\, \langle t|\,(-\mathrm{d}_t^2 + r^2)^{-1}\,|t\rangle \\
&\quad - \tfrac{1}{2}\int \mathrm{d}t_1 \mathrm{d}t_2\, \mu(t_1)\mu(t_2)\, \langle t_1|\,(-\mathrm{d}_t^2 + r^2)^{-1}\,|t_2\rangle\langle t_2|\,(-\mathrm{d}_t^2 + r^2)^{-1}\,|t_1\rangle + O(\mu^3).
\end{aligned}$$

We denote by $\mathcal{S}^{(2)}(\mu)$ the term quadratic in μ and write it as
$$\mathcal{S}^{(2)}(\mu) = -\frac{1}{2}\int \mathrm{d}t_1\mathrm{d}t_2\, \mu(t_1) K(t_1 - t_2)\mu(t_2).$$

Then,
$$K(t) = \frac{3N}{g}\delta(t) + \frac{N}{4}\Delta^2(t),$$
where $\Delta(t)$ is the gaussian two-point function (2.52):
$$\Delta(t) = \frac{1}{2r \sinh(r\beta/2)}\cosh\bigl(r(\beta/2 - |t|)\bigr).$$

To determine the spectrum of the kernel $K(t_1 - t_2)$, one can diagonalize it by Fourier transformation. To simplify, we consider only the ground state and, thus, take the limit $\beta \to \infty$. Then,
$$\tilde{K}(\kappa) = \int \mathrm{d}t\, \mathrm{e}^{i\kappa t}\, K(t) = \frac{3N}{g} + \frac{N}{8r}\frac{1}{4r^2 + \kappa^2}.$$

One notices that the eigenvalues of $\tilde{K}(\kappa)$ are strictly positive for all values of κ. Since μ is imaginary, the quadratic part of the action is positive and the saddle point is relevant. The gaussian integral then gives $(\det K)^{-1/2}$, which can be calculated explicitly.

3.5 Operator determinants

In many situations one has to calculate determinants of operators that can be cast into the form $\mathbf{M} = \mathbf{1} + \mathbf{K}$, if necessary after a few transformations. An example is provided by the determinant in expression (3.43) after the translation (3.50). Provided the traces of all powers of $\mathbf{K}$ exist, the following identity, valid for any matrix or operator $\mathbf{M}$,

$$\ln \det \mathbf{M} \equiv \operatorname{tr} \ln \mathbf{M}, \tag{3.51}$$

expanded in powers of the kernel $\mathbf{K}$:

$$\ln \det (\mathbf{1} + \mathbf{K}) = \sum_{p=1}^{\infty} \frac{(-1)^{p+1}}{p} \operatorname{tr} \mathbf{K}^p, \tag{3.52}$$

can be used. When the operators $\mathbf{M}$ and, thus, $\mathbf{K}$ are represented by the kernels $M(x,y) = \delta(x-y) + K(x,y)$, the successive traces take the explicit form

$$\operatorname{tr} \mathbf{K}^p = \int \mathrm{d}x_1 \cdots \mathrm{d}x_p \, K(x_1, x_2) K(x_2, x_3) \cdots K(x_p, x_1).$$

Remark. The determinant of a matrix of finite size $n \times n$, of the form $\mathbf{1} + \lambda \mathbf{K}$ is a polynomial in λ of degree n. The expansion (3.52) seems to generate an infinite series. Therefore, the traces of the powers of the matrix $\mathbf{K}$ satisfy identities that imply that all terms of degree larger than n vanish.

Application. We consider a quantum hamiltonian H_0 with a discrete spectrum of eigenvalues $E_n^{(0)}$ and eigenvectors $|n\rangle$ and a perturbed hamiltonian $H = H_0 + V$. We want to calculate the spectrum of H at second order in the perturbation V by using expansion (3.52). We set

$$G_0(E) = \frac{1}{H_0 - E},$$

and expand the eigenvalue E as

$$E_n = E_n^{(0)} + E_n^{(1)} + E_n^{(2)}.$$

One starts from the identities

$$\ln \det[(H-E)(H_0-E)^{-1}] = \operatorname{tr} \ln\bigl(1 + G_0(E)V\bigr) = \sum_{n=0} \ln\left(\frac{E_n - E}{E_n^{(0)} - E}\right).$$

One then expands the two expressions in powers of V. At first order,

$$E_n = E_n^{(0)} + E_n^{(1)} \Rightarrow \ln\left(\frac{E_n - E}{E_n^{(0)} - E}\right) = \frac{E_n^{(1)}}{E_n^{(0)} - E},$$

$$\operatorname{tr} \ln\bigl(1 + G_0(E)V\bigr) = \operatorname{tr} G_0(E)V = \sum_n \frac{\langle n|V|n\rangle}{E_n^{(0)} - E}.$$

Identifying the residues of the poles at $E = E_n^{(0)}$, one infers
$$E_n^{(1)} = \langle n | V | n \rangle.$$

At second order,
$$\begin{aligned} \operatorname{tr} \ln\bigl(1 + G_0(E)V\bigr) &= \operatorname{tr} G_0 V - \tfrac{1}{2} \operatorname{tr}(G_0 V)^2 \\ &= \sum_n \frac{\langle n | V | n \rangle}{E_n^{(0)} - E} - \frac{1}{2} \sum_{m,n} \frac{|\langle n | V | m \rangle|^2}{(E_m^{(0)} - E)(E_n^{(0)} - E)}. \end{aligned}$$

This expression must be compared with
$$\sum_n \ln\left(\frac{E_n - E}{E_n^{(0)} - E} \right) = \sum_n \frac{E_n^{(1)}}{E_n^{(0)} - E} + \sum_n \frac{E_n^{(2)}}{E_n^{(0)} - E} - \frac{1}{2} \left(\frac{E_n^{(1)}}{E_n^{(0)} - E} \right)^2.$$

One notices that the double pole contributions are identical on both sides. Again, one identifies the residues of the simple poles. Substituting
$$\frac{1}{(E_m^{(0)} - E)(E_n^{(0)} - E)} = \frac{1}{E_m^{(0)} - E_n^{(0)}} \left(\frac{1}{E_n^{(0)} - E} - \frac{1}{E_m^{(0)} - E} \right),$$

one recovers the classical result of second order perturbation theory:
$$E_n^{(2)} = \sum_{m \neq n} \frac{|\langle n | V | m \rangle|^2}{E_n^{(0)} - E_m^{(0)}}.$$

3.6 Hamiltonian: structure of the ground state

Uniqueness or, on the contrary, degeneracy of the ground state of a quantum hamiltonian plays an essential role for the structure of a theory. In particular, from the viewpoint of statistical physics, most phase transitions are associated with a transition between a situation where the ground state is unique to a situation where it is degenerate.

In this chapter as well as in Chapter 4, we consider only real quantum hamiltonians associated with a finite number of degrees of freedom. Then, the ground state is unique and phase transitions are impossible. This is the property we now prove for all hamiltonians of the form
$$H = \hat{\mathbf{p}}^2/2m + V(\hat{\mathbf{q}})$$

with potentials bounded on any finite interval (but the result can be generalized to potentials that are not too singular). Moreover, we assume that the ground state energy corresponds to an isolated point of the spectrum, in such a way that the eigenfunctions are square integrable.

The analysis is based on a variational principle: the ground state energy E_0 satisfies the inequality
$$E_0 \le \frac{\langle \psi | H | \psi \rangle}{\langle \psi | \psi \rangle} \qquad (3.53)$$
with
$$\langle \psi | \psi \rangle = \int d^d q\, \psi^2(q),$$
$$\langle \psi | H | \psi \rangle = \int d^d q \left[\tfrac{1}{2} (\nabla \psi(q))^2 + V(q) \psi^2(q) \right],$$
where the equality is possible only if $\psi(q)$ is the ground state eigenfunction.

To prove uniqueness, we first show that the ground state eigenfunction $\psi(q)$, which can be chosen real, does not vanish.

First, we consider the one-dimensional example ($d = 1$) and assume that the eigenfunction vanishes at a point that, for convenience, we choose to be the origin. Since the potential is bounded, from a local solution of the time-independent Schrödinger equation we infer
$$\psi(q) = \psi'_0 q + o(q^2).$$
We then consider the wave function
$$\varphi(q) = |\psi(q)|.$$
The function $\varphi(q)$ corresponds to the same energy E_0 since in (3.53) both the numerator and denominator remain unchanged. In the vicinity of the origin, $\varphi(q)$ behaves like $|\psi'_0||q|$. We now modify the function $\varphi(q)$ locally near the origin for $|q| \le \eta \ll 1$, replacing $|q|$ by $\sqrt{\varepsilon^2 + q^2}$, $0 < \varepsilon \ll \eta$. The variations of the denominator and numerator induced by this change are
$$\delta \langle \psi | \psi \rangle / \psi'^2_0 = \int_{-\eta}^{\eta} dq\, (q^2 + \varepsilon^2 - q^2) = O(\varepsilon^2),$$
$$\delta \langle \psi | H | \psi \rangle / \psi'^2_0 = 2\eta V(0) \varepsilon^2 + \int_{-\eta}^{\eta} dq \left(\frac{q^2}{q^2 + \varepsilon^2} - 1 \right) \sim -\pi \varepsilon.$$

The variation of the numerator is negative and of order ε, while the variation of the denominator is only of order ε^2. Therefore, the ratio (3.53) has decreased, in contradiction with the assumption that $\psi(q)$ is the ground state eigenfunction.

We conclude that the ground state eigenfunction can be chosen strictly positive.

If the ground state is degenerate, one thus finds two positive eigenfunctions, which can be chosen to be orthogonal since the hamiltonian H is hermitian. However, the scalar product of two positive functions is strictly positive, which is contradictory.

We conclude that the ground state is unique.

The argument generalizes to potentials less singular that $1/q^2$.

It also generalizes to any finite space dimension. If $\psi(q)$ vanishes at the origin, it can be expanded as
$$\psi(q) = \mathbf{q} \cdot \nabla \psi(0) + o(\mathbf{q}^2).$$
One then argues, using the component of the vector $\mathbf{q}$ on the vector $\nabla \psi(0)$.

Exercises

Remark. The path integral formalism must be used in all calculations. One assumes $m = \hbar = 1$.

Exercise 3.1

One considers the hamiltonian
$$H = \tfrac{1}{2}\hat{p}^2 + \tfrac{1}{2}\hat{q}^2 + \lambda \hat{q}^N,$$
where $\hat{p}$, $\hat{q}$ are the momentum and position operators:
$$[\hat{q}, \hat{p}] = i.$$
Write the partition function as a path integral. For $N = 6, 8$, infer the corrections of order λ to the *ground state* energy, and then to all states using Wick's theorem, generalizing the method of Section 3.

Solution. For $N = 6$ and $N = 8$ the energy eigenvalues are, respectively,
$$E_k = n + \tfrac{1}{2} + \tfrac{5}{2}\lambda(k + \tfrac{1}{2})\left(k^2 + k + \tfrac{3}{2}\right)$$
$$E_k = k + \tfrac{1}{2} + \tfrac{35}{8}\lambda\left(k^4 + 2k^3 + 5k^2 + 4k + \tfrac{3}{2}\right).$$

Exercise 3.2

We consider the hamiltonian
$$H = \tfrac{1}{2}(\hat{p}_1^2 + \hat{p}_2^2) + \tfrac{1}{2}(\hat{q}_1^2 + \hat{q}_2^2) + \tfrac{1}{4}g(\hat{q}_1^2 + \hat{q}_2^2)^2.$$
Express the partition function as a path integral, and infer the correction of order g to the *ground state* energy E_0 using Wick's theorem.

Solution.
$$E_0 = 1 + g/2 + O(g^2).$$

Exercise 3.3

Generalize the previous calculations to the example of the hamiltonian
$$H = \tfrac{1}{2}\hat{\mathbf{p}}^2 + \tfrac{1}{2}\hat{\mathbf{q}}^2 + \tfrac{1}{4}g(\hat{\mathbf{q}})^2$$
where $\mathbf{q}$ and $\mathbf{p}$ are N-component vectors (the preceding exercise corresponds to $N = 2$).

As a preparation, calculate, using Wick's theorem, the expectation value
$$I(N) = Z^{-1} \int d^N q \, (\mathbf{q}^2)^2 \, e^{-\mathbf{q}^2/2},$$
$$Z = \int d^N q \, e^{-\mathbf{q}^2/2},$$
where $\mathbf{q}$ is a N-component vector.

Calculate then the ground state energy E_0 at first order in g.

Solution.
$$I(N) = N(N+2),$$
$$E_0(g) = \tfrac{1}{2}N + \tfrac{1}{16}N(N+2)g + O(g^2).$$

Exercise 3.4

Calculate the *ground state* energy of the hamiltonian

$$H = \tfrac{1}{2}\hat{p}^2 + \tfrac{1}{2}\hat{q}^2 + \lambda\gamma\hat{q}^3 + \tfrac{1}{2}\lambda^2\hat{q}^4,$$

where γ is an arbitrary constant, at order λ^2.

Solution.
$$E_0 = \tfrac{1}{2} + \tfrac{1}{8}(3 - 11\gamma^2)\lambda^2 + O(\lambda^4).$$

Exercise 3.5

Calculate the energy eigenvalues E_k of the hamiltonian (note the difference in normalization of λ with respect to Section 3.1)

$$H = \tfrac{1}{2}\hat{p}^2 + \tfrac{1}{2}\hat{q}^2 + \lambda\hat{q}^4,$$

at order λ^2 by generalizing the method of Section 3.1.

Solution. Setting $\nu = k + \tfrac{1}{2}$, one finds

$$E = \nu + \tfrac{3}{2}\left(\nu^2 + \tfrac{1}{4}\right)\lambda - \tfrac{1}{4}\nu\left(17\nu^2 + \tfrac{67}{4}\right)\lambda^2 + O(\lambda^3).$$

Exercise 3.6

Generalize the calculation at order λ^2 to the potential

$$V(q) = \tfrac{1}{2}q^2 + \lambda v_1 q^4 + \lambda^2 v_2 q^6 + O(q^8),$$

where v_1, v_2 are two arbitrary constants. Invert the relation between ν and E.

Solution. One obtains an expansion that can be written more simply as

$$k + \tfrac{1}{2} = E - \lambda v_1 \left(\tfrac{3}{2}E^2 + \tfrac{3}{8}\right) + \lambda^2 \left[E^3 \left(\tfrac{35}{4}v_1^2 - \tfrac{5}{2}v_2\right) + E\left(\tfrac{85}{16}v_1^2 - \tfrac{25}{8}v_2\right)\right] + \cdots.$$

Exercise 3.7

Apply the variational method of Section 3.3 to the example of the hamiltonian (3.28) with the potential (3.29) when $\omega = 0$, that is,

$$H = \frac{1}{2}\hat{\mathbf{p}}^2 + \frac{1}{24}\frac{g}{N}(\hat{\mathbf{q}}^2)^2,$$

where $\mathbf{q}$ and $\mathbf{p}$ are N-component vectors. Infer an estimate of the ground state energy and compare it with the result (3.49) for $N \to \infty$.

Solution. One chooses an $O(N)$-symmetric gaussian ansatz with

$$S_0(\mathbf{q}) = \frac{1}{2}\int_0^\beta dt\left[\dot{\mathbf{q}}^2(t) + \omega^2 \mathbf{q}^2(t)\right],$$

where ω is a variational parameter. Then,

$$\int_0^\beta dt \left\langle \left(\mathbf{q}^2(t)\right)^2 \right\rangle_0 = \beta N(N+2)\Delta^2(0).$$

In the large β limit, one finds

$$E_{\text{var.}} = \frac{N\omega}{4} + \frac{(N+2)g}{96\omega^2},$$

and after minimization

$$E_{\text{var.}} = \frac{3N}{8}\left(\frac{g(N+2)}{12N}\right)^{1/3}.$$

4 CLASSICAL AND QUANTUM STATISTICAL PHYSICS

The main goal of this chapter is to provide a simple physical interpretation to the formal continuum limit that has led, from an integral over position variables corresponding to discrete times, to a path integral. We will show that the integral corresponding to discrete times can be considered as the partition function of a classical statistical system in one space dimension. The continuum limit, then, corresponds to a limit where the correlation length, which characterizes the decay of correlations at large distance, diverges. This limit has some universality properties in the sense that, as we have already pointed out, different discretized forms lead to the same path integral.

In this statistical framework, the correlation functions that have been introduced in Section 2.5, appear as continuum limits of the correlation functions of classical statistical models on a one-dimensional lattice.

The path integral, thus, allows us to exhibit a mathematical relation between classical statistical physics on a line and quantum statistical physics of a point-like particle at thermal equilibrium. This is the first example of more general relations that can be found between quantum statistical physics in D dimensions and classical statistical physics in $D+1$ dimensions.

Notice that this relation between quantum and classical physics is different from the classical limit of the quantum partition function that we have studied in Section 2.10, where the point-like quantum particle has the point-like classical particle as a limit. In fact, the semi-classical or high temperature limit appears as a kind of dimensional reduction in this context.

The specific form of the discretized path integral that we use in this chapter corresponds to a classical statistical model with nearest-neighbour interaction. To study such models, it is useful to introduce a transfer matrix, the direct analogue of the statistical operator (2.17) for short time intervals. The expressions of correlation functions in terms of the transfer matrix can then be used to investigate various issues like thermodynamic limit and large distance behaviour. Of particular interest is the two-point function, which we discuss more thoroughly.

Notice that in the whole chapter *the potential is time-independent.*

4.1 Classical partition function. Transfer matrix

We consider the one-dimensional lattice of points with integer coordinates. To each lattice point $k \in \mathbb{Z}$ is associated a stochastic variable q_k (e.g. the deviation of a particle from its equilibrium position in a crystal). To simplify the notation and many expressions, we restrict the study to one variable q_k per site.

We first define the model on a finite lattice $0 \le k < n$. Moreover, in the whole chapter we assume, for convenience, *periodic boundary conditions*. The lattice thus belongs to a circle and we identify $q_n = q_0$.

To a set of values of the variables q_k, we associate a Boltzmann weight $e^{-\mathcal{S}}$, exponential of a configuration energy

$$\mathcal{S}(q,\varepsilon) = \sum_{k=1}^{n} S(q_k, q_{k-1}), \tag{4.1}$$

$$S(q, q') = \left[\frac{1}{2} \frac{(q'-q)^2}{\varepsilon} + \frac{1}{2}\varepsilon \left(V(q') + V(q) \right) \right]. \tag{4.2}$$

This configuration energy defines a classical statistical model with nearest-neighbour interaction, as the Ising model in its simplest version. The control parameter ε plays a role, here, that is analogous to the temperature (it is directly related to the temperature in special cases, see Section 4.3).

One then notices that $\mathcal{S}(q,\varepsilon)$ is identical up to order ε, for $\varepsilon \to 0$, to the action (2.18) (for $t' = t + \varepsilon$) in the case $d = 1$, $m = 1$.

We also normalize the integration measure as $dq/\sqrt{2\pi\varepsilon}$.

The partition function of this classical model then reads

$$\mathcal{Z}(n,\varepsilon) = \int \prod_{k=1}^{n} \frac{dq_k}{\sqrt{2\pi\varepsilon}} \exp\left[-\mathcal{S}(q,\varepsilon)\right]. \tag{4.3}$$

In the case of periodic boundary conditions ($q_0 = q_n$), the classical partition function $\mathcal{Z}(n,\varepsilon)$ can also be considered as a discrete approximation, equivalent to the trace of the r.h.s. of equation (2.19), to the quantum partition function (2.23) $\mathcal{Z}(\beta) = \operatorname{tr} e^{-\beta H}$ corresponding to the quantum hamiltonian

$$H = \tfrac{1}{2}\hat{p}^2 + V(\hat{q}),$$

for $\hbar = 1$.

In expression (4.2), the two terms now have the following classical interpretation:

The potential term determines the distribution of the variables q_k at each site. In order for the distribution to be normalizable, the integral $\int dq\, e^{-\varepsilon V(q)}$ must converge for all $\varepsilon > 0$. This implies, in particular, that $V(q) \to +\infty$ for $|q| \to \infty$ and, thus, that the hamiltonian H has a discrete spectrum.

The kinetic term corresponds to an attractive interaction (it favours all q_k equal) between nearest neighbours on the lattice. A decrease of the parameter ε increases the interaction.

Notation. In what follows, it is convenient to introduce the usual bra and ket notation of quantum mechanics, and the complete basis in which the quantum position operator $\hat{q}$ is diagonal (equations (2.4, 2.5)):

$$\hat{q}\left|q\right\rangle = q\left|q\right\rangle, \quad \int dq \left|q\right\rangle \left\langle q\right| = 1. \tag{4.4}$$

Note that, in what follows, the notation $\langle \bullet \rangle$ means expectation value of $\bullet$ with respect to the statistical weight $e^{-\mathcal{S}}/\mathcal{Z}$, while $\langle q''|\mathbf{U}|q'\rangle$ means matrix element of the quantum operator $\mathbf{U}$ in the position basis.

Transfer matrix. The kernel

$$\mathcal{T}(q,q') \equiv \langle q'|\mathbf{T}(\varepsilon)|q\rangle = \frac{1}{\sqrt{2\pi\varepsilon}} \exp[-S(q,q')] \qquad (4.5)$$

defines, through its matrix elements in the position basis, a real symmetric operator $\mathbf{T}$, which is called the *transfer matrix* of the statistical model. Moreover, the assumption that the integral $\int dq\, e^{-\varepsilon V(q)}$ exists, implies that it has a discrete spectrum. In terms of the position and momentum quantum operators $\hat{q}$ and $\hat{p}$ (equation (2.10)), the transfer matrix reads

$$\mathbf{T}(\varepsilon) = e^{-\varepsilon V(\hat{q})/2}\, e^{-\varepsilon \hat{p}^2/2}\, e^{-\varepsilon V(\hat{q})/2}, \qquad (4.6)$$

a form that shows, in particular, that $\mathbf{T}$ is a positive operator.

Using the definition (4.5), it is straightforward to express the partition function $\mathcal{Z}(n,\varepsilon)$ of the statistical model in terms of the transfer matrix:

$$\mathcal{Z}(n,\varepsilon) = \int \prod_{k=0}^{n-1} dq_k\, \langle q_0|\mathbf{T}(\varepsilon)|q_1\rangle \langle q_1|\mathbf{T}(\varepsilon)|q_2\rangle \cdots \langle q_{n-1}|\mathbf{T}(\varepsilon)|q_0\rangle$$
$$= \operatorname{tr} \mathbf{T}^n(\varepsilon) \qquad (4.7)$$

and, thus, introducing the eigenvalues (discrete, real positive) $t_k(\varepsilon)$ of the transfer matrix $t_0 > |t_1| \geq \cdots$,

$$\mathcal{Z}(n,\varepsilon) = \sum_{k=0}^{\infty} t_k^n(\varepsilon). \qquad (4.8)$$

Thermodynamic limit. The thermodynamic limit is the limit in which the volume diverges, here $n \to \infty$. In this limit, the partition function is dominated by the largest eigenvalue of the transfer matrix (which can be shown to be non-degenerate):

$$\ln \mathcal{Z}(n,\varepsilon) \underset{n\to\infty}{\sim} n \ln t_0(\varepsilon).$$

In particular, one verifies that the free energy $\mathcal{W} = \ln \mathcal{Z}$ (omitting a temperature factor irrelevant in this context) is an extensive quantity, that is proportional to the volume n. The free energy density is given by

$$\frac{1}{n}\mathcal{W}(n,\varepsilon) = \frac{1}{n}\ln \mathcal{Z}(n,\varepsilon) \sim \ln t_0(\varepsilon).$$

4.2 Correlation functions

The correlation functions of the variables q_k are the moments of the distribution. The m-point correlation function is defined by

$$\langle q_{i_1} q_{i_2} \ldots q_{i_m} \rangle \equiv Z_n^{(m)}(i_1, i_2, \ldots, i_m)$$
$$= \mathcal{Z}^{-1}(n, \varepsilon) \int \left(\prod_{s=1}^n \frac{\mathrm{d}q_s}{\sqrt{2\pi\varepsilon}} \right) \exp\left[-\mathcal{S}(q, \varepsilon)\right] \prod_{\ell=1}^m q_{i_\ell}. \qquad (4.9)$$

4.2.1 Correlation functions and transfer matrix

In the special case of nearest-neighbour interactions, as generated by the configuration energy (4.1), correlation functions can be expressed in terms of the transfer matrix.

The one-point function

$$Z_n^{(1)}(k) \equiv \langle q_k \rangle = \mathcal{Z}^{-1}(n, \varepsilon) \int \left(\prod_{s=1}^n \frac{\mathrm{d}q_s}{\sqrt{2\pi\varepsilon}} \right) q_k \exp\left[-\mathcal{S}(q, \varepsilon)\right]$$

is the expectation value of q_k. The integral can be written as

$$\int \left(\prod_{s=1}^n \mathrm{d}q_s \right) q_k \, \mathrm{e}^{-\mathcal{S}(q,\varepsilon)} = \int q_k \mathrm{d}q_k \int \left(\prod_{s=1}^{k-1} \mathrm{d}q_s \right) \exp\left[-\sum_{l=1}^k S(q_l, q_{l-1})\right]$$
$$\times \int \left(\prod_{s=k+1}^n \mathrm{d}q_s \right) \exp\left[-\sum_{l=k+1}^n S(q_l, q_{l-1})\right].$$

Introducing the transfer matrix (4.5) and performing all integrals except that over q_k ($q_0 = q_n$), one obtains

$$Z_n^{(1)}(k) = \mathcal{Z}^{-1}(n, \varepsilon) \int \mathrm{d}q_k \, q_k \, \langle q_k | \mathbf{T}^n(\varepsilon) | q_k \rangle.$$

Introducing then the position operator $\hat{q}$ and using (equation (4.4))

$$\hat{q} |q_k\rangle = q_k |q_k\rangle,$$

one can rewrite the one-point function as

$$Z_n^{(1)}(k) = \mathcal{Z}^{-1}(n, \varepsilon) \int \mathrm{d}q_k \, \langle q_k | \mathbf{T}^n(\varepsilon) \hat{q} | q_k \rangle = \mathrm{tr}\left(\hat{q} \, \mathbf{T}^n(\varepsilon)\right) / \mathrm{tr}\, \mathbf{T}^n(\varepsilon).$$

The same method can be applied to the two-point correlation function, expectation value of the product of the variables q in two points k, l of the lattice:

$$Z_n^{(2)}(k, l) \equiv \langle q_k \, q_l \rangle = \mathcal{Z}^{-1}(n, \varepsilon) \int \int \left(\prod_{s=1}^n \frac{\mathrm{d}q_s}{\sqrt{2\pi\varepsilon}} \right) q_k q_l \exp\left[-\mathcal{S}(q, \varepsilon)\right].$$

Choosing, for example, $k \leq l$ and decomposing the interval $(1, n)$ into $(1, k)$, $(k+1, l)$, $(l+1, n)$ and integrating over all variables but q_k and q_l, one obtains

$$Z_n^{(2)}(k, l) = \mathcal{Z}^{-1}(n, \varepsilon) \int \mathrm{d}q_k \, \mathrm{d}q_l \, q_k q_l \, \langle q_k | \mathbf{T}^{l-k}(\varepsilon) | q_l \rangle \, \langle q_l | \mathbf{T}^{n+k-l}(\varepsilon) | q_k \rangle \, .$$

Using

$$\hat{q} | q_k \rangle = q_k | q_k \rangle \, , \quad \hat{q} | q_l \rangle = q_l | q_l \rangle \, ,$$

one can then perform the two last integrations and one finds

$$Z_n^{(2)}(k, l) = \operatorname{tr} \left(\hat{q} \, \mathbf{T}^{l-k}(\varepsilon) \, \hat{q} \, \mathbf{T}^{n+k-l}(\varepsilon) \right) / \operatorname{tr} \mathbf{T}^n(\varepsilon).$$

Note that while the function $Z_n^{(2)}(k, l)$ is symmetric in $k \leftrightarrow l$, this is not the case for the explicit expression in terms of the transfer matrix.

More generally, the same method leads to a representation of the m-point function, which in the sector $0 \leq i_1 \leq i_2 \ldots \leq i_m \leq n$, can be written as

$$\langle q_{i_1} q_{i_2} \ldots q_{i_m} \rangle = \operatorname{tr} \left[\mathbf{T}^{n-i_m+i_1} \hat{q} \, \mathbf{T}^{i_m-i_{m-1}} \hat{q} \ldots \mathbf{T}^{i_2-i_1} \hat{q} \right] / \operatorname{tr} \mathbf{T}^n \, . \tag{4.10}$$

4.2.2 Thermodynamic limit and large distance behaviour

We now denote by $|0\rangle, |1\rangle, \cdots$, the eigenvectors of $\mathbf{T}$ associated with the eigenvalues $t_0 > t_1 > \cdots$.

In the thermodynamic limit $n \to \infty$, $\mathbf{T}$ is dominated by the largest eigenvalues:

$$\mathbf{T}(\varepsilon) \underset{n \to \infty}{=} t_0^n(\varepsilon) \left[|0\rangle \langle 0| + |1\rangle \langle 1| \left(t_1(\varepsilon)/t_0(\varepsilon) \right)^n + O \left(\left(t_2(\varepsilon)/t_0(\varepsilon) \right)^n \right) \right] \, .$$

For $n \to \infty$, the one-point function thus has the limit

$$\lim_{n \to \infty} Z_n^{(1)}(k) = \langle q_k \rangle = \langle 0 | \hat{q} | 0 \rangle \, .$$

More generally, the m-point function in the sector $0 \leq i_1 \leq i_2 \ldots \leq i_m \leq n$ has the limit

$$\langle q_{i_1} q_{i_2} \ldots q_{i_m} \rangle = t_0^{i_1 - i_m} \langle 0 | \hat{q} \, \mathbf{T}^{i_m - i_{m-1}} \hat{q} \ldots \mathbf{T}^{i_2 - i_1} \hat{q} | 0 \rangle \, . \tag{4.11}$$

The two-point function at large distance. The two-point function in the thermodynamic limit becomes

$$Z^{(2)}(k, l) = t_0^{-|k-l|} \langle 0 | \hat{q} \mathbf{T}^{|k-l|} \hat{q} | 0 \rangle \, .$$

When the separation $|k - l|$ diverges, the two-point function is dominated by the first terms of the expansion of $\mathbf{T}^{|k-l|}$:

$$Z^{(2)}(k, l) = (\langle 0 | \hat{q} | 0 \rangle)^2 + \left(\frac{t_1(\varepsilon)}{t_0(\varepsilon)} \right)^{|k-l|} (\langle 0 | \hat{q} | 1 \rangle)^2 + O \left(\left(t_2(\varepsilon)/t_0(\varepsilon) \right)^{|k-l|} \right) \, .$$

The leading term is a constant, the square of the expectation value $\langle q_k \rangle = \langle 0|\hat{q}|0\rangle$. We then introduce the *connected* two-point function (see Section 1.4.2):

$$W^{(2)}(k,l) = Z^{(2)}(k,l) - Z^{(1)}(k)Z^{(1)}(l) = \langle (q_k - \langle q_k \rangle)(q_l - \langle q_l \rangle)\rangle. \tag{4.12}$$

The leading term cancels and

$$W^{(2)}(k,l) \underset{|k-l|\to\infty}{\sim} \left(\frac{t_1(\varepsilon)}{t_0(\varepsilon)}\right)^{|k-l|} (\langle 0|\hat{q}|1\rangle)^2.$$

(If $\langle 0|\hat{q}|1\rangle = 0$, the next term in the large $k - l$ expansion must be considered.) The connected two-point function thus decreases exponentially, for $|l - k| \to \infty$, as

$$W^{(2)}(k,l) \underset{|k-l|\to\infty}{\propto} e^{-|k-l|/\xi},$$

where we have introduced the *correlation length*

$$\xi = \frac{1}{\ln[t_0(\varepsilon)/t_1(\varepsilon)]}. \tag{4.13}$$

This is a first example of a more general property. Connected correlation functions satisfy a *cluster property*: namely, they vanish asymptotically when the distance between two non-empty complementary sets of points goes to infinity.

Remark. The relations (4.12) show that the connected two-point function is identical to the two-point function of a shifted variable q'_k obtained by removing the expectation value:

$$q'_k = q_k - \langle q_k \rangle \Rightarrow \langle 0|\hat{q}'|0\rangle = \langle q'_k \rangle = 0.$$

The expectation value $\langle q_k \rangle$ has, in general, no special physical meaning since it corresponds simply to a choice of coordinate. However, in some situations the point $q = 0$ is distinguished, for example, when a system is invariant under the reflection $q \mapsto -q$. Then, a non-vanishing expectation value has the meaning of a *spontaneous breaking of a symmetry*. By going through the preceding arguments, one verifies that this implies that the vector $|0\rangle$ is not invariant under the same symmetry. Therefore, the symmetric of $|0\rangle$ is an eigenvector with the same eigenvalue and thus the eigenvalue t_0 is degenerate. Generalizing the result derived in Section 3.6, it is possible to prove that such a situation cannot occur in quantum systems with a finite number of degrees of freedom.

4.3 Classical model at low temperature: an example

We have shown that the discrete approximation to the path integral has a natural interpretation in terms of a classical statistical system on a one-dimensional lattice. We illustrate now this idea with a simple example. We consider the partition function

$$\mathcal{Z}_n(T) = \int \left(\prod_{k=1}^{n} \mathrm{d}\rho(q_k)\right) \exp\left[-E(q_i)/T\right], \tag{4.14}$$

where q_i characterizes the configuration at point i of a one-dimensional lattice (for example the deviation of a particle from its equilibrium position), n the size of the lattice, T the temperature, $\mathrm{d}\rho(q)$ the distribution of the variable q and $E(q_i)$ the configuration energy, which we choose of the special form (nearest-neighbour interaction)

$$E(q) = \sum_{i=1}^{n} \tfrac{1}{2} J (q_i - q_{i-1})^2, \quad J > 0, \tag{4.15}$$

(J characterizes the strength of the interaction). Moreover, we impose, for convenience, periodic boundary conditions: $q_n = q_0$.

We set

$$\mathrm{d}\rho(q) = \mathrm{e}^{-v(q)}\,\mathrm{d}q.$$

The simplest example corresponds to a function $v(q)$ with a unique minimum at $q = 0$ where it is regular,

$$v(q) = \tfrac{1}{2} v_2 q^2 + O\left(q^3\right).$$

One can interpret this term as an effective representation of a force that recalls a particle to its equilibrium position, and which has a negligible dependence on the temperature.

For $T \to 0$, the leading configuration is obtained by minimizing the energy (4.14) and corresponds to all q_i equal. Thus, for $n \to \infty$, only configurations close to the configuration $q_i = 0$, which is the most probable, contribute to the partition function. In these limits, after the change of variables $q_i \mapsto T^{1/4} q_i$, the partition function is given by

$$\mathcal{Z}_n(T) \propto \int \prod_{k=1}^{n} \mathrm{d}q_k \, \exp\left[-\mathcal{S}(q)\right] \tag{4.16}$$

with

$$\mathcal{S}(q) = \frac{1}{2} \sum_{i=1}^{n} \left[J(q_i - q_{i-1})^2/\sqrt{T} + v_2 \sqrt{T} q_i^2\right]. \tag{4.17}$$

Comparing with expression (2.39) for $\hbar = \omega = 1$, one infers that, for $T \to 0$,

$$\mathcal{Z}_n(T) \propto \mathrm{tr}\, \mathrm{e}^{-n\varepsilon \hat{H}},$$

where $\hat{H}$ is the hamiltonian of the harmonic oscillator:

$$\hat{H} = \frac{1}{2m}\hat{p}^2 + \frac{m}{2}\hat{q}^2 \tag{4.18}$$

with

$$\varepsilon = \sqrt{v_2 T/J} + O(T^{3/2}), \quad m = \sqrt{v_2 J} + O(T).$$

The continuum limit, which leads to the path integral, corresponds to $\varepsilon \propto \sqrt{T} \to 0$, that is, to a low temperature limit.

The behaviour of the two-point function at large distance is related to the difference between the two first eigenvalues of $\hat{H}$. Defining the correlation length ξ by

$$\langle q_k q_0 \rangle \underset{k \to \infty}{\propto} e^{-k/\xi},$$

one infers from equation (4.13) that for $T \to 0$:

$$\xi \sim 1/\varepsilon = \sqrt{J/v_2 T}.$$

The path integral is thus obtained in a limit in which the correlation length diverges, a property that is discussed in more general terms in Section 4.4.1.

4.4 Continuum limit and path integral

We examine now, more generally, the behaviour of the partition function (4.3) and the correlation functions (4.9) in the limit ε goes to zero. In the discussion of Section 2.3, ε is a time step. In the limit $\varepsilon \to 0$, integrals over variables associated with discrete time steps formally yield path integrals. We show that the existence of this continuum limit relies on the divergence of the correlation length in the corresponding classical statistical model.

4.4.1 Continuum limit

The representation (4.6) shows that

$$\ln \mathbf{T}(\varepsilon) \underset{\varepsilon \to 0}{\sim} -\varepsilon H,$$

a behaviour consistent with the discussion of Section 2.2, which leads to

$$\mathcal{T}(q, q') \underset{\varepsilon \to 0}{\sim} \langle q' | e^{-\varepsilon H} | q \rangle.$$

These relations translate into a relation between eigenvalues of the transfer matrix and the eigenvalues E_k of H (which has a discrete spectrum and whose ground state is not degenerate, see Section 3.6):

$$\ln t_k(\varepsilon) \underset{\varepsilon \to 0}{\sim} -\varepsilon E_k.$$

Thus, in the limit $\varepsilon \to 0$, the differences between eigenvalues of the transfer matrix vanish.

Divergence of the correlation length. In the limit $\varepsilon \to 0$, the correlation length (4.13) becomes

$$\xi \sim \frac{1}{\varepsilon(E_1 - E_0)}.$$

4.4 Classical and quantum statistical physics

The limit $\varepsilon \to 0$, which leads to the path integral with a continuous time, in the statistical framework thus corresponds to a limit where the correlation length diverges. In this limit, a length scale large with respect to the microscopic scale, here the lattice spacing, is generated dynamically.

Partition function in the continuum limit. The classical partition function (4.7),

$$\mathcal{Z}(n,\varepsilon) = \operatorname{tr} \mathbf{T}^n(\varepsilon),$$

can be considered as a discretized approximation of the quantum partition function $\mathcal{Z}(\beta) = \operatorname{tr} e^{-\beta H}$ as given by the path integral representation (2.23). It follows from the discussion of Section 2.3.1, that the quantum function is obtained in the double limit $n \to \infty$, $\varepsilon \to 0$ with $n\varepsilon = \beta$ fixed:

$$\lim_{\substack{n \to \infty \\ \varepsilon \to 0}} \mathcal{Z}(n,\varepsilon) = \mathcal{Z}(\beta) = \operatorname{tr} e^{-\beta H}.$$

Even though the volume n diverges, this is not the thermodynamic limit, which corresponds, instead, to $n \to \infty$ with ε fixed. In the framework of the classical statistical model, we can now give a physical interpretation to this limit, called *continuum limit* because the lattice has disappeared. Since

$$\frac{n}{\xi} = \beta(E_1 - E_0),$$

the double limit $n \to \infty$, $\varepsilon \to 0$ at $\beta = n\varepsilon$ fixed, is a limit where the size n of the system diverges but the ratio of the size and the correlation length remains fixed; β can be considered as the size of the system measured in the scale of the correlation length. The thermodynamic limit of the classical system then is obtained for $\beta \to \infty$, that is in the zero temperature limit of the quantum system. In this limit

$$\ln \mathcal{Z}(\beta) \underset{\beta \to \infty}{\sim} -\beta E_0.$$

The free energy of the classical system $W = \ln \mathcal{Z}(\beta)$ is extensive: it increases as the 'volume' when the volume diverges, and is proportional to the ground state energy of the quantum system.

In the classical model, the dependence on β is called a finite size effect.

Continuum limit and universality. The existence of a continuum limit relies on the divergence of the correlation length ξ and the emergence of a non-trivial large distance physics, which does not depend on the initial lattice structure and of most details of the microscopic model (in particular, as the discussion of Section 2.2 has shown, we could have chosen a form more general than expression (4.2)). This independence of large distance continuum properties from the detailed initial structure of the microscopic model is called *universality*.

The phenomenon that we observe here has a deeper meaning: quite generally, in a statistical model a non-trivial continuum limit exists only if the correlation length diverges. Such a divergence occurs also at a continuous or second order phase transition. This phenomenon explains why continuous phase transitions (where the correlation length diverges) can be described by statistical (euclidean) field theories. In one space dimension with short range interactions, phase transitions are impossible and the correlation length can diverge only in the zero temperature limit, which explains the role of the limit $\varepsilon \to 0$.

4.4.2 Correlation functions and continuum limit

We have shown that the m-point function, in the sector $0 \le i_1 \le i_2 \cdots \le i_m \le n$, can be written as (equation (4.10))

$$\langle q_{i_1} q_{i_2} \ldots q_{i_m} \rangle = \operatorname{tr} \mathbf{T}^{n-i_m+i_1} \hat{q}\, \mathbf{T}^{i_m - i_{m-1}} \hat{q} \ldots \mathbf{T}^{i_2 - i_1} \hat{q} / \operatorname{tr} \mathbf{T}^n .$$

We introduce the variables

$$t_k = \varepsilon i_k \;\Rightarrow\; i_k/\xi = t_k(E_1 - E_0),$$

and the trajectory $q(t)$ such that $q(k\varepsilon) = q_k$ (see Section 2.3.1) and, thus, $q(t_k) = q_{i_k}$. The continuum limit is now a limit $\varepsilon \to 0$ at β and t_k fixed: distances are fixed in the scale of the correlation length, instead of the lattice spacing, which is the microscopic scale. In this limit, the transfer matrix can be expressed in terms of the quantum hamiltonian:

$$\langle q(t_1) q(t_2) \ldots q(t_m) \rangle_\beta$$
$$= \operatorname{tr} \mathrm{e}^{-(\beta - t_m + t_1)H} \hat{q}\, \mathrm{e}^{-(t_m - t_{m-1})H} \hat{q} \ldots \mathrm{e}^{-(t_2 - t_1)H} \hat{q} / \operatorname{tr} \mathrm{e}^{-\beta H}. \qquad (4.19)$$

Simultaneously, the multiple integrals converge toward path integrals:

$$\langle q(t_1) q(t_2) \ldots q(t_m) \rangle_\beta = \mathcal{Z}^{-1}(\beta) \int [\mathrm{d}q]\, q(t_1) \ldots q(t_m)\, \mathrm{e}^{-\mathcal{S}(q)} . \qquad (4.20)$$

In the continuum limit, correlation functions of the statistical model converge towards correlation functions defined in terms of the path integral. In this limit, correlation functions also exhibit universality: they can be expressed in terms of path integrals, where any reference to the initial lattice has disappeared.

Thermodynamic limit. The thermodynamic limit of the classical one-dimensional model nows corresponds to $\beta \to \infty$, that is to the zero temperature limit of a quantum statistical model in zero dimension (one point-like particle). In this limit, correlation functions become

$$\langle q(t_1) q(t_2) \ldots q(t_m) \rangle = \langle 0 | \hat{q}\, \mathrm{e}^{-(t_m - t_{m-1})(H - E_0)} \hat{q} \ldots \mathrm{e}^{-(t_2 - t_1)(H - E_0)} \hat{q} | 0 \rangle, \qquad (4.21)$$

where $|0\rangle$ is the ground state of the hamiltonian.

The limit is a ground state expectation value of a product of operators, which in the framework of the many-body theory (see Chapters 6 and 7) or quantum field theory is also called a *vacuum expectation value*.

Quantum and classical statistical models. We have shown how a quantum statistical model with one particle is obtained as the continuum limit of a one-dimensional classical statistical model. This relation has generalizations in higher space dimensions (D dimensions for the quantum model, $D+1$ dimensions for the classical model), the quantum character being reflected classically by an additional dimension of the size of the inverse temperature β, with periodic boundary conditions.

Note, however, that the analogy between quantum and classical statistical models is incomplete, in the sense that only equal-time correlation functions have a direct quantum interpretation, as thermal expectation values of powers of the position operator.

In contrast, in higher dimensions one can find classical equal-time correlation functions that have quantum analogues.

Finally, after analytic continuation to real time $t \mapsto it$, (but not for β!), classical correlation functions transform into time-dependent, thermal, quantum correlation functions.

Remarks.

(i) Other stochastic processes lead to path integrals, where the corresponding correlation functions are physical observables. For example, the simplest random walk corresponds to a free quantum hamiltonian. The solution of the Fokker–Planck equation (Section 5.5), which describes more general stochastic processes, can be written as a path integral.

(ii) When the euclidean action is not real (as in the case of a hamiltonian in a magnetic field, see Section 5.1), expectation values of the form (4.20) no longer have a simple statistical interpretation. For reasons that originate from the study of real-time quantum evolution, they remain useful quantities to consider and, for convenience, we will still call them correlation functions.

Generating functional. Perturbation theory. We have shown in Section 2.8 how to calculate a path integral perturbatively in terms of the gaussian path integral (2.47) for all hamiltonians of the form $p^2/2m + V(q)$. The argument immediately generalizes to the corresponding correlation functions. Moreover, by adding to the potential a term corresponding to an arbitrary external force, one obtains a generating functional of correlation functions (see Section 2.5). Defining

$$\mathcal{Z}(b) = \int [\mathrm{d}q(t)] \exp\left[-\mathcal{S}(q) + \int \mathrm{d}t\, q(t) b(t)\right], \tag{4.22}$$

one verifies

$$\langle q(t_1) \ldots q(t_n)\rangle = \mathcal{Z}^{-1}(b=0) \left(\frac{\delta}{\delta b(t_1)} \cdots \frac{\delta}{\delta b(t_n)}\right) \mathcal{Z}(b)\bigg|_{b=0}.$$

One can then combine this representation and the perturbative expansion.

Finally, the functional $\mathcal{W}(b) = \ln \mathcal{Z}(b)$ is the generating functional of connected correlation functions (see Section 1.4.2).

Quantum equation of motion. From representation (4.22), one can infer relations between correlation functions, by expressing that the integral over whole space of a total derivative vanishes:

$$\int [\mathrm{d}q(t)] \frac{\delta}{\delta q(\tau)} \exp\left[-\mathcal{S}(q) + \int \mathrm{d}t\, q(t)b(t)\right] = 0,$$

or more explicitly:

$$\int [\mathrm{d}q(t)] \left[b(\tau) - \frac{\delta \mathcal{S}(q)}{\delta q(\tau)}\right] \exp\left[-\mathcal{S}(q) + \int \mathrm{d}t\, q(t)b(t)\right] = 0.$$

One then notices that $q(t)$ inside the integral can be replaced by a function differentiation with respect to $b(t)$, which can be taken out of the integral, yielding the equation

$$\left[b(\tau) - \frac{\delta \mathcal{S}(\delta/\delta b(t))}{\delta q(\tau)}\right] \int [\mathrm{d}q(t)] \exp\left[-\mathcal{S}(q) + \int \mathrm{d}t\, q(t)b(t)\right] = 0.$$

In the remaining integral one recognizes the generating functional $\mathcal{Z}(b)$. The equation can thus be rewritten as

$$\left[b(\tau) - \frac{\delta \mathcal{S}(\delta/\delta b(t))}{\delta q(\tau)}\right] \mathcal{Z}(b) = 0. \tag{4.23}$$

This functional equation satisfied by $\mathcal{Z}(b)$, called Schwinger–Dyson equation in quantum field theory, has the initial path integral for solution. If one expands the functional equation in powers of b, one generates an infinite set of equations relating different correlation functions.

4.5 The two-point function: perturbative expansion, spectral representation

The two-point correlation function plays an important role in quantum field theory and in statistical physics. In the framework of a simple quantum model, we first calculate it perturbatively to first order beyond the gaussian approximation, a calculation that illustrates some aspects of the analysis of previous sections. We then establish the existence and the properties of its spectral representation.

4.5.1 Perturbative calculation

We now calculate the two-point function $\langle q(t)q(u)\rangle_\lambda$ corresponding to the action

$$\mathcal{S}(q) = \int_{-\beta/2}^{\beta/2} \mathrm{d}t \left[\tfrac{1}{2}\dot{q}^2(t) + \tfrac{1}{2}q^2(t) + \frac{1}{4!}\lambda q^4(t)\right], \tag{4.24}$$

at order λ, and in the limit $\beta \to \infty$. We assume periodic boundary conditions.

The algebra is the same as in Section 1.4.1. The expansion to order λ can be written as

$$Z^{(2)}(t,u) \equiv \langle q(t)q(u)\rangle_\lambda$$

$$= \frac{\mathcal{Z}(\beta,0)}{\mathcal{Z}(\beta,\lambda)}\left[\Delta(t-u) - \tfrac{1}{24}\lambda \int_{-\beta/2}^{\beta/2} d\tau\, \langle q(t)q(u)q^4(\tau)\rangle_0\right] + O(\lambda^2),$$

where the partition function $\mathcal{Z}(\beta,\lambda)$ has been calculated at this order in Section 3.1 and $\Delta(t)$ is given in (2.52).

An application of Wick's theorem leads to

$$\langle q(t)q(u)q^4(\tau)\rangle_0 = 3\Delta(t-u)\Delta^2(0) + 12\Delta(t-\tau)\Delta(u-\tau)\Delta(0).$$

We recognize in the first contribution the product of the gaussian two-point function by the correction to the partition function. This term cancels in the ratio

$$Z^{(2)}(t,u) = \left[1 - \tfrac{1}{8}\lambda\beta\Delta^2(0)\right]^{-1}$$

$$\times \left[\Delta(t-u)\left(1 - \tfrac{1}{8}\lambda\beta\Delta^2(0)\right) - \tfrac{1}{2}\lambda\Delta(0)\int_{-\beta/2}^{\beta/2} d\tau\, \Delta(t-\tau)\Delta(u-\tau)\right]$$

and, thus, only two contributions remain, which are displayed in Fig. 1.3:

$$Z^{(2)}(t,u) = \Delta(t-u) - \tfrac{1}{2}\lambda\Delta(0)\int_{-\beta/2}^{\beta/2} d\tau\, \Delta(t-\tau)\Delta(\tau-u) + O(\lambda^2).$$

In the limit $\beta \to \infty$ one finds, in particular,

$$Z^{(2)}(t,u) = \tfrac{1}{2}e^{-|t-u|}\left[1 - \tfrac{1}{8}\lambda(1+|t-u|)\right] + O(\lambda^2). \tag{4.25}$$

We have shown (see the discussion of Section 4.2) that for $|t-u| \to \infty$

$$Z^{(2)}(t,u) \underset{|t-u|\to\infty}{\sim} A\, e^{-(E_1-E_0)|t-u|}.$$

The energy eigenvalues E_k have been calculated in Section 3.1. The difference between the two lowest eigenvalues is

$$E_1 - E_0 = 1 + \tfrac{1}{8}\lambda + O(\lambda^2).$$

One thus identifies the term proportional to $|t-u|e^{-|t-u|}$ with a contribution to the expansion of the exponential

$$e^{-(1+\lambda/8)|t-u|} = e^{-|t-u|}\left(1 - \lambda|t-u|/8\right) + O(\lambda^2).$$

The two-point function, at this order, can thus be written as

$$Z^{(2)}(t,u) = \tfrac{1}{2}(1 - \tfrac{1}{8}\lambda)\, e^{-(E_1-E_0)|t-u|} + O(\lambda^2).$$

One verifies, in particular, that this expression satisfies up to order λ the relation (4.28) ($\hbar = m = 1$)

$$\lim_{t\to 0+}\frac{d}{dt}Z^{(2)}(t,0) = -\tfrac{1}{2}(1 - \tfrac{1}{8}\lambda)(E_1 - E_0) = -\tfrac{1}{2} + O(\lambda^2).$$

4.5.2 Spectral representation

In the zero-temperature limit $\beta \to \infty$ (or the thermodynamic limit of the classical model), the two-point correlation function has the representation (4.21):

$$Z^{(2)}(t) \equiv \langle q(0) q(t) \rangle = \langle 0 | \hat{q} \, e^{-|t|(H - E_0)} \, \hat{q} | 0 \rangle.$$

We assume that the hamiltonian H is hermitian, bounded from below and, to simplify notation, has a discrete spectrum. In the basis in which H is diagonal, the two-point function can be rewritten as

$$Z^{(2)}(t) = \sum_{n \geq 0} |\langle 0 | \hat{q} | n \rangle|^2 \, e^{-(\varepsilon_n - \varepsilon_0)|t|}, \qquad (4.26)$$

where the quantities $|n\rangle$ and ε_n are, respectively, the eigenfunctions and eigenvalues of H. The hermiticity of H implies that its eigenvalues are real and the exponentials in the r.h.s. have positive coefficients. The Fourier transform of the two-point function

$$\widetilde{Z}^{(2)}(\omega) = \int dt \, Z^{(2)}(t) \, e^{i\omega t}$$

then has the representation

$$\widetilde{Z}^{(2)}(\omega) = 2\pi \, |\langle 0 | \hat{q} | 0 \rangle|^2 \, \delta(\omega) + 2 \sum_{n > 0} \frac{(\varepsilon_n - \varepsilon_0) \, |\langle 0 | \hat{q} | n \rangle|^2}{[\omega^2 + (\varepsilon_n - \varepsilon_0)^2]}. \qquad (4.27)$$

Two properties of the Fourier transform of the two-point function follow: except for a possible distribution at $\omega = 0$, it is an analytic function of ω^2 with all poles located on the real negative axis. Moreover, the residues of the poles are all positive. As a consequence, $\widetilde{Z}^{(2)}(\omega)$ cannot decrease faster than $1/\omega^2$ for $\omega^2 \to \infty$.

A more precise result can even be proved. Let us calculate the limit, when $t \to 0+$, of the derivative of

$$Z^{(2)}(t) = \frac{1}{\mathcal{Z}(\beta)} \, \text{tr}[e^{-(\beta - t)H} \, \hat{q} \, e^{-tH} \, \hat{q}]$$

at β finite. Then,

$$\lim_{t \to 0+} \frac{d}{dt} Z^{(2)}(t) = \frac{1}{\mathcal{Z}(\beta)} \, \text{tr}[e^{-\beta H}(H \hat{q}^2 - \hat{q} H \hat{q})].$$

Using the cyclic property of the trace, one infers

$$2 \, \text{tr}\{e^{-\beta H}(H \hat{q}^2 - \hat{q} H \hat{q})\} = \text{tr}\{e^{-\beta H}(H \hat{q}^2 + \hat{q}^2 H - 2 \hat{q} H \hat{q})\}$$
$$= \text{tr}\{e^{-\beta H} [\hat{q}[\hat{q}, H]]\}.$$

For any hamiltonian quadratic in the momentum operator of the form $H = \frac{1}{2m} \hat{p}^2 + O(\hat{p})$, the commutator can be evaluated explicitly:

$$[\hat{q}\,[\hat{q}, H]] = -\frac{\hbar^2}{m} \quad \Rightarrow \quad \lim_{t \to 0+} \frac{d}{dt} Z^{(2)}(t) = -\frac{\hbar^2}{2m}. \qquad (4.28)$$

Applying this result in the limit $\beta \to \infty$ to expression (4.26), one finds

$$\frac{\hbar^2}{2m} = \sum_{n \geq 0} |\langle 0 | \hat{q} | n \rangle|^2 \, (\varepsilon_n - \varepsilon_0).$$

One concludes

$$\widetilde{Z}^{(2)}(\omega) \underset{\omega \to \infty}{\sim} \frac{\hbar^2}{m\omega^2}.$$

This useful result is not entirely surprising. The behaviour for $\omega \to \infty$ is related to short time evolution, and we have seen that the most singular contribution to the matrix elements of the statistical operator is determined by the kinetic term in the hamiltonian.

Finally, when the spectrum of H has a continuous part, the sum in (4.27) is replaced by an integral, poles are replaced by a cut with a positive discontinuity and the other conclusions remain unchanged. The relativistic generalization of the representation (4.27) is called Källen–Lehmann representation.

Regularity of paths contributing to path integrals. The preceding calculation implies

$$\left\langle \bigl(q(t+\tau) - q(t)\bigr)^2 \right\rangle = -2 \left\langle q(t)\bigl(q(t+\tau) - q(t)\bigr) \right\rangle$$

$$\underset{\tau \to 0}{\sim} \frac{\hbar}{m} \hbar |\tau|,$$

which generalizes the result (2.59) (here t has the dimension of an inverse energy instead of time). This result confirms that the regularity properties of the paths that contribute to path integrals are those of the brownian paths and are independent of the potential.

4.6 Operator formalism. Time-ordered products

Using the definition of path integrals as limits of integrals with discrete times, we have already shown that correlation functions, in the continuum limit, have a path integral representation (expression (4.20)), but can also be expressed in terms of quantum operators (expression (4.19)). Let us verify, here, that this result can be directly derived in the continuum.

We order n times $t_1, t_2, \ldots, t_n$ as

$$0 \leq t_1 \leq t_2 \leq \cdots \leq t_n \leq \beta. \tag{4.29}$$

We then decompose the interval $(0, \beta)$ into $n+1$ sub-intervals: $(0, t_1)$, (t_1, t_2), ..., (t_n, β). The total action is the sum of the corresponding contributions:

$$\mathcal{S}(q) = \sum_{i=1}^{n+1} \int_{t_{i-1}}^{t_i} \left[\tfrac{1}{2} m \dot{q}^2 + V(q)\right] dt \quad \text{with } t_0 = 0, \ t_{n+1} = \beta. \tag{4.30}$$

We rewrite the integral (4.20) with the help of the identity

$$\prod_{i=1}^{n} q(t_i) = \int \prod_{i=1}^{n} dq_i\, \delta\left[q(t_i) - q_i\right] q_i\,.$$

The path integral then factorizes into a product of path integrals corresponding to the different time sub-intervals. Returning to the very definition of the path integral (equations (2.21, 2.22)), one notices that the numerator of expression (4.20) with the order (4.29) is exactly (taking into account the order (2.1))

$$\mathcal{Z}(\beta) \langle q(t_1)\ldots q(t_n)\rangle_\beta = \operatorname{tr}\left[e^{-(\beta-t_n)H}\,\hat{q}\,e^{-(t_n-t_{n-1})H}\,\hat{q}\ldots e^{-(t_2-t_1)H}\,\hat{q}\,e^{-t_1 H}\right],$$

in agreement with equations (4.19, 4.20).

Introducing the Heisenberg representation of the operator $\hat{q}$ (after analytic continuation $it \mapsto t$)

$$Q(t) = e^{tH}\,\hat{q}\,e^{-tH}, \tag{4.31}$$

(for t real, $Q(t)$ does not necessarily exists and the definition then is somewhat formal), one can rewrite the correlation function as

$$\langle q(t_1)\ldots q(t_n)\rangle_\beta = \mathcal{Z}^{-1}(\beta)\operatorname{tr}\left[e^{-\beta H}\,Q(t_n)\ldots Q(t_1)\right]. \tag{4.32}$$

Time-ordered products. To express correlation functions in terms of operators in this formalism, it is necessary to choose an order between times, even though the functions are symmetric. A convenient formalism, based on the introduction of time-ordered products of operators, allows restoring the symmetry between times.

We introduce the operation T that orders times in operator products: to the operators $A_1(t_1)$, ..., $A_l(t_l)$, considered as functions of time, it associates the time-ordered product (T-product) of these operators. For example, for $l = 2$,

$$\operatorname{T}\left[A_1(t_1)A_2(t_2)\right] = A_1(t_1)A_2(t_2)\theta(t_1 - t_2) + A_2(t_2)A_1(t_1)\theta(t_2 - t_1).$$

Irrespective of the order between the times $t_1, \ldots, t_n$, expression (4.32) can then be rewritten as

$$\langle q(t_1)\ldots q(t_n)\rangle = \mathcal{Z}^{-1}(\beta)\operatorname{tr}\left\{e^{-\beta H}\,\operatorname{T}\left[Q(t_1)Q(t_2)\ldots Q(t_n)\right]\right\}. \tag{4.33}$$

In particular, if H has a unique and isolated ground state $|0\rangle$, in the limit $\beta \to \infty$, correlation functions have the representation (4.21), which can be rewritten as

$$\langle q(t_1)\ldots q(t_n)\rangle \underset{\beta\to\infty}{=} \langle 0 | \operatorname{T}\left[Q(t_1)Q(t_2)\ldots Q(t_n)\right] | 0 \rangle. \tag{4.34}$$

These time-ordered products are the analytic continuations to imaginary times of the time-ordered products that one introduces in the real time formulation of quantum field theory. After analytic continuation, they become the Green functions from which, for example, scattering amplitudes can be calculated. However, all physically acceptable theories from the viewpoint of Green functions, do not necessarily correspond to real euclidean actions and, thus, their analytic continuation may lead to correlation functions only in a rather formal sense.

Generating functional. In terms of time-ordered products, the generating functional (4.22) can also be written as

$$\mathcal{Z}(b,\beta) = \mathcal{Z}^{-1}(\beta)\,\mathrm{tr}\left\{e^{-\beta H}\,\mathrm{T}\left[\exp\int dt\,Q(t)b(t)\right]\right\}. \tag{4.35}$$

This representation is directly related to expression (9.55) of the perturbative expansion in Section 9.7.

Exercises

One intends to apply the method of the transfer matrix to the partition function (4.16) of the gaussian model of Section 4.3, rewritten as

$$\mathcal{Z}_n(T) = \int \prod_{i=1}^{n} \frac{dq_i}{\sqrt{2\pi T}}\,\exp[-\mathcal{E}(q)]$$

with

$$\mathcal{E}(q) = \frac{1}{2}\sum_{i=1}^{n}\left[(q_i - q_{i-1})^2/T + vq_i^2\right] \tag{4.36}$$

and periodic boundary conditions.

Exercise 4.1

Partition function and eigenvalues. Verify that the transfer matrix $\mathcal{T}(q,q')$ of the model can be cast into the form

$$\mathcal{T}(q,q') = \sqrt{b/2\pi}\,\exp\left[-\tfrac{1}{2}a\left(q^2 + q'^2\right) + bqq'\right] \tag{4.37}$$

with

$$a = \frac{1}{T} + \frac{v}{2},\ b = \frac{1}{T}. \tag{4.38}$$

It is also convenient to parametrize the coefficients a,b in terms of parameters ρ,θ:

$$\cosh\theta = a/b = 1 + vT/2,\ \rho = \sqrt{a^2 - b^2} = \sqrt{(v/T(1 + vT/4))}.$$

Verify by explicit calculation that the transfer matrix then satisfies

$$\mathbf{T}(\theta)\mathbf{T}(\theta') = \mathbf{T}(\theta + \theta'),$$

where $\mathbf{T}(\theta)$ is the operator corresponding to the kernel $\mathcal{T}(q,q')$ with parameter θ. Infer the partition function $\mathcal{Z}_n(T)$ and the eigenvalues of the transfer matrix.

Solution.

$$\mathcal{Z}_n(T) = \frac{1}{2\sinh(n\theta/2)} = \sum_{k=0}^{\infty} e^{-n(k+1/2)\theta}.$$

Th eigenvalues t_k thus are

$$t_k = e^{-(k+1/2)\theta},$$

where for $T \to 0$:

$$\theta \sim \sqrt{vT}.$$

Exercise 4.2

Eigenvectors of the gaussian transfer matrix. The preceding expressions show that the spectrum of the transfer matrix is related to the spectrum of an harmonic oscillator. One can obtain its eigenvectors by similar algebraic methods, that is by constructing annihilation and creation type operators.

In terms of momentum $\hat{p}$ and position $\hat{q}$ operators,

$$\hat{p} = \frac{1}{i}\frac{d}{dq} \quad \Rightarrow \quad [\hat{q}, \hat{p}] = i, \qquad (4.39)$$

one defines two operators of annihilation and creation type:

$$\mathbf{A} = i\hat{p} + \frac{\sinh\theta}{T}\hat{q}, \quad \mathbf{A}^\dagger = -i\hat{p} + \frac{\sinh\theta}{T}\hat{q}. \qquad (4.40)$$

Determine the commutation relation of $\mathbf{A}$ and $\mathbf{A}^\dagger$ and prove

$$\mathbf{AT} = e^{-\theta}\,\mathbf{TA}, \quad \mathbf{A}^\dagger\mathbf{T} = e^{\theta}\,\mathbf{TA}^\dagger. \qquad (4.41)$$

Solution. The commutation relation of $\mathbf{A}$ and $\mathbf{A}^\dagger$ is

$$[\mathbf{A}, \mathbf{A}^\dagger] = \frac{2}{T}\sinh\theta.$$

Acting then on an arbitrary vector $|\psi\rangle$, one infers

$$\mathbf{AT}|\psi\rangle \equiv \int dq' \left(\frac{d}{dq} + \frac{\sinh\theta}{T}q\right) T(q,q')\psi(q')$$
$$= \frac{1}{T}\int dq' \left(q' - q e^{-\theta}\right) T(q,q')\psi(q'),$$

and

$$\mathbf{TA}|\psi\rangle \equiv \int dq'\, T(q,q') \left(\frac{d}{dq'} + \frac{\sinh\theta}{T}q'\right) \psi(q')$$
$$= \int dq'\, \psi(q') \left(\frac{\sin\theta}{T}q' - \frac{d}{dq'}\right) T(q,q')$$
$$= \frac{1}{T}\int dq' \left(q' e^{\theta} - q\right) T(q,q')\psi(q').$$

The commutation relations (4.41) follow (the second one being implied by hermitian conjugation).

Exercise 4.3

Recover directly the eigenvalues and determine the eigenvectors of the transfer matrix.

Solution. The commutation relations allow recovering the spectrum and determining the eigenvectors. We denote by $|m\rangle$ the eigenvectors and τ_m the corresponding eigenvalues. From the commutation relations, one infers

$$\mathbf{A}^\dagger \mathbf{T} |m\rangle = \tau_m \mathbf{A}^\dagger |m\rangle = \mathrm{e}^\theta \, \mathbf{T}\mathbf{A}^\dagger |m\rangle.$$

This implies that $\mathbf{A}^\dagger |m\rangle$ is an eigenvector of $\mathbf{T}$ with eigenvalue $\mathrm{e}^{-\theta} \tau_m = \tau_{m+1}$. The same argument applies to $\mathbf{A}$. Thus,

$$\mathbf{A}^\dagger |m\rangle \propto |m+1\rangle, \quad \mathbf{A} |m\rangle \propto |m-1\rangle.$$

Since no eigenvector can have an eigenvalue larger than τ_0, the second relation applied for $m = 0$ can only generate a vanishing vector:

$$\mathbf{A} |0\rangle = 0.$$

The eigenvalues are determined up to a factor $\tau_m = \mathrm{e}^{-m\theta} \tau_0$, which can determined by calculating $\mathrm{tr}\,\mathbf{T}$.

Noting that $\hat{q}$ is proportional to $\mathbf{A} + \mathbf{A}^\dagger$, one infers $\hat{q} |0\rangle \propto |1\rangle$. Using the representations of $\mathbf{A}, \hat{\mathbf{Q}}$ in the basis $|q\rangle$, one finds explicitly

$$\langle q|0\rangle \propto \mathrm{e}^{-q^2 \sinh\theta/2T}, \quad \langle q|1\rangle \propto q\,\mathrm{e}^{-q^2 \sinh\theta/2T}.$$

Exercise 4.4

Generalize the calculation of the two-point function of Section 4.5.1 to

$$W_{ij}(t_1, t_2) \equiv \langle q_i(t_1) q_j(t_2) \rangle = \lim_{\beta\to\infty} \frac{1}{\mathcal{Z}(\beta)} \int [\mathrm{d}q(t)] q_i(t_1) q_j(t_2)\,\mathrm{e}^{-\mathcal{S}(\mathbf{q})},$$

where $\mathbf{q}$ is an N-component vector $(i, j = 1, \cdots, N)$, $\mathbf{q}(-\beta/2) = \mathbf{q}(\beta/2)$, and $\mathcal{Z}(\beta)$ is the partition function corresponding to the action

$$\mathcal{S}(\mathbf{q}) = \int_{-\beta/2}^{\beta/2} \mathrm{d}t \left(\frac{1}{2}(\dot{\mathbf{q}})^2 + \frac{1}{2}\mathbf{q}^2 + \frac{\lambda}{4!}(\mathbf{q}^2)^2 \right).$$

Note that the action is symmetric under transformations of the $O(N)$ group (rotations and reflections in N-dimensional space). Infer the difference between the two first energy eigenvalues.

Solution. Symmetry under the $O(N)$ group implies that the two-point function must have the form

$$W_{ij}(t) = W(t)\delta_{ij}.$$

The expansions of function $W(t)$ for $N = 1$ and general N are then quite similar. Only the coefficient in front of the contribution of order λ is affected. One obtains

$$E_1 - E_0 = 1 + \tfrac{N+2}{24}\lambda + O(\lambda^2).$$

Exercise 4.5

Complete the calculation of the two-point function of Section 4.5.1 for β finite. Show, by expanding $Z^{(2)}$ in a Fourier series, that the result can be written as

$$Z^{(2)}(t,u) = \frac{1}{2\omega \sinh(\omega\tau/2)} \cosh\bigl(\omega(\tau/2 - |t|)\bigr) + O(\lambda^2)$$

with

$$\omega = E_1 - E_0 = 1 + \tfrac{1}{8}\lambda.$$

Exercise 4.6

Obtain, by functional differentiation of equation (4.23), a relation between the two- and four-point functions in the case of the action (4.24).

Solution. First,

$$\frac{\delta S}{\delta q(t)} = -\ddot{q}(t) + q(t) + \frac{\lambda}{3!} q^3(t).$$

Equation (4.23) then can be written as

$$\left\{\left[-\left(\frac{\mathrm{d}}{\mathrm{d}t}\right)^2 + 1\right]\frac{\delta}{\delta b(t)} + \frac{\lambda}{3!}\left(\frac{\delta}{\delta b(t)}\right)^3\right\} \mathcal{Z}(b) = b(t).$$

Differentiating the equation with respect to $b(u)$ and taking the limit $b \equiv 0$, one finds

$$\left[-(\mathrm{d}_t)^2 + 1\right] Z^{(2)}(t,u) + \frac{\lambda}{3!} Z^{(4)}(t,t,t,u) = \delta(t-u).$$

Exercise 4.7

Recover the two-point function (4.25) from the preceding equation.

Solution. To leading order, one finds $Z^{(2)}(t,u) = \Delta(t-u)$. Next order requires the gaussian four-point function, which is directly given by Wick's theorem:

$$Z^{(4)}(t_1, t_2, t_3, t_4) = \Delta(t_1 - t_2)\Delta(t_3 - t_4) + \Delta(t_1 - t_3)\Delta(t_2 - t_4) \\ + \Delta(t_1 - t_4)\Delta(t_3 - t_2),$$

and, thus,

$$Z^{(4)}(t,t,t,u) = 3\Delta(0)\Delta(t-u).$$

The problem is reduced to the solution of the differential equation

$$\left[-(\mathrm{d}_t)^2 + 1\right] Z^{(2)}(t,u) = \delta(t-u) - \frac{\lambda}{2}\Delta(0)\Delta(t-u)$$

with the condition $\Delta(t) \to 0$ for $|t| \to \infty$. The equation can then be solved by Fourier transformation.

5 PATH INTEGRALS AND QUANTIZATION

In Chapter 2, we have constructed a path integral representation of the matrix elements of the operator $\mathrm{e}^{-\beta H}$ in the case of hamiltonians of the form $p^2/2m + V(q)$. Such hamiltonians share the property that they are the sum of a function of p and a function of q. In this situation, the transition from a classical to a quantum hamiltonian follows from the correspondence 'principle': one replaces the real classical variables p and q by the corresponding quantum operators. In this chapter, we discuss potentials linear in the momentum variables, producing additional contributions to the action linear in the velocities. We examine two examples, a quantum system coupled to a magnetic field, and diffusion as described by the Fokker–Planck equation. In both examples, the hamiltonian contains products of the position and momentum operators. A quantization problem then arises since these operators do not commute, and the correspondence principle is no longer sufficient to determine the quantum hamiltonian. The order between operators is determined by additional conditions, such as hermiticity or conservation of probabilities. We show that the calculation of the corresponding path integral then suffers from ambiguities, directly related to this quantization problem. The continuum limit is no longer unique, but depends on the limiting process.

In the last part of the chapter, we examine a situation where space has a non-trivial topology, in this case a circle, and show how this influences the calculation of the path integral.

5.1 Gauge transformations

Before discussing quantization in a magnetic field, we briefly recall a few relevant properties of classical lagrangian mechanics.

Classical gauge transformations. Classical equations of motion that are obtained by varying an action, which is the time integral of a lagrangian $\mathcal{L}$,

$$\mathcal{A}(\mathbf{q}) = \int_{t'}^{t''} dt\, \mathcal{L}(\mathbf{q}, \dot{\mathbf{q}}),$$

are insensitive to the addition of a total derivative to the lagrangian:

$$\mathcal{L}(\mathbf{q}, \dot{\mathbf{q}}) \mapsto \mathcal{L}_\Omega = \mathcal{L}(\mathbf{q}, \dot{\mathbf{q}}) + \dot{\mathbf{q}} \cdot \nabla \Omega(\mathbf{q}). \tag{5.1}$$

Indeed, the corresponding action becomes

$$\mathcal{A}_\Omega(\mathbf{q}) = \int_{t'}^{t''} dt\, \mathcal{L}_\Omega = \mathcal{A}(\mathbf{q}) + \Omega\bigl(\mathbf{q}(t'')\bigr) - \Omega\bigl(\mathbf{q}(t')\bigr)$$

and, thus, the equation of motion is not modified:

$$\frac{\delta \mathcal{A}}{\delta \mathbf{q}(\tau)} = \frac{\delta \mathcal{A}_\Omega}{\delta \mathbf{q}(\tau)}.$$

This addition to the lagrangian affects, however, the conjugate momentum:

$$\mathbf{p} = \frac{\partial \mathcal{L}}{\partial \dot{\mathbf{q}}} \mapsto \mathbf{p} = \frac{\partial \mathcal{L}_\Omega}{\partial \dot{\mathbf{q}}} = \frac{\partial \mathcal{L}}{\partial \dot{\mathbf{q}}} + \nabla \Omega(\mathbf{q}), \tag{5.2}$$

and thus the hamiltonian

$$H(\mathbf{p},\mathbf{q}) = \mathbf{p} \cdot \dot{\mathbf{q}} - \mathcal{L},$$

$$\mapsto H_\Omega(\mathbf{p},\mathbf{q}) = \mathbf{p} \cdot \dot{\mathbf{q}} - \mathcal{L}_\Omega = \frac{\partial \mathcal{L}}{\partial \dot{\mathbf{q}}} \cdot \dot{\mathbf{q}} - \mathcal{L} = H(\mathbf{p} - \nabla \Omega(\mathbf{q}), \mathbf{q}). \tag{5.3}$$

The new hamiltonian is equal to the initial hamiltonian in which the variable $\mathbf{p}$ has been replaced by $\mathbf{p} - \nabla \Omega(\mathbf{q})$. This transformation, which does not affect physics but only its description, is called a gauge transformation. Note, however, that for a system invariant under space translations, one choice is the simplest, which corresponds directly to the conserved quantities. An example is provided by the free hamiltonian $H = \mathbf{p}^2/2m$.

Quantum gauge transformations. To the addition of a total derivative to the classical lagrangian corresponds, in quantum mechanics, a unitary transformation, which again does not affect physical properties.

In what follows, we denote by $\hat{\mathbf{q}}, \hat{\mathbf{p}}, \hat{H}$ the quantum operators corresponding to the classical quantities $\mathbf{q}, \mathbf{p}, H$.

To the classical transformations (5.2, 5.3) corresponds, in quantum mechanics, a unitary transformation generated by the operator

$$\mathbf{\Omega} = e^{i\Omega(\hat{\mathbf{q}})/\hbar}. \tag{5.4}$$

Indeed,

$$\mathbf{\Omega} \hat{\mathbf{p}} \mathbf{\Omega}^\dagger = \hat{\mathbf{p}} - \nabla \Omega(\hat{\mathbf{q}}), \quad \mathbf{\Omega} \hat{\mathbf{q}} \mathbf{\Omega}^\dagger = \hat{\mathbf{q}}. \tag{5.5}$$

The new conjugate momentum is obtained by a unitary transformation that leaves the position operator $\hat{\mathbf{q}}$, and the commutation relations between $\hat{\mathbf{q}}$ and $\hat{\mathbf{p}}$, invariant. Applying the unitary transformation (5.5) to a hamiltonian $\hat{H}$, one finds

$$\mathbf{\Omega} \hat{H}(\hat{\mathbf{p}},\hat{\mathbf{q}}) \mathbf{\Omega}^\dagger = \hat{H}(\hat{\mathbf{p}} - \nabla \Omega, \hat{\mathbf{q}}) = \hat{H}_\Omega(\hat{\mathbf{p}},\hat{\mathbf{q}}),$$

where $\hat{H}_\Omega(\hat{\mathbf{p}},\hat{\mathbf{q}})$ is a quantized form of the transformed classical hamiltonian (5.3).

In this unitary transformation, wave functions $\psi(\mathbf{q})$ are multiplied by a function of the position $\mathbf{q}$, of modulus 1 (an element of the $U(1)$ group):

$$\psi_\Omega(\mathbf{q}) = e^{i\Omega(\mathbf{q})/\hbar} \psi(\mathbf{q}). \tag{5.6}$$

5.2 Path integrals and quantization

Moreover, the operator $U(t'', t')$, solution of equation (2.2), transforms like

$$U_\Omega = \mathbf{\Omega} \, U \, \mathbf{\Omega}^\dagger,$$

and, thus, its matrix elements become

$$\langle \mathbf{q}'' | U_\Omega(t'', t') | \mathbf{q}' \rangle = \mathrm{e}^{i(\Omega(\mathbf{q}'') - \Omega(\mathbf{q}'))/\hbar} \langle \mathbf{q}'' | U(t'', t') | \mathbf{q}' \rangle. \tag{5.7}$$

In quantum mechanics also, the transformation that we have described, and which is called *gauge transformation*, has no physical implication, and all choices of conjugate momenta $\hat{\mathbf{p}}$ are *a priori* equivalent. However, in the case of lagrangians of the form $\tfrac{1}{2} m \dot{\mathbf{q}}^2 - V(\mathbf{q})$, one choice is clearly simpler.

5.2 Coupling to a magnetic field: gauge symmetry

We now recall how the introduction of a magnetic field generates a particular symmetry called a gauge symmetry.

5.2.1 Classical gauge invariance

The action corresponding to a particle in a potential coupled to a magnetic field, is the integral of a lagrangian that can be written as

$$\mathcal{L}(\mathbf{q}, \dot{\mathbf{q}}) = \tfrac{1}{2} m \dot{\mathbf{q}}^2 - e \mathbf{A}(\mathbf{q}) \cdot \dot{\mathbf{q}} - V(\mathbf{q}), \tag{5.8}$$

where $\mathbf{A}(\mathbf{q})$ is a vector potential, from which the magnetic field $\mathbf{B}$ derives, and e a charge that is introduced here only to follow usual conventions. The corresponding equation of motion is

$$m \ddot{q}_\mu = -e \sum_\nu F_{\mu\nu} \dot{q}_\nu - \partial_\mu V,$$

where

$$F_{\mu\nu} = \partial_\mu A_\nu - \partial_\nu A_\mu \tag{5.9}$$

is the magnetic tensor and we have used the notation $\partial_\mu \equiv \partial/\partial x_\mu$.

In dimension 3, the antisymmetric tensor $F_{\mu\nu}$ can be parametrized in terms of an axial vector, the magnetic field $\mathbf{B}$, as

$$F_{\mu\nu} = \sum_\rho \epsilon_{\mu\nu\rho} B_\rho,$$

where $\epsilon_{\mu\nu\rho}$ is the tensor completely antisymmetric in its three indices and $\epsilon_{123} = 1$. Then,

$$m \ddot{q}_\mu = -e \sum_{\nu\rho} \epsilon_{\mu\nu\rho} B_\rho \dot{q}_\nu - \partial_\mu V,$$

which, in a vector notation, can also be written as

$$m \ddot{\mathbf{q}} = e \mathbf{B} \times \dot{\mathbf{q}} - \nabla V$$

with
$$\mathbf{B} = \nabla \times \mathbf{A}.$$

The equations of motion are still insensitive to the addition of a total derivative to the lagrangian:
$$\mathcal{L}(\mathbf{q}, \dot{\mathbf{q}}) \mapsto \mathcal{L}_\Omega = \mathcal{L}(\mathbf{q}, \dot{\mathbf{q}}) + \dot{\mathbf{q}} \cdot \nabla \Omega(\mathbf{q}).$$

But now a total derivative can be cancelled by the corresponding *gauge transformation* of the vector potential:
$$\mathbf{A}(\mathbf{q}) \mapsto \mathbf{A}_\Omega(\mathbf{q}) = \mathbf{A}(\mathbf{q}) + \frac{1}{e} \nabla \Omega(\mathbf{q}). \tag{5.10}$$

In the presence of a vector potential, the gauge invariance of the equations of motion has a new implication: since the equations of motion are invariant under the gauge transformations (5.1), they are also invariant under the transformation (5.10), a property that can be immediately verified since the magnetic tensor (5.9) is gauge invariant. The transformation (5.10) thus defines equivalence classes of vector potentials.

The corresponding classical hamiltonian is
$$H = \frac{1}{2m} [\mathbf{p} + e\mathbf{A}(\mathbf{q})]^2 + V(\mathbf{q}). \tag{5.11}$$

Conversely, one notes that the addition of a vector potential solves the problem of constructing a lagrangian invariant under the addition of a total derivative or a hamiltonian invariant under the transformation (5.3):
$$\mathbf{p} \mapsto \mathbf{p} - \nabla \Omega(\mathbf{q}).$$

In the hamiltonian formalism, gauge invariance is implemented by the substitution
$$\mathbf{p} \mapsto \mathbf{p} + e\mathbf{A}(\mathbf{q}).$$

This symmetry principle has received the name of *gauge symmetry* and, quite generally, leads to the introduction of a *gauge field* (mathematically also called a connection), here the vector potential.

Of course, in classical mechanics, only the equations of motion are physical, and the vector potential appears only as a convenient mathematical quantity. However, in quantum mechanics the situation is somewhat different.

5.2.2 Quantum gauge invariance and quantization

When classical variables are replaced by quantum operators in the classical hamiltonian (5.11), the problem of the ordering of operators in the product $\mathbf{p} \cdot \mathbf{A}(\mathbf{q})$ arises. In the case of a magnetic field, the hermiticity condition determines the quantum hamiltonian and leads to
$$\hat{H} = \frac{1}{2m} [\hat{\mathbf{p}}^2 + e\mathbf{A}(\hat{\mathbf{q}}) \cdot \hat{\mathbf{p}} + e\hat{\mathbf{p}} \cdot \mathbf{A}(\hat{\mathbf{q}}) + e^2 \mathbf{A}^2(\hat{\mathbf{q}})] + V(\hat{\mathbf{q}}),$$
$$= \frac{1}{2m} [\hat{\mathbf{p}} + e\mathbf{A}(\hat{\mathbf{q}})]^2 + V(\hat{\mathbf{q}}). \tag{5.12}$$

It is interesting to realize that the same form is obtained by implementing *gauge invariance*. Gauge invariance implies that hamiltonians that correspond to vector potentials $\mathbf{A}(\hat{\mathbf{q}})$ related by a gauge transformation (5.10), are unitary equivalent. From gauge symmetry, thus, it follows that the hamiltonian can depend on the operator $\hat{\mathbf{p}}$ only through the combination $\hat{\mathbf{p}} + e\mathbf{A}(\hat{\mathbf{q}})$. Indeed, the transformations (5.5, 5.10) lead to

$$\Omega \left[\hat{\mathbf{p}} + e\mathbf{A}_\Omega(\hat{\mathbf{q}})\right] \Omega^\dagger = \hat{\mathbf{p}} + e\mathbf{A}(\hat{\mathbf{q}}).$$

Both the hermiticity and gauge invariance conditions lead to the same quantum hamiltonian.

Another choice of quantization would result into the addition of a term proportional to the commutator

$$e[\hat{\mathbf{p}}, \mathbf{A}(\hat{\mathbf{q}})] = -ie\hbar \nabla \cdot \mathbf{A}$$

to the hamiltonian, which is a quantum correction that has the form of an imaginary potential violating both hermiticity and gauge invariance.

Finally, note that the operator

$$\mathbf{D} = \frac{i}{\hbar}\left(\hat{\mathbf{p}} + e\mathbf{A}(\hat{\mathbf{q}})\right) = \nabla_q + \frac{ie}{\hbar}\mathbf{A}(\hat{\mathbf{q}}),$$

which is a differential operator when acting on wave functions, leads to *covariant derivatives*. Its components satisfy the commutation relation

$$[\mathrm{D}_\mu, \mathrm{D}_\nu] = \frac{ie}{\hbar} F_{\mu\nu}.$$

Remark. The condition that an operator of the form $\hat{\mathbf{\Pi}} = \hat{\mathbf{p}} + e\mathbf{A}(\hat{\mathbf{q}})$ satisfies the canonical commutation relations, reduces to

$$[\hat{\Pi}_\mu, \hat{\Pi}_\nu] = 0 \iff F_{\mu\nu} = 0.$$

In simple situations (with trivial topologies), this implies that $\mathbf{A}(\hat{\mathbf{q}})$ is a gradient or pure gauge.

5.2.3 Gauge invariance and path integral

We now show that the principle of gauge symmetry determines also, to a large extent, the general form of the path integral giving the matrix elements of the statistical operator $U(t'', t')$ (equation (2.2)).

Indeed, these elements transform by multiplication as indicated by equation (5.7). The phase can then be written as the integral of a total derivative:

$$\Omega(\mathbf{q}'') - \Omega(\mathbf{q}') = \int_{t'}^{t''} dt \, \dot{\mathbf{q}} \cdot \nabla\Omega(\mathbf{q}(t)).$$

In the path integral (2.22), the transformation (5.7) has the interpretation of a modification of the action (2.21):

$$\mathcal{S}(\mathbf{q}) \mapsto \mathcal{S}_\Omega(\mathbf{q}) = \int_{t'}^{t''} dt \left[\tfrac{1}{2}m\dot{\mathbf{q}}^2(t) + V\big(\mathbf{q}(t),t\big) - i\dot{\mathbf{q}}(t) \cdot \nabla\Omega\big(\mathbf{q}(t)\big)\right].$$

Such a variation can then be cancelled by adding to the lagrangian a term proportional to a vector potential $\mathbf{A}(\mathbf{q})$,

$$\mathcal{S}(\mathbf{q}) = \int_{t'}^{t''} dt \left[\tfrac{1}{2}m\dot{\mathbf{q}}^2 + ie\mathbf{A}(\mathbf{q}) \cdot \dot{\mathbf{q}} + V(\mathbf{q})\right], \tag{5.13}$$

which transforms like (equation (5.10))

$$\mathbf{A}_\Omega(\mathbf{q}) = \mathbf{A}(\mathbf{q}) + (1/e)\nabla\Omega(\mathbf{q}).$$

The resulting euclidean action (5.13) is the integral of the lagrangian (5.8).

A direct calculation, based on the solution of the short time evolution equation with the hamiltonian (5.12) (see Section 5.4), confirms that, like in the case without magnetic field, the path integral involves only the classical euclidean action (5.13):

$$\langle \mathbf{q}'' | U(t'',t') | \mathbf{q}' \rangle = \int_{\mathbf{q}(t')=\mathbf{q}'}^{\mathbf{q}(t'')=\mathbf{q}''} [d\mathbf{q}(t)] \exp\left[-\mathcal{S}(\mathbf{q})/\hbar\right]. \tag{5.14}$$

Note that the resulting euclidean action is no longer real and, thus, no longer defines a positive measure. This is a direct consequence of a property of the hamiltonian: in the presence of a magnetic field, the hamiltonian is still hermitian but no longer real symmetric. Also, from the viewpoint of a continuation from a real time lagrangian to an euclidean lagrangian, we observe that a term linear in $\dot{q}$ is multiplied by an additional factor i compared to the other terms.

Note also that the (imaginary) magnetic term has a remarkable property: it depends only on the geometric trajectory, but not on the explicit motion. Indeed,

$$\int dt\, \mathbf{A}(\mathbf{q}) \cdot \dot{\mathbf{q}} = \oint d\mathbf{q} \cdot \mathbf{A}(\mathbf{q}). \tag{5.15}$$

This property has various consequences in quantum mechanics, such as the quantization of the charge of a magnetic monopole (if a monopole exists) or, more generally, quantization properties when space is not simply connected and, thus, the quantity (5.15) multivalued.

5.3 Quantization and path integrals

The path integral (5.14) superficially depends only on the classical action and, nevertheless, determines quantum observables. One may thus wonder whether it implies some special choice of quantization. The answer is negative, as an explicit calculation will show. Without an additional constraint, the path integral is ill-defined because ambiguous quantities appear in its calculation.

5.3.1 Discrete times and continuum limit

The nature of the ambiguities can be understood by returning to discrete times and short time evolution, $t'' - t' = \varepsilon \to 0$. Then, for continuous paths, for $\varepsilon \to 0$,

$$\int_{t'}^{t''=t'+\varepsilon} dt\, \mathbf{A}(\mathbf{q}(t)) \cdot \dot{\mathbf{q}}(t) \sim (\mathbf{q}'' - \mathbf{q}') \cdot \mathbf{A}(\mathbf{q}'') \sim (\mathbf{q}'' - \mathbf{q}') \cdot \mathbf{A}(\mathbf{q}').$$

A priori, it would seem that the two discretized terms, which differ only by the argument, $\mathbf{q}'$ or $\mathbf{q}''$, of $\mathbf{A}(\mathbf{q})$, correspond to equivalent discretized actions since they have the same continuum limit when $\varepsilon \to 0$. This is not the case: indeed, the difference between $\mathbf{A}(\mathbf{q}'')$ and $\mathbf{A}(\mathbf{q}')$ is of order $\mathbf{q}'' - \mathbf{q}'$ and, thus, taking into account the factor $\mathbf{q}'' - \mathbf{q}'$, it induces a modification of the discretized action of order

$$(\mathbf{q}'' - \mathbf{q}') \cdot \mathbf{A}(\mathbf{q}') - (\mathbf{q}'' - \mathbf{q}') \cdot \mathbf{A}(\mathbf{q}'') \sim -\sum_{\mu\nu} (q'_\mu - q''_\mu)(q'_\nu - q''_\nu) \partial_\nu A_\mu. \quad (5.16)$$

This modification is of order ε because the typical paths are brownian and thus $|\mathbf{q}'' - \mathbf{q}'| = O(\sqrt{\varepsilon})$. But as we have shown in Section 2.3.1, the terms of order ε in the discretized action contribute to the path integral in the continuum limit. In contrast with the case without a field, one can thus define a one-parameter family of continuum limits that are only determined by the specific form of the discretized action. Note, however, that the concept of continuum limit remains meaningful in the sense that only one parameter survives in the limit.

This subtlety is no longer apparent in the formal expression of the path integral, which thus corresponds to any continuum limit. The formal definition of the path integral must be supplemented by an additional condition.

One then notices that the operator $\mathrm{e}^{-\varepsilon \hat{H}/\hbar}$ is hermitian because $\hat{H}$ is hermitian. Thus, its matrix elements are invariant under the transformation: complex conjugation combined with an exchange $\mathbf{q}' \leftrightarrow \mathbf{q}''$. For the magnetic term, this condition is only satisfied if one chooses a symmetric function of $\mathbf{q}'', \mathbf{q}'$ like $\mathbf{A}((\mathbf{q}'' + \mathbf{q}')/2)$.

The precise choice is then irrelevant because two symmetric functions differ only at order $(\mathbf{q}'' - \mathbf{q}')^2$, which leads to a negligible difference of order $\varepsilon^{3/2}$ in the action.

Alternatively, the condition of gauge invariance is satisfied if one takes the continuum action and approximates the path $\mathbf{q}(t)$ by a free motion. One then obtains an equivalent symmetric form. We show in Section 5.4 that this is the form that emerges from a direct calculation of the matrix elements of the statistical operator corresponding to the hamiltonian (5.12), in the short time interval limit.

In the presence of a magnetic field, it is then possible to fix the ambiguity in the definition of the path integral by requesting that the matrix elements of the statistical operator have the correct properties: gauge covariance or hermiticity.

We are confronted here with a special example of a more general problem: the difficulties generated by the choice of quantization can often be solved by requesting that the quantum theory preserves some symmetries. We will meet this problem again in Section 5.5, in Sections 6.2, 7.8 or in Chapter 10.

5.3.2 Ambiguity and perturbative calculation

To illustrate the discussion of Section 5.3.1, we now examine how the ambiguity in the definition of the path integral shows up in perturbative calculations, and how the problem can be solved.

We consider the example of the calculation of the ground state energy corresponding to the action (we set here $\hbar = 1$)

$$\mathcal{S}(\mathbf{q}) = \int_{-\beta/2}^{\beta/2} dt \left[\tfrac{1}{2} m \dot{\mathbf{q}}^2 + ie\mathbf{A}(\mathbf{q}) \cdot \dot{\mathbf{q}} + \tfrac{1}{2}\omega^2 \mathbf{q}^2 \right],$$

where we assume that $\mathbf{A}(\mathbf{q})$ is a polynomial.

For this purpose, we expand the partition function in powers of the charge e. At first order, one finds (see Section 2.8)

$$\mathcal{Z}(\beta) = \mathcal{Z}_0(\beta) \left[1 - ie \int_{-\beta/2}^{\beta/2} dt \left\langle \dot{q}(t) \cdot \mathbf{A}(q(t)) \right\rangle \right] + O(e^2), \tag{5.17}$$

where $\langle \bullet \rangle$ means expectation value with respect to the gaussian measure associated with the harmonic oscillator. For each monomial contributing to $\mathbf{A}$, one must evaluate

$$\sum_i \sum_{j_1, j_2, \ldots, j_p} \left\langle \dot{q}_i(t) A_{i, j_1 j_2 \ldots j_p} q_{j_1}(t) \ldots q_{j_p}(t) \right\rangle.$$

The expectation value can be calculated with the help of Wick's theorem (2.58). The factor $\dot{q}$ must be paired in all possible ways with a factor q. All pairings give the same result, which yields a factor p. Then, the remaining q factors must be paired. One verifies that the result can then be written as

$$\sum_i \sum_{j_1, j_2, \ldots, j_p} \left\langle \dot{q}_i(t) A_{i, j_1 j_2 \ldots j_p} q_{j_1}(t) \ldots q_{j_p}(t) \right\rangle$$

$$= \sum_{i,j} \left\langle \dot{q}_i(t) q_j(t) \right\rangle \frac{\partial}{\partial q_j(t)} \sum_{j_1, j_2, \ldots, j_p} \left\langle A_{i, j_1 j_2 \ldots j_p} q_{j_1}(t) \ldots q_{j_p}(t) \right\rangle.$$

Summing all contributions, one concludes that the expectation value in (5.17) can be cast into the form (see also equation (1.20))

$$-ie \int dt \sum_{i,j} \left\langle \dot{q}_i(t) q_j(t) \right\rangle \left\langle \frac{\partial A_i}{\partial q_j} \right\rangle.$$

The gaussian two-point function, in the limit $\beta \to \infty$, is given by equation (2.53):

$$\Delta_{ij}(t) = \langle q_i(t) q_j(0) \rangle = \frac{\delta_{ij}}{2\omega} e^{-\omega|t|}.$$

A problem arises because one needs the expectation value $\langle \dot{q}_i(t) q_j(t) \rangle$. Indeed, by differentiating $\Delta_{ij}(t - t')$, one finds

$$\langle \dot{q}_i(t) q_j(t') \rangle = \frac{\mathrm{d}\Delta_{ij}(t - t')}{\mathrm{d}t} = -\tfrac{1}{2}\delta_{ij}\, \mathrm{sgn}(t - t')\, \mathrm{e}^{-\omega|t - t'|},$$

where $\mathrm{sgn}(t) = 1$ for $t > 0$, $\mathrm{sgn}(t) = -1$ for $t < 0$. Clearly $\mathrm{sgn}(0)$ is not defined. The calculation in the continuum is ambiguous and the ambiguity, at first order in e, is proportional to

$$\tfrac{1}{2} i e\, \mathrm{sgn}(0) \int \mathrm{d}t\, \langle \nabla \cdot \mathbf{A}(q(t)) \rangle$$

and, thus, corresponds to the addition of a commutator term to the action, exactly as one expects from the discussion of Section 5.2.2. Note, however, that

$$\frac{\mathrm{d}}{\mathrm{d}t} \langle q_i(t) q_j(t) \rangle = \frac{\mathrm{d}}{\mathrm{d}t} \frac{\delta_{ij}}{2\omega} = 0$$

is defined. If one insists that differentiation and expectation value should commute, one finds, for $i = j$,

$$\frac{\mathrm{d}}{\mathrm{d}t} \langle q_i(t) q_i(t) \rangle = 2 \langle \dot{q}_i(t) q_i(t) \rangle = 0,$$

a result compatible only with $\mathrm{sgn}(0) = 0$. This is precisely the commutation property in *equal-time products* that is required in order to prove directly gauge invariance of the term of order e in the perturbative expression (5.17). Indeed, in the gauge transformation (5.10), the contribution transforms like

$$\text{variation of } ie \int_{-\beta/2}^{\beta/2} \mathrm{d}t\, \langle \dot{\mathbf{q}}(t) \cdot \mathbf{A}(\mathbf{q}(t)) \rangle = i \int_{-\beta/2}^{\beta/2} \mathrm{d}t\, \langle \dot{\mathbf{q}}(t) \cdot \nabla \Omega(\mathbf{q}) \rangle$$

$$= i \int_{-\beta/2}^{\beta/2} \mathrm{d}t\, \left\langle \frac{\mathrm{d}}{\mathrm{d}t} \Omega(\mathbf{q}(t)) \right\rangle.$$

If differentiation and expectation value commute, the variation can be calculated:

$$i \int_{-\beta/2}^{\beta/2} \mathrm{d}t\, \left\langle \frac{\mathrm{d}}{\mathrm{d}t} \Omega(\mathbf{q}(t)) \right\rangle = i \left[\Omega(\mathbf{q}(\beta/2)) - \Omega(\mathbf{q}(-\beta/2)) \right] = 0,$$

as a consequence of periodic boundary conditions.

Moreover, another choice of quantization would lead to the addition of a term proportional to $i\nabla \cdot \mathbf{A}$ to the hamiltonian, which is an imaginary potential. All other terms in the classical action such that $i\dot{\mathbf{q}} \cdot \mathbf{A}$ or the real potential are invariant under the double transformation, complex conjugation and time reversal, unlike $i\nabla \cdot \mathbf{A}$. In particular, hermiticity is implemented more easily if $\mathrm{sgn}(t)$ itself is symmetric by time reversal and one sets $\mathrm{sgn}(0) = \tfrac{1}{2}(\mathrm{sgn}(0_+) + \mathrm{sgn}(0_-)) = 0$.

Note that often, instead of the sign function $\mathrm{sgn}(t)$, one introduces the step function $\theta(t)$: $\mathrm{sgn}(t) = 2\theta(t) - 1$. The choice $\mathrm{sgn}(0) = 0$ then corresponds to $\theta(0) = \tfrac{1}{2}$.

5.4 Magnetic field: direct calculation

For completeness, we now directly calculate (for $m = \hbar = 1$) the matrix elements of the operator $U(t) = \mathrm{e}^{-tH}$ with

$$H = \tfrac{1}{2}(\mathbf{p} + e\mathbf{A}(\mathbf{q}))^2,$$

for $t \to 0$, from a generalized form of equation (2.11), in d dimensions:

$$\partial_t \langle \mathbf{q}|\, U(t)\, |\mathbf{q}'\rangle = \tfrac{1}{2} \sum_{\mu=1}^{d} (\partial_\mu + ieA_\mu)^2 \langle \mathbf{q}|\, U(t)\, |\mathbf{q}'\rangle, \qquad (5.18)$$

where $\partial_\mu \equiv \partial/\partial q_\mu$. One sets

$$\langle \mathbf{q}|\, U(t)\, |\mathbf{q}'\rangle = \mathrm{e}^{-\sigma(\mathbf{q},\mathbf{q}';t)}. \qquad (5.19)$$

Equation (5.18) then becomes

$$\nabla_q^2 \sigma - \sum_\mu \left[ie\partial_\mu A_\mu + (\partial_\mu \sigma - ieA_\mu)^2 \right] = 2\partial_t \sigma. \qquad (5.20)$$

The function σ has, for $t \to 0$, an expansion of the form

$$\sigma = \frac{1}{2t}(\mathbf{q} - \mathbf{q}')^2 + \frac{d}{2}\ln 2\pi t + \sigma_0 + \sigma_1 t + O(t^2). \qquad (5.21)$$

Thus,

$$\partial_t \sigma = -\frac{1}{2t^2}(\mathbf{q} - \mathbf{q}')^2 + \frac{d}{2t} + \sigma_1 + O(t),$$

$$\partial_\mu \sigma = \frac{1}{t}(q - q')_\mu + \partial_\mu \sigma_0 + \partial_\mu \sigma_1 t + O(t^2).$$

One infers

$$\sum_\mu (\partial_\mu \sigma)^2$$
$$= \frac{1}{t^2}(\mathbf{q} - \mathbf{q}')^2 + \sum_\mu \left[\frac{2}{t}(q - q')_\mu \partial_\mu \sigma_0 + (\partial_\mu \sigma_0)^2 + 2(q - q')_\mu \partial_\mu \sigma_1 \right] + O(t),$$

$$\nabla_q^2 \sigma = \frac{d}{t} + \nabla_q^2 \sigma_0 + t\nabla_q^2 \sigma_1 + O(t^2).$$

The order t^{-1} yields the first non-trivial equation:

$$\sum_\mu (q - q')_\mu (\partial_\mu \sigma_0 - ieA_\mu(\mathbf{q})) = 0.$$

The solution is

$$\sigma_0(\mathbf{q},\mathbf{q}') = ie\int_0^1 ds \sum_\mu (q-q')_\mu A_\mu(\mathbf{q}' + s(\mathbf{q}-\mathbf{q}')).$$

It has the expected gauge properties:

$$\delta A_\mu(\mathbf{q}) = (1/e)\partial_\mu \Omega(\mathbf{q}) \Rightarrow \delta\sigma_0(\mathbf{q},\mathbf{q}') = i(\Omega(\mathbf{q}) - \Omega(\mathbf{q}')).$$

One infers

$$\partial_\mu \sigma_0 - ieA_\mu(\mathbf{q}) = ie\int_0^1 ds\, s \sum_\nu (q-q')_\nu F_{\mu\nu}(\mathbf{q}' + s(\mathbf{q}-\mathbf{q}')), \quad (5.22)$$

where

$$F_{\mu\nu} = \partial_\mu A_\nu - \partial_\nu A_\mu,$$

is the magnetic tensor.

The term of order t^0 yields

$$2\sigma_1 + \sum_\mu 2(q-q')_\mu \partial_\mu \sigma_1 = \sum_\mu \partial_\mu(\partial_\mu \sigma_0 - ieA_\mu(\mathbf{q})) - (\partial_\mu \sigma_0 - ieA_\mu(\mathbf{q}))^2.$$

Since $|\mathbf{q}-\mathbf{q}'| = O(\sqrt{t})$, one needs only $\sigma_1(\mathbf{q},\mathbf{q})$.

Equation (5.22) shows that $\partial_\mu \sigma_0 - ieA_\mu(\mathbf{q})$, which is of order $\mathbf{q} - \mathbf{q}'$, can now be neglected. Moreover,

$$\sum_\mu \partial_\mu(\partial_\mu \sigma_0 - ieA_\mu(\mathbf{q})) = ie\int_0^1 ds\, s^2 \sum_{\mu,\nu}(q-q')_\nu \partial_\mu F_{\mu\nu}(\mathbf{q}' + s(\mathbf{q}-\mathbf{q}'))$$

and, thus, it vanishes also for $\mathbf{q} = \mathbf{q}'$. One concludes

$$\sigma_1(\mathbf{q},\mathbf{q}) = 0.$$

Thus, in terms of the free motion

$$\mathbf{q}(\tau) = \mathbf{q}' + \tau(\mathbf{q}-\mathbf{q}')/t,$$

the additional magnetic contribution to the action can be written as

$$ie\int_0^t d\tau\, \dot{\mathbf{q}}(\tau) \cdot A(\mathbf{q}(\tau)),$$

which is an expression symmetric in $\mathbf{q}$ and $\mathbf{q}'$, in agreement with the discussion of Section 5.2.2.

5.5 Diffusion, random walk, Fokker–Planck equation

We now consider markovian processes describing classical diffusion phenomena, random walks or brownian motion. We consider only processes in continuous space and time, but the equations that we study describe also processes on space lattices in discrete time in the asymptotic limit of large distances and large times.

We denote by $P(\mathbf{q}'', \mathbf{q}'; t'', t')$ the probability density that the random variable $\mathbf{q}$ that takes the value $\mathbf{q}'$ at time t', takes the value $\mathbf{q}''$ at a later time t''. A process is markovian if the probability distribution for the variable $\mathbf{q}$ at time t depends only on its value at time $t - \mathrm{d}t$ and not on prior history. To express this property, it is convenient to introduce the operator $\mathbf{P}$ whose matrix elements are the probability densities:

$$\langle \mathbf{q}'' | \mathbf{P}(t'', t') | \mathbf{q}' \rangle \equiv P(\mathbf{q}'', \mathbf{q}'; t'', t').$$

The Markov property then takes the form

$$\mathbf{P}(t_3, t_2)\mathbf{P}(t_2, t_1) = \mathbf{P}(t_3, t_1), \quad t_3 \geq t_2 \geq t_1. \tag{5.23}$$

Equation (5.23) is identical to the relation (2.1) satisfied by the statistical operator.

However, P satisfies two additional constraints. Probabilities are positive:

$$P(\mathbf{q}, \mathbf{q}'; t, t') \geq 0.$$

Moreover, the total probability is conserved:

$$\int \mathrm{d}^d q \, P(\mathbf{q}, \mathbf{q}'; t, t') = 1. \tag{5.24}$$

One can verify that these conditions are consistent with Markov's property (5.23).

We assume, in addition, that $\mathbf{P}$ is a differentiable function of time and set

$$\mathbf{P}(t + \varepsilon, t) = \mathbf{1} - \varepsilon \mathbf{H}(t) + O(\varepsilon^2).$$

Taking the limit $t_3 \to t_2$ in equation (5.23), one derives an equation of type (2.2):

$$\frac{\partial}{\partial t} \mathbf{P}(t, t') = -\mathbf{H}(t) \mathbf{P}(t, t'), \tag{5.25}$$

where, in analogy with quantum mechanics, we call $\mathbf{H}$ the Fokker–Planck hamiltonian.

5.5.1 A simple example: random walk or brownian motion

We consider a stochastic process of the random walk type, time-independent, space-isotropic and translation-invariant. This implies that the probability distribution has the form

$$P(\mathbf{q}'', \mathbf{q}'; t'', t') = p(|\mathbf{q}' - \mathbf{q}'|; t'' - t')$$

and, thus, the operator $\mathbf{H}$ in equation (5.25) is time-independent. Formally,
$$\mathbf{P}(t'',t') = e^{-(t''-t')\mathbf{H}}.$$
Moreover, we assume a short range dynamics, that is that $p(|\mathbf{q}|, t)$ decreases at least exponentially for $|\mathbf{q}| \to \infty$. This is a *locality condition*, which plays the role, in this context, of the locality of quantum hamiltonians discussed in Section 2.1.2.

To exploit space translation invariance, it is natural to introduce the Fourier transform of the distribution $p(|\mathbf{q}|, t)$:
$$\tilde{p}(\mathbf{k},t) = \int d^d q \, e^{-i\mathbf{q}\cdot\mathbf{k}} p(|\mathbf{q}|;t).$$
In this basis, the operators $\mathbf{P}$ and $\mathbf{H}$ are diagonal and, thus,
$$\tilde{p}(\mathbf{k},t) = e^{-t\tilde{h}(\mathbf{k})},$$
where we have denoted by $\tilde{h}(\mathbf{k})$ the matrix elements of $\mathbf{H}$.

The function $\tilde{p}$ is a real function that depends only on $|\mathbf{k}|$. Moreover, the positivity of $p(|\mathbf{q}|, t)$ and its normalization imply
$$|\tilde{p}(\mathbf{k},t)| \leq \tilde{p}(0,t) = 1. \tag{5.26}$$
In addition, the upper-bound is reached only for $\mathbf{k} = 0$. One infers
$$\tilde{h}(\mathbf{k}) \geq 0, \ \tilde{h}(0) = 0.$$
The locality condition implies that $\tilde{p}(\mathbf{k}, t)$ is an analytic function of $\mathbf{k}$. Together with the normalization condition (5.26), it implies that $\tilde{h}(\mathbf{k})$ is analytic at $\mathbf{k} = 0$ and, thus,
$$\tilde{h}(\mathbf{k}) \underset{\mathbf{k}\to 0}{\sim} \tfrac{1}{2} D \mathbf{k}^2,$$
where D is a diffusion constant. The distribution is obtained by inverting the Fourier transformation:
$$p(\mathbf{q};t) = \frac{1}{(2\pi)^d} \int d^d p \, \exp\left[i\mathbf{k}\cdot\mathbf{q} - t\tilde{h}(\mathbf{k})\right].$$
For $t \to \infty$, the integral is dominated by the unique minimum of $\tilde{h}$ at $\mathbf{k} = 0$. One can thus replace $\tilde{h}$ by its leading term as $\mathbf{k} \to 0$:
$$p(\mathbf{q};t) \underset{t\to\infty}{\sim} \frac{1}{(2\pi)^d} \int d^d k \, \exp\left[i\mathbf{k}\cdot\mathbf{q} - \tfrac{1}{2} t D \mathbf{k}^2\right]$$
$$= \frac{1}{(2\pi Dt)^{d/2}} e^{-\mathbf{q}^2/2Dt}.$$
One recognizes the asymptotic behaviour of the simple random walk. The result follows from arguments that also lead to the central limit theorem.

The asymptotic form of $p(\mathbf{q};t)$ is analogous to the form (2.9). Therefore, for large times, $p(\mathbf{q};t)$ has a path integral representation with the action
$$\mathcal{S}(\mathbf{q}) = \int_0^t d\tau \, \left(\dot{\mathbf{q}}(\tau)\right)^2/2D,$$
which is analogous to the action of a free particle in quantum mechanics.

5.5.2 General diffusion equation

As in the case of the statistical operator, the rule (5.23) allows deriving $\mathbf{P}(t'',t')$ from its evaluation $\mathbf{P}(t+\varepsilon,t)$ at short times, $\varepsilon \to 0$. We still assume that brownian motion dominates diffusion for short times, but we now allow an additional, position-dependent, anisotropy in such a way that for short times $P(\mathbf{q},\mathbf{q}';t,t')$ takes the form

$$\log[(2\pi D\varepsilon))^{d/2} P(\mathbf{q},\mathbf{q}';t'+\varepsilon,t')] = -\frac{(\mathbf{q}-\mathbf{q}')^2}{2D\varepsilon} + (\mathbf{q}-\mathbf{q}') \cdot \mathbf{A}(\mathbf{q}') + B(\mathbf{q}'). \tag{5.27}$$

Conservation of probability then translates into the equation

$$(2\pi D\varepsilon))^{d/2} = \int d^d q \; e^{-(\mathbf{q}-\mathbf{q}')^2/(2D\varepsilon)+(\mathbf{q}-\mathbf{q}')\cdot\mathbf{A}(\mathbf{q}')+B(\mathbf{q}')}.$$

The integration over $\mathbf{q}$ yields

$$\exp(\tfrac{1}{2} D\varepsilon \mathbf{A}^2 + B) = 1 \;\Rightarrow\; B = -\tfrac{1}{2}\varepsilon D \mathbf{A}^2.$$

Thus,

$$P(\mathbf{q},\mathbf{q}';t'+\varepsilon,t') \underset{\varepsilon\to 0}{\sim} \frac{1}{(2\pi D\varepsilon)^{d/2}} \exp\left[-\frac{(\mathbf{q}-\mathbf{q}'-D\varepsilon \mathbf{A}(\mathbf{q}'))^2}{2D\varepsilon}\right]. \tag{5.28}$$

Using the same strategy as in Chapter 2, one derives the path integral representation

$$P(\mathbf{q}'',\mathbf{q}';t'',t') = \int [d\mathbf{q}(t)] \, e^{-\mathcal{S}(\mathbf{q})} \tag{5.29}$$

with

$$\mathcal{S}(\mathbf{q}) = \int_{t'}^{t''} dt \, \tfrac{1}{2D} \left[\dot{\mathbf{q}}(t) - D\mathbf{A}(\mathbf{q})\right]^2, \tag{5.30}$$

and the boundary conditions $\mathbf{q}(t')=\mathbf{q}'$ and $\mathbf{q}(t'')=\mathbf{q}''$.

The path integral provides a representation of the probability distribution P that is obviously positive and satisfies Markov's property (5.23).

Quantization ambiguities. The form of the path integral suggests that the distribution P in this more general framework still satisfies a Schrödinger-type equation, called Fokker–Planck equation with a non-hermitian hamiltonian and a potential linear in the momentum operator. To derive the equation, we Fourier transform expression (5.28) with respect to $\mathbf{q}$:

$$\int d^d q \, e^{-i\mathbf{q}\cdot\mathbf{p}} P(\mathbf{q},\mathbf{q}';\varepsilon) = e^{-D\varepsilon \mathbf{p}^2/2 - i\varepsilon D\mathbf{p}\cdot\mathbf{A}(\mathbf{q}')} \, e^{-i\mathbf{p}\cdot\mathbf{q}'}. \tag{5.31}$$

The term of order ε in the expansion of the r.h.s. directly yields the matrix elements of the corresponding hamiltonian in the mixed position-momentum representation:

$$\langle \mathbf{p} | \mathbf{H} | \mathbf{q}' \rangle = \tfrac{1}{2} D \left[\mathbf{p}^2 + 2i\mathbf{p}\cdot\mathbf{A}(\mathbf{q}')\right] e^{-i\mathbf{p}\cdot\mathbf{q}'}.$$

5.5 Path integrals and quantization

After Fourier transformation with respect to **p**, one obtains the matrix elements of the hamiltonian in the position basis. One concludes that $P(\mathbf{q}, \mathbf{q}'; t)$ satisfies an equation analogous to equation (2.11):

$$\dot{P}(\mathbf{q}, \mathbf{q}'; t) = \tfrac{1}{2} D \sum_i \partial_i \left[\partial_i - 2 A_i(\mathbf{q}) \right] P(\mathbf{q}, \mathbf{q}'; t),$$

called the Fokker–Planck equation. The r.h.s. is a divergence and this guarantees conservation of probabilities.

If one knows only the 'classical' form of the Fokker–Planck hamiltonian, one is confronted with a problem of quantization. In the present context, the form of the 'quantum' hamiltonian is fixed by the condition of conservation of probabilities, which implies that $\dot{P}$ is given by a divergence.

Again, one expects to meet an ambiguity problem in the definition of the path integral. The nature of the problem has already been exhibited in the expansions (5.16) and (5.17). If in equation (5.27), $(\mathbf{q}-\mathbf{q}') \cdot \mathbf{A}(\mathbf{q}')$ is replaced by $(\mathbf{q}-\mathbf{q}') \cdot \mathbf{A}(\mathbf{q})$, two expressions that have the same continuum limit, the limit of the integral over discrete times is modified because, for brownian motion, the difference between the two terms is of order ε, as we have learned from the expansion (5.16).

Different choices of quantization correspond to different values of sgn(0). To resolve this difficulty, it is useful to return to the discretized form, which is not ambiguous. In particular, if one insists on using a formalism that is symmetric under time reflection (thus with sgn(0) = 0), one finds out that one must insert in the path integral the half-sum of the normal and anti-normal ordered hamiltonian (for a discussion see Section 10.2.1). This leads to the continuum action

$$\mathcal{S}(\mathbf{q}) = \int_{t'}^{t''} dt \left[\tfrac{1}{2D} \left(\dot{\mathbf{q}}(t) - D \mathbf{A}(\mathbf{q}) \right)^2 + \tfrac{1}{2} D \nabla \cdot \mathbf{A} \right]. \tag{5.32}$$

This expression differs from the form (5.30) because it assumes a different convention for sgn(0).

Dissipative Fokker–Planck equation. If the vector **A** is itself a gradient,

$$\mathbf{A} = -\tfrac{1}{2} \nabla E(\mathbf{q})/D,$$

the Fokker–Planck hamiltonian is equivalent to a hermitian positive operator. Moreover, $e^{-E(\mathbf{q})/D}$ is a time-independent solution of the Fokker–Planck equation. If $e^{-E(\mathbf{q})/D}$ is a normalizable distribution, it is the limiting distribution for $t \to \infty$, and is called *equilibrium distribution*. In the action (5.32), the term linear in the velocity is then a total derivative and can be integrated. The action becomes

$$\mathcal{S}(\mathbf{q} = \int_{t'}^{t''} dt \left[\tfrac{1}{2D} \left(\dot{\mathbf{q}}^2(t) + \tfrac{1}{4} (\nabla E(\mathbf{q}))^2 \right) - \tfrac{1}{4} \nabla^2 E(\mathbf{q}) \right] + \tfrac{1}{2D} \left[E(\mathbf{q}'') - E(\mathbf{q}') \right]. \tag{5.33}$$

However, this integration is allowed inside the path integral only if differentiation and expectation value commute, which, as we have seen, implies the convention sgn(0) = 0 and thus is valid only for expression (5.32).

Finally, the equivalent hermitian hamiltonian can be directly inferred from the action (5.33).

5.6 The spectrum of the $O(2)$ rigid rotator

To illustrate the effect of a non-trivial topology of space on the calculation of path integrals, we now discuss the spectrum of the quantum two-dimensional rigid rotator, a system invariant under the $O(2)$ group (rotations-reflection in the plane), first directly using the Schrödinger equation, then using the path integral formalism. We recall that the abelian (or commutative) rotation group in the plane is denoted by $SO(2)$. It is subgroup of the orthogonal group $O(2)$ (rotations-reflection).

The determination of the spectrum $O(2)$ rigid rotator using the time-independent Schrödinger equation is straightforward. What is more remarkable is that the path integral can also be calculated exactly. Moreover, its calculation provides an example of the influence on path integrals of a space that, like the circle, has a non-trivial topology. It also illustrates explicitly the new quantization problem that arises in the case of hamiltonians that are general quadratic functions of momenta and that will be discussed more generally in Section 10.4.

5.6.1 The rigid rotator: hamiltonian and spectrum

We consider a planar, rotation invariant system that can be completely characterized by an angle $\theta(t)$. In this parametrization, the classical lagrangian then corresponds to free motion on the circle:

$$\mathcal{L}(\theta, \dot\theta) = \tfrac{1}{2} mR^2 \dot\theta^2.$$

It has the form of the lagrangian for free motion on the line, a difference arising only from the angular character of the variable θ, which is defined only modulo 2π. The variable conjugate to θ is the angular momentum

$$L = \frac{\partial \mathcal{L}}{\partial \dot\theta} = mR^2 \dot\theta,$$

and the corresponding classical hamiltonian reads

$$H = L\dot\theta - \mathcal{L} = L^2/2mR^2.$$

With this parametrization of the circle, quantization is straightforward and leads to the quantum hamiltonian

$$\hat{H} = \frac{\hat{L}^2}{2mR^2} = -\frac{\hbar^2}{2mR^2} \frac{\partial^2}{(\partial\theta)^2}. \tag{5.34}$$

The $SO(2)$ symmetry is explicit since the angular momentum operator

$$\hat{L} = \frac{\hbar}{i} \frac{\partial}{\partial\theta},$$

which is a generator of the Lie algebra of the $SO(2)$ group, commutes with the hamiltonian (5.34) (reflection corresponds to $\theta \mapsto -\theta$).

The quantum hamiltonian has also the form of a free hamiltonian. However, since θ is an angular variable, the wave functions $\psi(\theta)$ are periodic with period 2π and, thus, have a Fourier series expansion:

$$\psi(\theta) = \sum_{\ell \in \mathbb{Z}} e^{i\ell\theta} \psi_\ell,$$

instead of a Fourier transform. This leads to a quantization of the spectrum:

$$\hat{L} e^{i\ell\theta} = \hbar\ell\, e^{i\ell\theta} \Rightarrow \hat{H} e^{i\ell\theta} = E_\ell e^{i\ell\theta}, \quad E_\ell = \frac{\ell^2 \hbar^2}{2mR^2}.$$

5.6.2 Path integral

Matrix elements of the statistical operator $\mathrm{e}^{-\beta H}$ are given by the path integral

$$\langle \theta''| \mathrm{e}^{-\beta H} |\theta'\rangle = \int_{\theta(-\beta/2)=\theta'}^{\theta(\beta/2)=\theta''} [\mathrm{d}\theta(t)] \exp\left[-\frac{\sigma}{2}\int_{-\beta/2}^{\beta/2} \dot\theta^2(t)\mathrm{d}t\right] \quad (5.35)$$

with $\sigma = mR^2/\hbar^2$.

Note that this path integral also describes brownian motion on the circle.

It is a gaussian path integral and one expects to be able to calculate it exactly. As usual, one first solves the classical equation of motion to determine the dependence on the boundary conditions, that is, on the angles θ' and θ''. The cyclic character of the variable θ then has the following consequence: since θ' and θ'' are angles, one finds an *infinite number* of trajectories that go from θ' to θ'' in a time β. They are given by

$$\theta_n(t) = \tfrac{1}{2}(\theta' + \theta'') + (\theta'' - \theta' + 2\pi n)t/\beta, \quad \text{with } n \in \mathbb{Z}.$$

The number of turns n (called the winding number) has a *topological* meaning in the sense that continuous deformations of a trajectory cannot change the number of turns. The integral over the fluctuations around a trajectory with a given n, thus, cannot include the contributions of trajectories corresponding to other values of n. In contrast to free motion on the line, the path integral (5.35) is given by a sum of an infinite number of contributions.

To calculate the contribution for a given value of n, one changes variables, $\theta(t) \mapsto u(t)$, setting

$$\theta(t) = \theta_n(t) + u(t), \quad u(0) = u(\beta) = 0. \quad (5.36)$$

Then, the sum of contributions can be written as

$$\langle \theta''| \mathrm{e}^{-\beta H} |\theta'\rangle = \mathcal{N} \sum_{n=-\infty}^{+\infty} \exp\left[-\frac{\sigma}{2\beta}(\theta'' - \theta' + 2\pi n)^2\right], \quad (5.37)$$

where we have factorized the normalization

$$\mathcal{N}(\beta/\sigma) = \int_{u(-\beta/2)=0}^{u(\beta/2)=0} [\mathrm{d}u(t)] \exp\left[-\tfrac{1}{2}\sigma \int_{-\beta/2}^{\beta/2} \dot{u}^2(t)\mathrm{d}t\right]$$

because it is independent of n. The normalization is also independent of θ', θ'' and for dimensional reasons depends only on the ratio β/σ. Since the integration over $u(t)$ sums the fluctuations around the classical trajectory, one expects that the angular character of $u(t)$ plays no role and, thus,

$$\mathcal{N}(\beta/\sigma) = \sqrt{2\pi\sigma/\beta} = \hbar^{-1}\sqrt{2\pi m/\beta}, \tag{5.38}$$

but this is not completely obvious because the integration is restricted to configurations in the zero topological sector. Thus, one cannot immediately exclude corrections of order $\mathrm{e}^{-2\pi^2\sigma/\beta}$. Nevertheless, with this assumption one finds

$$\langle \theta'' | \mathrm{e}^{-\beta H} | \theta' \rangle = \frac{1}{\hbar}\sqrt{\frac{2\pi m}{\beta}} \sum_{n=-\infty}^{+\infty} \exp\left[-\frac{\sigma}{2\beta}(\theta'' - \theta' + 2\pi n)^2\right]. \tag{5.39}$$

Because all trajectories θ_n have been taken into account, the expression is periodic in θ' and θ'', as it should.

5.6.3 Partition function and spectrum

Expression (5.39) is only function of the difference $\theta'' - \theta'$, as a consequence of the $SO(2)$ symmetry, which implies translation invariance on the circle. Thus, it can be expanded in a Fourier series of the form

$$\langle \theta'' | \mathrm{e}^{-\beta H} | \theta' \rangle = \frac{1}{2\pi} \sum_{\ell=-\infty}^{+\infty} \mathrm{e}^{i\ell(\theta''-\theta')} \mathrm{e}^{-\beta E_\ell}. \tag{5.40}$$

We have identified the coefficients of the Fourier series with the eigenvalues of $\mathrm{e}^{-\beta H}$ because the functions $\mathrm{e}^{i\ell\theta}/(2\pi)^{1/2}$ form an orthonormal basis and, thus, the operator $\mathrm{e}^{-\beta H}$ is directly diagonalized.

The coefficients of the Fourier series can then be calculated explicitly from expression (5.39), by using Poisson's formula.

Poisson's formula. We consider a function $g(\theta)$ continuous and decreasing sufficiently fast for $|\theta| \to \infty$. To $g(\theta)$, we associate a periodic function, with period 2π,

$$f(\theta) = \sum_{n=-\infty}^{+\infty} g(\theta + 2n\pi).$$

A periodic function can be expanded into a Fourier series:

$$f(\theta) = \sum_{\ell=-\infty}^{+\infty} \mathrm{e}^{i\ell\theta} f_\ell,$$

where the coefficients are given by

$$f_\ell = \frac{1}{2\pi} \int_0^{2\pi} \mathrm{d}\theta\, \mathrm{e}^{-i\ell\theta} f(\theta) = \frac{1}{2\pi} \int_0^{2\pi} \mathrm{d}\theta\, \mathrm{e}^{-i\ell\theta} \sum_n g(\theta + 2n\pi).$$

Inverting sum over n and integration, and changing variables, $\theta + 2n\pi \mapsto \theta$, one then obtains the relations

$$f_\ell = \sum_{n=-\infty}^{+\infty} \int_{2n\pi}^{2(n+1)\pi} \mathrm{d}\theta\, \mathrm{e}^{-i\ell\theta}\, g(\theta) = \frac{1}{2\pi} \int_{-\infty}^{+\infty} \mathrm{d}\theta\, \mathrm{e}^{-i\ell\theta}\, g(\theta). \tag{5.41}$$

These relations can be summarized by the identity

$$\sum_{n=-\infty}^{+\infty} g(\theta + 2n\pi) = \frac{1}{2\pi} \sum_{\ell=-\infty}^{+\infty} \int_{-\infty}^{+\infty} \mathrm{d}\theta'\, \mathrm{e}^{i\ell(\theta-\theta')}\, g(\theta').$$

Application. We now apply Poisson's formula (5.41) to expression (5.39) ($\theta \equiv \theta'' - \theta'$) with

$$g(\theta) = \mathcal{N}(\beta)\, \mathrm{e}^{-\sigma\theta^2/2\beta}.$$

Comparing with the expansion (5.40), one finds

$$\exp(-\beta E_\ell) = \mathcal{N}(\beta/\sigma) \int_{-\infty}^{+\infty} \mathrm{d}\theta\, \mathrm{e}^{-i\ell\theta - \sigma\theta^2/2\beta} = \mathcal{N}(\beta/\sigma) \sqrt{\frac{2\pi\beta}{\sigma}}\, \mathrm{e}^{-\ell^2\beta/2\sigma}. \tag{5.42}$$

This result confirms the normalization (5.38) and yields the spectrum

$$E_\ell = \frac{1}{2\sigma}\ell^2 = \frac{\hbar^2\ell^2}{2mR^2}, \tag{5.43}$$

which is, indeed, the exact result. An exact calculation is possible because the $SO(2)$ rotation group is abelian. The discussion of the general rigid rotator with $O(N)$ symmetry (the orthogonal group of rotations-reflections in N-dimensional space), or free motion on the sphere S_{N-1}, is more complicated (see Section 10.5.2).

5.6.4 Other parametrizations

Since the parametrization of the circle by an angle is, in some sense, redundant, one could think about using other parametrizations. Let us introduce a periodic function $q(\theta)$, monotonous in $[0, 2\pi[$ and, thus, discontinuous at $\theta = 2n\pi$, to describe the rigid rotator. Then, the classical lagrangian becomes

$$\mathcal{L}(q, \dot{q}) = \tfrac{1}{2} g(q) \dot{q}^2$$

with

$$g(q) = mR^2 (\theta'(q))^2.$$

An example of a function $q(\theta)$ is $2R\tan(\theta/2)$, which is invertible on $]-\pi, \pi[$ but singular at $\theta = \pm\pi$ (a singularity is unavoidable). Alternatively, for example, the function $R\sin\theta$, is regular and adequate near $\theta = 0$, but not globally invertible on $[-\pi, \pi[$.

With such parametrizations, the classical hamiltonian takes the general form
$$H = \tfrac{1}{2}p^2/g(q)$$
and, thus, a quantization problem arises. In the case of the rigid rotator, the form of the quantum hamiltonian $\hat H$ is, to a large extent, determined by requiring $SO(2)$ invariance and thus by the condition that $\hat H$ must commute with the angular momentum operator $\hat L$, generator of the Lie algebra of $SO(2)$. In this general parametrization
$$\hat L = \frac{\hbar}{i}\frac{\partial}{\partial\theta} = \frac{\hbar}{i}\frac{\partial q}{\partial\theta}\frac{\partial}{\partial q} = \frac{\hbar}{i}R\sqrt{m/g(q)}\frac{\partial}{\partial q}.$$
Commutation then implies
$$\hat H = \hat L^2/2mR^2 + \text{const.}.$$
The quantum hamiltonian is thus determined up to an additive constant.

From the viewpoint of the path integral, the change of variables $\theta \mapsto q(\theta)$ yields an integration measure that for discrete times takes the form
$$\mathrm{d}\theta_k = \frac{\mathrm{d}\theta_k}{\mathrm{d}q_k}\mathrm{d}q_k \equiv \sqrt{g(q_k)}\,\mathrm{d}q_k\,,$$
and, thus,
$$\prod_k \sqrt{g(q_k)}\,\mathrm{d}q_k = \exp\left[\frac{1}{2}\sum_k \ln\bigl(g(q(t_k))\bigr)\right]\prod_k \mathrm{d}q_k\,.$$
This has the effect of adding a term to the action, but with a coefficient that diverges in the continuum limit:
$$\frac{1}{2}\sum_k \ln\bigl(g(q(t_k))\bigr) = \frac{1}{2\varepsilon}\sum_k \varepsilon\ln\bigl(g(q(t_k))\bigr) \underset{\varepsilon\to 0}{\sim} \frac{1}{2\varepsilon}\int \mathrm{d}t\,\ln\bigl(g(q(t))\bigr).$$

This term has the form of a quantum correction since it has no $1/\hbar$ factor. Correspondingly, in a perturbative calculation of the path integral, the second derivative of the propagator (2.53) at $t=0$ appears, which is divergent. This divergence cancels formally the divergence coming from the measure, but the finite part is then ambiguous. One verifies that the condition of rotation invariance constrains the finite parts. This problem will be discussed in more detail in Section 10.4.

One concludes that the initial choice of parametrization of the plane rigid rotator, for which the action of the $SO(2)$ is trivial and no problem of ordering of operators arises, is much simpler. Such a parametrization does not exist in higher dimensions because spheres have a local curvature (in the sense of riemannian geometry).

Non-linear changes of variables. This discussion also illustrates the problem of non-linear changes of variables in path integrals. The effect of such changes are ill-defined in the continuum limit and can only be discussed by returning to some discretized form (the path integral has be regularized, in the language of quantum field theory.)

Exercises

Exercise 5.1

Use expression (5.13) to calculate the matrix elements of the statistical operator $U(\beta) = e^{-\beta H}$ for $\mathbf{A}(\mathbf{q}) = \frac{1}{2}\mathbf{B} \times \mathbf{q}$, $\mathbf{B}$ constant, that is for a constant magnetic field and $V(\mathbf{q}) = 0$.

Solution. We first consider the effect of a translation on the statistical operator. In the path integral we change variables, setting
$$\mathbf{q}(t) \mapsto \mathbf{q}(t) + \mathbf{a},$$
where $\mathbf{a}$ is a constant vector. The boundary conditions then become
$$\mathbf{q}(-\beta/2) = \mathbf{q}' - \mathbf{a}, \quad \mathbf{q}(\beta/2) = \mathbf{q}'' - \mathbf{a}, \quad \beta = t'' - t'.$$
The effect of a translation is to add to the magnetic term of the action the contribution $\frac{1}{2}ie\mathbf{B} \times \mathbf{a}(\mathbf{q}'' - \mathbf{q}')$, which leads to the relation
$$\langle \mathbf{q}'' | U(\beta) | \mathbf{q}' \rangle = \exp[\tfrac{1}{2}ie(\mathbf{B} \times \mathbf{a})(\mathbf{q}'' - \mathbf{q}')] \langle \mathbf{q}'' - \mathbf{a} | U(\beta) | \mathbf{q}' - \mathbf{a} \rangle.$$
This relation can be used for $\mathbf{a} = \mathbf{q}'$ to simplify the calculation. The phase factor then becomes $\frac{1}{2}ie\mathbf{B} \cdot (\mathbf{q}' \times \mathbf{q}'')$.

One has still to evaluate $\langle \mathbf{q}'' - \mathbf{q}' | U(\beta) | \mathbf{0} \rangle$. One thus solves the equation of the classical motion. The motion parallel to $\mathbf{B}$ being free, the problem can be restricted to the plane perpendicular to $\mathbf{B}$. One finds
$$\mathcal{S}_{\mathrm{c}} = \frac{eB}{4\tanh(eB\beta/2m)} (\mathbf{q}'' - \mathbf{q}')^2.$$
One has then to calculate the determinant resulting from the gaussian integration. It can be inferred from the eigenvalues of the differential operator
$$-m\frac{\mathrm{d}^2}{(\mathrm{d}t)^2}\delta_{ij} + ieB_k \epsilon_{kij}\frac{\mathrm{d}}{\mathrm{d}t}$$
with the boundary conditions that eigenfunctions vanish at $\pm\beta/2$. After a Fourier transformation, the operator takes the form
$$\begin{pmatrix} m\omega^2 & -\omega eB \\ \omega eB & m\omega^2 \end{pmatrix},$$
where ω is the corresponding frequency. The eigenvalues follow:
$$\lambda_n = n^2\pi^2/\beta^2 + e^2B^2/4m^2,$$
each eigenvalue being twice degenerate. The product of eigenvalues divided by its value in zero field can be calculated using the identity
$$\det\left[\left(-\frac{\mathrm{d}^2}{\mathrm{d}t^2} + \omega^2\right) \bigg/ \left(-\frac{\mathrm{d}^2}{\mathrm{d}t^2}\right)\right] = \prod_{n>0}\left(1 + \frac{\omega^2\beta^2}{n^2\pi^2}\right) = \frac{\sinh\omega\beta}{\omega\beta}. \quad (5.44)$$
Finally, normalizing with respect to the free hamiltonian, one obtains
$$\langle \mathbf{q}'' | U(\beta) | \mathbf{q}' \rangle = \frac{eB}{4\pi\sinh(\beta eB/2m)} \exp\left[\tfrac{1}{2}ie\mathbf{B} \cdot (\mathbf{q}' \times \mathbf{q}'')\right] \exp[-\mathcal{S}_{\mathrm{c}}].$$

Exercise 5.2

Add to the preceding action the contribution of the harmonic potential

$$V(\mathbf{q}) = \tfrac{1}{2}m\omega^2 q^2$$

and calculate, again, the matrix elements of the statistical operator.

Solution. The classical action then becomes

$$S_c = \frac{m\omega'}{2\sinh\beta\omega'}\left[(\mathbf{q}'^2 + \mathbf{q}''^2)\cosh\beta\omega' - 2\mathbf{q}'\mathbf{q}''\cosh\left(\frac{eB\beta}{2m}\right)\right.$$
$$\left. - 2i\sinh\left(\frac{eB\beta}{2m}\right)\hat{\mathbf{B}}\cdot(\mathbf{q}'\times\mathbf{q}'')\right],$$

where

$$\hat{\mathbf{B}} = \mathbf{B}/B, \qquad \omega' = \sqrt{\omega^2 + e^2 B^2/4m^2}.$$

The gaussian integration generates the factor

$$\mathcal{N} = \frac{eB}{4\pi\sinh(\beta\omega')}.$$

Exercise 5.3

The Fokker–Planck equation

1. Using the path integral (5.29) with the action (5.33), calculate the probability distribution $P(q,\tau;q_0,0)$ solution of the Fokker–Planck equation

$$\frac{\partial}{\partial t}P(q,t;q_0,t_0) = \frac{1}{2}D\frac{\partial}{\partial q}\left[\frac{\partial}{\partial q}P + \frac{1}{D}\frac{\partial E}{\partial q}P\right],$$

with the function

$$E(q) = q^2.$$

One may relate the problem to the harmonic oscillator and use the corresponding result. Infer, then, the limiting distribution for $\tau \to +\infty$.

2. Using the path integral, calculate $P(q,\tau;q_0,0)$ for $q_0 = 0$ in the example of the function

$$E(q) = -q^2.$$

What happens in the limit $\tau \to +\infty$?

Solutions.
1. In the case of the function

$$E(q) = q^2,$$

using the corresponding result for the harmonic oscillator, one finds

$$P(q,\tau;q_0,0) = \frac{e^{-[(q^2+q_0^2)\cosh\tau - 2q_0 q]/2D\sinh\tau}\, e^{-(q^2-q_0^2)/2D}}{\sqrt{2\pi D(1-e^{-2\tau})}}.$$

The limiting (or equilibrium) distribution for $\tau \to +\infty$ then is

$$P(q,\tau;q_0,0) \underset{\tau\to\infty}{\sim} \frac{1}{\sqrt{\pi D}} e^{-q^2/D} \propto e^{-E(q)/D}.$$

2. In the case of the function

$$E(q) = -q^2,$$

for $q_0 = 0$, one infers from the path integral representation

$$P(q,\tau;q_0,0) = \frac{1}{\sqrt{\pi D(e^{2\tau}-1)}} e^{-q^2/D(e^{2\tau}-1)}.$$

In the limit $\tau \to +\infty$, the distribution becomes uniform and converges to zero everywhere. This result reflects the property that the distribution $e^{-E(q)/D}$ is not normalizable and, thus, is not a suitable equilibrium distribution.

6 PATH INTEGRALS AND HOLOMORPHIC FORMALISM

In this chapter, we introduce a description of quantum mechanics in terms of a Hilbert space of square integrable analytic functions, called the holomorphic representation. We then construct the corresponding path integral representation of the statistical operator.

The holomorphic formalism is specially well adapted to a study of the harmonic oscillator and, more generally, of perturbed harmonic oscillators. Nevertheless, one may wonder whether it is really useful to construct another representation of the harmonic oscillator, which the usual path integral formalism already allows discussing quite thoroughly. Actually, the main motivation comes from quantum N-particle problems (one speaks also of N-body problems). It is based on a characteristic property of the harmonic oscillator, the additivity of its spectrum: the energy of the level N of the harmonic oscillator is the sum of the ground state energy and N times the splitting between two neighbouring levels. This energy thus can also be considered as the sum of a vacuum energy and the total energy of N identical and independent particles. The perturbations to the harmonic oscillator then correspond to interactions between particles.

This gives an intuitive understanding of the role of generalized holomorphic path integrals (field integrals) in representations of the partition function of the quantum Bose gas and, in quantum field theory, of boson scattering or S-matrices. As an illustration, we apply the formalism to the Bose–Einstein condensation.

In Chapter 7, we will show how to construct a parallel formalism for fermions. In the latter problem complex variables must be replaced by Grassmann, that is anticommuting variables, elements of an antisymmetric algebra that allows reproducing the statistical properties of fermions.

However, before describing the holomorphic formalism, it is useful to recall some properties of a certain type of integrals over complex variables.

6.1 Complex integrals and Wick's theorem

In the case of integrals over $2n$ real variables $\{x_\alpha\}$ and $\{y_\alpha\}$, $\alpha = 1, \ldots, n$, where the integrand is invariant under a simultaneous identical rotation in all planes (x_α, y_α), it is natural to introduce, in each plane, a complex parametrization on which rotations act by multiplication.

Let us first discuss the nature of this change of variables in the case of one pair (x, y). Quite generally, we consider an integral of the form

$$I = \int_{\mathbb{R}^2} dx\, dy\, f(x, y).$$

If f is a holomorphic function, one can embed the $\mathbb{R}^2$ plane into complex $\mathbb{C}^2$, and consider the variables x, y as complex variables, the initial domain of integration being defined by $\operatorname{Im} x = \operatorname{Im} y = 0$. In $\mathbb{C}^2$, one can then perform a linear change of variables, $(x, y) \mapsto (z, z')$, defined in terms of a unitary matrix:

$$z = (x + iy)/\sqrt{2}, \ z' = (x - iy)/\sqrt{2} \Rightarrow dz' dz = i dx\, dy. \tag{6.1}$$

In the new variables, the domain of integration is then defined by $z' = \bar{z}$, where $\bar{z}$ denotes the complex conjugate of z, and the integral becomes

$$I = -i \int_{z'=\bar{z}} dz\, dz'\, f\bigl(x(z, z'), y(z, z')\bigr).$$

Moreover, a rotation of angle θ acting on the vector (x, y) is then represented by

$$z \mapsto z\, e^{i\theta}, \ z' \mapsto e^{-i\theta}\, z'.$$

In particular, if a function f is rotation invariant, it is a function only of $x^2 + y^2$ and, thus, after the change of variables, it is a function only of the product zz'.

Formal complex conjugation. Notation. We now introduce a standard, convenient, but somewhat dangerous notation: we denote the variable z' by $\bar{z}$. We stress that the complex variables z and $\bar{z}$ remain *independent integration variables* and are complex conjugate only in a formal sense. Indeed, the integration contours over the variables x and y could be deformed in complex space and, then, these variables would also take complex values. The symbol $dz d\bar{z}$ thus corresponds to an integration over a surface of real dimension 2, embedded in $\mathbb{C}^2$.

However, the notation is motivated by the important role played by formal complex conjugation, which corresponds to a reflection $y \mapsto -y$ combined with a complex conjugation of coefficients

$$f_{mn} x^m y^n \mapsto \bar{f}_{mn} x^m (-y)^n.$$

In the variables $z, \bar{z}$, formal complex conjugation has a very simple form

$$f(z, \bar{z}) \mapsto \bar{f}(z, \bar{z}).$$

6.1.1 Gaussian integrals

The simplest gaussian integral that has an integrand invariant under rotations in the plane is

$$\frac{1}{2\pi} \int_{\mathbb{R}^2} dx\, dy\, e^{-a(x^2+y^2)/2} = \int \frac{dz d\bar{z}}{2i\pi} e^{-a\bar{z}z} = \frac{1}{a},$$

where, in the initial variables (x, y), one integrates over the real plane and $\operatorname{Re} a > 0$.

More generally, we consider gaussian integrals of the form

$$\mathcal{Z}(\mathbf{A}) = \int \left(\prod_{i=1}^n \frac{dz_i d\bar{z}_i}{2i\pi} \right) e^{-A(\bar{\mathbf{z}}, \mathbf{z})},$$

where $A(\bar{\mathbf{z}}, \mathbf{z})$ is the quadratic form

$$A(\bar{\mathbf{z}}, \mathbf{z}) = \sum_{i,j=1}^{n} \bar{z}_i A_{ij} z_j. \tag{6.2}$$

We assume that the determinant of the complex matrix $\mathbf{A}$ with elements A_{ij} does not vanish. Because terms of the form zz and $\bar{z}\bar{z}$ are absent, the integral is left invariant by the linear change of variables

$$z_i \mapsto z_i \, \mathrm{e}^{i\theta}, \quad \bar{z}_i \mapsto \bar{z}_i \, \mathrm{e}^{-i\theta}. \tag{6.3}$$

If $\mathbf{A}$ is a hermitian positive matrix, it can be diagonalized by a unitary transformation:

$$\mathbf{A} = \mathbf{U}\mathbf{D}\mathbf{U}^{\dagger},$$

where $\mathbf{U}$ is a unitary matrix and $\mathbf{D}$ is the diagonal matrix with eigenvalues $a_i > 0$. The change of variables

$$z_i = \sum_j U_{ij} z'_j, \quad \bar{z}_i = \sum_j \bar{U}_{ij} \bar{z}'_j$$

has a jacobian $|\det \mathbf{U}|^2 = 1$ and, thus, the integral becomes

$$\mathcal{Z}(\mathbf{A}) = \prod_i \int \frac{\mathrm{d}z'_i \mathrm{d}\bar{z}'_i}{2i\pi} \mathrm{e}^{-a_i \bar{z}'_i z'_i} = \prod_i \frac{1}{a_i} = \frac{1}{\det \mathbf{A}}.$$

Remark. In contrast with the real gaussian integral (1.4), the result is a rational function of the elements of the matrix and can thus be extended by analytic continuation to any complex matrix. Actually, from a purely algebraic viewpoint, it is possible to perform an asymmetric change of variables like $z_i \mapsto z'_i = \sum_j A_{ij} z_j$, leaving the variables $\bar{z}$ unchanged, which has for jacobian $1/\det \mathbf{A}$. One then obtains

$$\mathcal{Z}(\mathbf{A}) = \frac{1}{\det \mathbf{A}} \int \left(\prod_{i=1}^n \frac{\mathrm{d}z_i \mathrm{d}\bar{z}_i}{2i\pi} \right) \exp\left[-\sum_i \bar{z}_i z_i \right] = \frac{1}{\det \mathbf{A}} \left[\int \frac{\mathrm{d}z \mathrm{d}\bar{z}}{2i\pi} \mathrm{e}^{-\bar{z}z} \right]^n,$$

which is consistent with

$$\mathcal{Z}(\mathbf{A}) = \int \left(\prod_{i=1}^n \frac{\mathrm{d}z_i \mathrm{d}\bar{z}_i}{2i\pi} \right) \mathrm{e}^{-A(\bar{\mathbf{z}}, \mathbf{z})} = \frac{1}{\det \mathbf{A}}. \tag{6.4}$$

6.1.2 General gaussian integral

The general gaussian integral

$$\mathcal{Z}(\mathbf{A}; \mathbf{b}, \bar{\mathbf{b}}) = \int \left(\prod_{i=1}^{n} \frac{\mathrm{d}z_i \mathrm{d}\bar{z}_i}{2i\pi} \right) \exp\left[-A(\bar{\mathbf{z}}, \mathbf{z}) + \bar{\mathbf{b}} \cdot \mathbf{z} + \bar{\mathbf{z}} \cdot \mathbf{b} \right], \tag{6.5}$$

where $\mathbf{b}$ and $\bar{\mathbf{b}}$ correspond to two independent sets of variables, generates all expectation values with the weight $\mathrm{e}^{-A(\bar{\mathbf{z}}, \mathbf{z})}$. As usual, the integral can be calculated by shifting $\mathbf{z}$ and $\bar{\mathbf{z}}$. One solves the equations

$$\frac{\partial}{\partial z_i} \left[A(\bar{\mathbf{z}}, \mathbf{z}) - \bar{\mathbf{b}} \cdot \mathbf{z} - \bar{\mathbf{z}} \cdot \mathbf{b} \right] = \frac{\partial}{\partial \bar{z}_i} \left[A(\bar{\mathbf{z}}, \mathbf{z}) - \bar{\mathbf{b}} \cdot \mathbf{z} - \bar{\mathbf{z}} \cdot \mathbf{b} \right] = 0.$$

The solutions can be conveniently expressed in terms of the inverse matrix

$$\Delta = \mathbf{A}^{-1}.$$

One finds

$$\sum_j z_j A_{ji} = \bar{b}_i \Rightarrow \bar{z}_i = \sum_j \bar{b}_j \Delta_{ji}, \quad \sum_j A_{ij} z_j = b_i \Rightarrow z_i = \sum_j \Delta_{ij} b_j.$$

The change of variables $\{z_i, \bar{z}_i\} \mapsto \{v_i, \bar{v}_i\}$:

$$z_i = v_i + \sum_j \Delta_{ij} b_j, \quad \bar{z}_i = \bar{v}_i + \sum_j \bar{b}_j \Delta_{ji}, \tag{6.6}$$

eliminates the terms linear in z_i and $\bar{z}_i$. The remaining gaussian integral over the variables $v, \bar{v}$ is simply the integral (6.4) and, thus,

$$\mathcal{Z}(\mathbf{A}; \mathbf{b}, \bar{\mathbf{b}}) = (\det \mathbf{A})^{-1} \exp\left[\sum_{i,j=1}^{n} \bar{b}_i \Delta_{ij} b_j \right]. \tag{6.7}$$

Gaussian expectation values. This result allows calculating expectation values with the gaussian weight $\mathrm{e}^{-A}/\mathcal{Z}(\mathbf{A})$:

$$\langle \bar{z}_{i_1} \ldots \bar{z}_{i_p} z_{j_1} \ldots z_{j_q} \rangle = \det \mathbf{A} \int \left(\prod_{i=1}^{n} \frac{\mathrm{d}z_i \mathrm{d}\bar{z}_i}{2i\pi} \right) \bar{z}_{i_1} \ldots \bar{z}_{i_p} z_{j_1} \ldots z_{j_q} \, \mathrm{e}^{-A(\bar{\mathbf{z}}, \mathbf{z})}. \tag{6.8}$$

As a direct consequence of the symmetry (6.3), only the monomials with an equal number of z and $\bar{z}$ factors have a non-vanishing gaussian expectation value. Indeed, in the change of variables $z_i \mapsto \mathrm{e}^{i\theta} z_i$, $\bar{z}_i \mapsto \mathrm{e}^{i\theta} \bar{z}_i$, the jacobian is 1 and $A(\bar{\mathbf{z}}, \mathbf{z})$ is invariant. Thus,

$$\langle \bar{z}_{i_1} \ldots \bar{z}_{i_p} z_{j_1} \ldots z_{j_q} \rangle = \mathrm{e}^{i(q-p)\theta} \langle \bar{z}_{i_1} \ldots \bar{z}_{i_p} z_{j_1} \ldots z_{j_q} \rangle = 0 \quad \text{if} \quad p \neq q.$$

Then, differentiating expression (6.5) and using the result (6.7), one first obtains the basic quantity for gaussian expectation values:

$$\langle \bar{z}_k z_\ell \rangle = \frac{\partial}{\partial b_k} \frac{\partial}{\partial \bar{b}_\ell} \exp\left[\sum_{i,j=1}^n \bar{b}_i \Delta_{ij} b_j \right]\bigg|_{b=\bar{b}=0} = \Delta_{\ell k}.$$

Wick's theorem. More generally, by differentiating the integral (6.5) with respect to b and $\bar{b}$ and using the result (6.7), one obtains a general expression for all gaussian expectation values (6.8):

$$\langle \bar{z}_{i_1} z_{j_1} \ldots \bar{z}_{i_p} z_{j_p} \rangle = \frac{\partial}{\partial b_{i_1}} \frac{\partial}{\partial \bar{b}_{j_1}} \cdots \frac{\partial}{\partial b_{i_p}} \frac{\partial}{\partial \bar{b}_{j_p}} \exp\left[\sum_{i,j=1}^n \bar{b}_i \Delta_{ij} b_j \right]\bigg|_{b=\bar{b}=0}.$$

Contributions that do not vanish in the limit $b = \bar{b} = 0$, are obtained by pairing each differentiation with respect to b with a differentiation with respect to $\bar{b}$ in all possible ways. From this observation, one infers the complex Wick theorem:

$$\langle \bar{z}_{i_1} z_{j_1} \ldots \bar{z}_{i_p} z_{j_p} \rangle = \sum_{\substack{\text{all permutations} \\ P \text{ of } \{j_1,\ldots,j_p\}}} \Delta_{j_{P_1} i_1} \Delta_{j_{P_2} i_2} \ldots \Delta_{j_{P_p} i_p} \tag{6.9a}$$

$$= \sum_{\substack{\text{all permutations} \\ P \text{ of } \{j_1,\ldots,j_p\}}} \langle \bar{z}_{i_1} z_{j_{P_1}} \rangle \langle \bar{z}_{i_2} z_{j_{P_2}} \rangle \cdots \langle \bar{z}_{i_p} z_{j_{P_p}} \rangle. \tag{6.9b}$$

For example,

$$\langle \bar{z}_i z_j \bar{z}_k z_l \rangle = \langle \bar{z}_i z_j \rangle \langle \bar{z}_k z_l \rangle + \langle \bar{z}_i z_l \rangle \langle \bar{z}_k z_j \rangle.$$

6.1.3 Perturbative expansion

The previous results allows expanding perturbed gaussian integrals around the gaussian approximation. For example, the integral

$$Z(\lambda) = \int \left(\prod_{i=1}^n \frac{\mathrm{d}z_i \mathrm{d}\bar{z}_i}{2i\pi} \right) \exp\left[-A(\bar{\mathbf{z}}, \mathbf{z}) - \frac{1}{4}\lambda \sum_{i=1}^n \bar{z}_i^2 z_i^2 \right], \tag{6.10}$$

can be expanded in powers of the parameter λ. The few first terms can be calculated using Wick's theorem. Denoting by $\langle \bullet \rangle$ gaussian expectation values, one finds

$$Z(\lambda)/Z(0) = 1 - \frac{\lambda}{4} \sum_i \langle \bar{z}_i^2 z_i^2 \rangle_0 + \frac{\lambda^2}{2!\, 4^2} \sum_{i,j} \langle \bar{z}_i^2 z_i^2 \bar{z}_j^2 z_j^2 \rangle_0 + O(\lambda^3)$$

$$= 1 - \frac{\lambda}{2} \sum_i \Delta_{ii}^2 + \lambda^2 \sum_{i,j} \left(\tfrac{1}{8}\Delta_{ij}^2 \Delta_{ji}^2 + \tfrac{1}{2}\Delta_{ii}\Delta_{ij}\Delta_{jj}\Delta_{ji} \right)$$

$$+ \frac{\lambda^2}{2!\, 4^2} \langle \bar{z}_i^2 z_i^2 \rangle_0 \langle \bar{z}_j^2 z_j^2 \rangle_0 + O(\lambda^3).$$

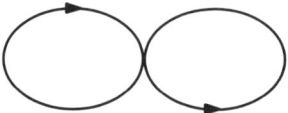

Fig. 6.1 Feynman diagram: the contribution $\langle \bar{z}^2 z^2 \rangle_0$ at order λ.

As expected, the last non-connected contribution cancels in $\ln Z$. Note also that, since $\langle \bar{z}_i z_j \rangle$ is not symmetric in ij, a faithful representation of each contribution in terms of Feynman diagrams involves oriented lines, which go, for instance, from z to $\bar{z}$ (see Fig. 6.1).

To give another example, we expand the expectation value $\langle \bar{z}_k z_\ell \rangle_\lambda$ corresponding to the normalized measure induced by the integrand (6.10). To order λ^2, the expansion is (diagrams of Figs 1.3 and 1.4 with oriented lines)

$$\langle \bar{z}_k z_\ell \rangle_\lambda = \frac{Z(0)}{Z(\lambda)} \left[\Delta_{\ell k} - \frac{\lambda}{4} \sum_i \langle \bar{z}_k z_\ell \bar{z}_i^2 z_i^2 \rangle_0 + \frac{\lambda^2}{32} \sum_{i,j} \langle \bar{z}_k z_\ell \bar{z}_i^2 z_i^2 \bar{z}_j^2 z_j^2 \rangle_0 \right] + O(\lambda^3)$$

$$= \Delta_{\ell k} - \lambda \sum_i \Delta_{\ell i} \Delta_{ii} \Delta_{ik} + \lambda^2 \sum_{i,j} (\Delta_{\ell j} \Delta_{jj} \Delta_{ji} \Delta_{ii} \Delta_{ik}$$

$$+ \Delta_{\ell i} \Delta_{ij} \Delta_{jj} \Delta_{ji} \Delta_{ik} + \tfrac{1}{2} \Delta_{\ell j} \Delta_{ji} \Delta_{ij} \Delta_{ji} \Delta_{ik}) + O(\lambda^3).$$

6.2 Holomorphic representation

The holomorphic formalism originates from the idea of associating (complex) classical variables $z, \bar{z}$ to the creation and annihilation operators $a^\dagger, a$ that, for example, are introduced in the simplest algebraic solution of the harmonic oscillator.

6.2.1 Hilbert space of analytic functions

One considers the complex vector space of analytic entire functions endowed with a scalar product: the scalar product of two entire functions f and g is defined by

$$(g, f) = \int \frac{dz\, d\bar{z}}{2i\pi} \, e^{-z\bar{z}} \, \overline{g(z)} \, f(z), \tag{6.11}$$

where $\bar{z}$ is the variable formally complex conjugate to z and the integration is defined as in Section 6.1. The scalar product (f, f) defines a positive norm $\|f\| = (f, f)^{1/2}$. Entire functions with a finite norm form a Hilbert space, which we denote below by $\mathfrak{H}$.

One can then define operators acting on this Hilbert space, and, thus, construct a representation of quantum mechanics, called the holomorphic representation.

Notice that a necessary condition for the norm of a function f to be finite is that it satisfies the bound

$$|f(z)| \leq C \, e^{|z|^2/2}, \tag{6.12}$$

6.2 Path integrals and holomorphic formalism

where C is a positive constant.

Orthonormal basis. The monomials $z^n/\sqrt{n!}$ form an orthonormal basis of the Hilbert space $\mathfrak{H}$ since

$$\int \frac{\mathrm{d}z\mathrm{d}\bar{z}}{2i\pi}\, \mathrm{e}^{-z\bar{z}}\, \bar{z}^n z^m = n!\, \delta_{mn}\,. \tag{6.13}$$

This result can easily be verified, either by applying Wick's theorem (6.9), or, directly, by changing variables $z, \bar{z} \mapsto \rho, \theta$, where ρ and θ are the modulus and argument of the complex variable z, respectively. Indeed,

$$\int \frac{\mathrm{d}z\mathrm{d}\bar{z}}{2i\pi}\, \mathrm{e}^{-z\bar{z}}\, \bar{z}^n z^m = \int \rho\mathrm{d}\rho \frac{\mathrm{d}\theta}{\pi}\, \mathrm{e}^{-\rho^2}\, \rho^{n+m}\, \mathrm{e}^{i\theta(m-n)}$$

$$= \delta_{mn} \int_0^\infty 2\rho\mathrm{d}\rho\, \rho^{2n}\, \mathrm{e}^{-\rho^2}\,.$$

Then, the norm of a function expanded on this basis:

$$f(z) = \sum_{n \geq 0} f_n z^n,$$

is finite if

$$(f,f) = \sum_n |f_n|^2 n! < \infty\,, \tag{6.14}$$

which also implies the bound (6.12).

δ function. In the holomorphic formalism, the role of Dirac's δ function is played by the function

$$\delta(z) \equiv \frac{1}{2i\pi} \int \mathrm{d}\bar{z}\, \mathrm{e}^{-\bar{z}z}\,. \tag{6.15}$$

Indeed, as a consequence of the orthogonality relations,

$$\int \frac{\mathrm{d}z\mathrm{d}\bar{z}}{2i\pi}\, \mathrm{e}^{-z\bar{z}}\, f(z) = \int \frac{\mathrm{d}z\mathrm{d}\bar{z}}{2i\pi}\, \mathrm{e}^{-z\bar{z}} \sum_{n=0} \frac{z^n}{n!} f^{(n)}(0) = f(0)\,.$$

This remark will be useful in Section 6.3.

6.2.2 Harmonic oscillator and holomorphic representation

We now construct a representation of the operators of quantum mechanics acting on the Hilbert space $\mathfrak{H}$. For this purpose, we show that the operators z, acting by multiplication, and $\mathrm{d}/\mathrm{d}z$ represent the creation and annihilation operators $a^\dagger, a$ that allows diagonalizing the hamiltonian of the harmonic oscillator

$$H_0 = \tfrac{1}{2}\hat{p}^2 + \tfrac{1}{2}\omega^2 \hat{q}^2, \tag{6.16}$$

by algebraic methods.

Creation and annihilation operators. We recall that the creation and annihilation operators $a^\dagger$ and a, associated with the hamiltonian (6.16), are defined by (in this chapter $\hbar = 1$ except in Sections 6.8 and 6.9)

$$\hat{p} - i\omega\hat{q} = -i\sqrt{2\omega}\,a\,, \quad \hat{p} + i\omega\hat{q} = i\sqrt{2\omega}\,a^\dagger \Rightarrow [a, a^\dagger] = 1\,. \tag{6.17}$$

Expressed in terms of $a, a^\dagger$, the hamiltonian takes the simple form

$$H_0 = \omega a^\dagger a + \tfrac{1}{2}\omega$$

with $\omega > 0$, a *sign convention* that we adopt throughout this chapter.

The eigenvectors $|k\rangle$ of the hamiltonian are then generated by acting repeatedly on the ground state $|0\rangle$, which satisfies

$$a\,|0\rangle = 0\,,$$

with the creation operator $a^\dagger$:

$$|k\rangle = \frac{1}{\sqrt{k!}}(a^\dagger)^k\,|0\rangle\,.$$

The corresponding eigenvalues then are $E_k = (k + 1/2)\omega$. In what follows we omit, in general, the global energy shift $\omega/2$, translating H_0 by $-\omega/2$.

Holomorphic representation. The operator $a^\dagger$ is represented by a multiplication by the complex variable z:

$$a^\dagger f \mapsto z f(z). \tag{6.18}$$

The operator a is then represented by the differential operator $\mathrm{d}/\mathrm{d}z$

$$a f \mapsto \mathrm{d}f(z)/\mathrm{d}z\,. \tag{6.19}$$

First, this choice is consistent with the commutation relation (6.17) since

$$[\mathrm{d}/\mathrm{d}z, z] = 1\,.$$

Then, let us consider the scalar product of a function $g(z)$ by a function $f(z)$ on which acts the operator $\mathrm{d}/\mathrm{d}z$:

$$(g, \mathrm{d}f/\mathrm{d}z) = \int \frac{\mathrm{d}z\,\mathrm{d}\bar{z}}{2i\pi}\,\mathrm{e}^{-z\bar{z}}\,\overline{g(z)}\,\frac{\mathrm{d}}{\mathrm{d}z}f(z).$$

After integration by parts, one finds

$$(g, \mathrm{d}f/\mathrm{d}z) = -\int \frac{\mathrm{d}z\,\mathrm{d}\bar{z}}{2i\pi}\,f(z)\overline{g(z)}\,\frac{\mathrm{d}}{\mathrm{d}z}\,\mathrm{e}^{-z\bar{z}}$$

$$= \int \frac{\mathrm{d}z\,\mathrm{d}\bar{z}}{2i\pi}\,\overline{zg(z)}\,f(z).$$

This identity proves that z and $\mathrm{d}/\mathrm{d}z$ are hermitian conjugate operators with respect to the scalar product (6.11), in the same way as the operators $a^\dagger$ and a. As a direct consequence, hermitian conjugation is preserved in the representation (6.18, 6.19).

The holomorphic representation of quantum mechanics is thus isomorphic to the more usual representation in terms of momentum and position, or creation and annihilation operators.

Harmonic oscillator. The hamiltonian $H_0 = \omega a^\dagger a$ is then represented by

$$H_0 = \omega z \frac{\mathrm{d}}{\mathrm{d}z}. \tag{6.20}$$

The holomorphic representation is particularly well adapted to studying the harmonic oscillator. Indeed,

$$H_0 z^n = \omega z \frac{\mathrm{d}}{\mathrm{d}z} z^n = \omega n\, z^n$$

and, thus, the monomials z^n are the eigenfunctions of the hamiltonian. We have already shown that the monomials z^n are orthogonal (equation (6.13)) with respect to the scalar product (6.11), a property consistent with the hermiticity of H_0.

6.3 Kernel of operators

To the functions $\{z^n/\sqrt{n!}\}$, which form an orthonormal basis, is associated a representation of the identity operator in the form of a kernel:

$$\mathcal{I}(z,\bar{z}) = \sum_{n=0}^\infty \frac{\bar{z}^n}{\sqrt{n!}} \frac{z^n}{\sqrt{n!}} = \mathrm{e}^{z\bar{z}}, \tag{6.21}$$

a result that can be verified by a direct calculation. Indeed,

$$\int \frac{\mathrm{d}z'\mathrm{d}\bar{z}'}{2i\pi} \mathrm{e}^{-z'\bar{z}'} \mathcal{I}(z,\bar{z}') f(z') = \int \frac{\mathrm{d}z'\mathrm{d}\bar{z}'}{2i\pi} \mathrm{e}^{-(z'-z)\bar{z}'} f(z') = f(z), \tag{6.22}$$

as a consequence of the representation (6.15) of the δ function.

Identity (6.22) leads to a new representation of any operator function of a and $a^\dagger$. Using the commutation relation (6.17), one can expand any operator on a basis of monomials written in a canonical form with all operators a at the right of all operators $a^\dagger$ (*normal order*)

$$a^{\dagger m} a^n \mapsto z^m \left(\frac{\mathrm{d}}{\mathrm{d}z}\right)^n.$$

Then, acting with the operator in its holomorphic representation on equation (6.22), one finds

$$\begin{aligned}z^m \left(\frac{\mathrm{d}}{\mathrm{d}z}\right)^n f(z) &= \int \frac{\mathrm{d}z'\mathrm{d}\bar{z}'}{2i\pi} \mathrm{e}^{-z'\bar{z}'} z^m \left(\frac{\mathrm{d}}{\mathrm{d}z}\right)^n \mathcal{I}(z,\bar{z}') f(z') \\ &= \int \frac{\mathrm{d}z'\mathrm{d}\bar{z}'}{2i\pi} \mathrm{e}^{-z'\bar{z}'} \left[z^m \bar{z}'^n \mathrm{e}^{z\bar{z}'}\right] f(z').\end{aligned}$$

To any operator $\mathcal{O}$ written in normal form as
$$\mathcal{O} = \sum_{m,n} \mathcal{O}_{mn} a^{\dagger m} a^n,$$
is thus associated the kernel
$$\mathcal{O}(z, \bar{z}) = O(z, \bar{z}) \, e^{z\bar{z}} \quad \text{with} \tag{6.23a}$$
$$O(z, \bar{z}) = \sum_{m,n} \mathcal{O}_{mn} z^m \bar{z}^n. \tag{6.23b}$$

To give to these expressions a more suggestive form, we introduce a formal notation of matrix elements with bras and kets:
$$\mathcal{O}(z, \bar{z}) \equiv \langle z | \mathcal{O} | \bar{z} \rangle, \tag{6.24}$$
without trying to define very precisely the corresponding vectors.

The action of an operator $\mathcal{O}$ with kernel $\mathcal{O}(z, \bar{z})$ on a vector $f(z)$ is then given by
$$(\mathcal{O}f)(z) = \int \frac{\mathrm{d}z' \mathrm{d}\bar{z}'}{2i\pi} \langle z | \mathcal{O} | \bar{z}' \rangle \, e^{-z'\bar{z}'} f(z').$$

The kernel associated with the product of two operators follows:
$$\int \frac{\mathrm{d}\bar{z}' \mathrm{d}z'}{2i\pi} \langle z | \mathcal{O}_2 | \bar{z}' \rangle \, e^{-z'\bar{z}'} \langle z' | \mathcal{O}_1 | \bar{z} \rangle = \langle z | \mathcal{O}_2 \mathcal{O}_1 | \bar{z} \rangle. \tag{6.25}$$

Moreover, the trace of the operator $\mathcal{O}$ is given by
$$\operatorname{tr} \mathcal{O} = \int \frac{\mathrm{d}\bar{z} \mathrm{d}z}{2i\pi} \, e^{-z\bar{z}} \langle z | \mathcal{O} | \bar{z} \rangle. \tag{6.26}$$

Using the relation (6.25), one verifies that this definition satisfies the cyclic condition $\operatorname{tr} \mathcal{O}_1 \mathcal{O}_2 = \operatorname{tr} \mathcal{O}_2 \mathcal{O}_1$.

Remarks.

(i) From the property that z and $\mathrm{d}/\mathrm{d}z$ are hermitian conjugate, follows that hermitian conjugation acting on kernels is represented by formal complex conjugation:
$$\mathcal{O} \mapsto \mathcal{O}^{\dagger} \quad \Rightarrow \quad \mathcal{O}(z, \bar{z}) \mapsto \overline{\mathcal{O}}(z, \bar{z}), \tag{6.27}$$
as one can also directly verify by comparing the scalar products $(f, \mathcal{O}g)$ and $(\mathcal{O}^{\dagger} f, g)$.

(ii) To an operator $\mathcal{O}$ that has matrix elements $\mathcal{O}_{mn}$ in the harmonic oscillator basis, is associated the kernel $\sum_{m,n} \mathcal{O}_{mn}(z^m/\sqrt{m!})(\bar{z}^n/\sqrt{n!})$. From the definition (6.26) of the trace follows the expected identity
$$\operatorname{tr} \mathcal{O} = \int \frac{\mathrm{d}\bar{z} \mathrm{d}z}{2i\pi} \, e^{-z\bar{z}} \sum_{m,n} \mathcal{O}_{mn} \frac{\bar{z}^n}{\sqrt{n!}} \frac{z^m}{\sqrt{m!}} = \sum_n \mathcal{O}_{nn}.$$

(iii) Matrix elements of operators in the holomorphic representation are analogous to matrix elements in the mixed position–momentum representation. The latter can be obtained from the matrix elements $\langle q|\mathcal{O}|q'\rangle$ of an operator in the position basis by a partial Fourier transformation on the r.h.s. argument:

$$\langle q|\mathcal{O}|p\rangle = \int dq'\, e^{ipq'/\hbar}\, \langle q|\mathcal{O}|q'\rangle.$$

In the classical limit, the complex variables $(z,\bar z)$ provide a parametrization of phase space alternative to the parametrization in terms of the real momentum and position variables (p,q) (see Chapter 10).

Hamiltonian and statistical operator. The matrix elements of the hamiltonian (6.20) are

$$\langle z|H_0|\bar z\rangle = \omega z\bar z\, e^{z\bar z}. \tag{6.28}$$

The matrix elements $\langle z|U_0(t)|\bar z\rangle$ of the statistical operator

$$U_0(t) = e^{-H_0 t}$$

satisfy

$$\frac{\partial}{\partial t}\langle z|U_0(t)|\bar z\rangle = -\langle z|H_0 U_0(t)|\bar z\rangle = -\omega z\frac{\partial}{\partial z}\langle z|U_0(t)|\bar z\rangle.$$

The solution of this differential equation has the general form

$$\langle z|U_0(t)|\bar z\rangle = \mathcal{U}(e^{-\omega t}z,\bar z).$$

The boundary condition

$$\langle z|U_0(0)|\bar z\rangle = \mathcal{I}(z,\bar z) = e^{z\bar z}$$

then implies

$$\langle z|U_0(t)|\bar z\rangle = \mathcal{I}(e^{-\omega t}z,\bar z) = e^{z\bar z\, e^{-\omega t}}. \tag{6.29}$$

Note that the action of the quantum statistical operator on a function $f(z)$ is given by

$$[U_0 f](z) = \int \frac{dz'\, d\bar z'}{2i\pi} e^{z\bar z'\, e^{-\omega t}} e^{-z'\bar z'} f(z') = f(e^{-\omega t}z), \tag{6.30}$$

where the equation (6.15) has been used.

Finally, using property (6.27), one verifies directly that H_0 and $U_0(t)$ are hermitian.

The partition function $\mathcal{Z}_0(\beta) = \operatorname{tr} e^{-\beta H_0}$ is the trace of $U_0(\beta)$. Using equations (6.26, 6.29), one recovers the partition function of a shifted harmonic oscillator:

$$\mathcal{Z}_0(\beta) = \operatorname{tr} U_0(\beta) = \int \frac{d\bar z\, dz}{2i\pi} e^{-z\bar z} e^{z\bar z\, e^{-\omega\beta}} = \frac{1}{1 - e^{-\omega\beta}}. \tag{6.31}$$

6.4 Path integral: the harmonic oscillator

We now construct a path integral representation of the quantum statistical operator, based on the holomorphic formalism. As in the case of real coordinates, we first discuss the harmonic oscillator and, then, more general systems.

Path integral. The matrix elements of the statistical operator $U_0(t)$ are given in (6.29) and, thus, at first order in t,

$$\langle z | U_0(t) | \bar{z} \rangle = \exp\left[\bar{z}z(1 - \omega t) + O\left(t^2\right)\right]. \tag{6.32}$$

The semi-group property $U_0(t) = [U_0(t/n)]^n$, then, allows calculating $U_0(t)$ at finite time. Using equation (6.25) iteratively to calculate the matrix elements of the product, one obtains

$$\langle z'' | U_0(t'', t') | \bar{z}' \rangle = \lim_{n \to \infty} \int \prod_{k=1}^{n-1} \frac{dz_k d\bar{z}_k}{2i\pi} \exp\left[\bar{z}_0 z_1 - \mathcal{S}_\varepsilon(z, \bar{z})\right] \tag{6.33}$$

with

$$\mathcal{S}_\varepsilon(z, \bar{z}) = -\sum_{k=1}^{n-1} \bar{z}_k (z_{k+1} - z_k) + \omega\varepsilon \sum_{k=0}^{n-1} \bar{z}_k z_{k+1}, \tag{6.34}$$

where $\varepsilon = (t'' - t')/n$ is the time interval and the boundary conditions are

$$\bar{z}_0 = \bar{z}', \quad z_n = z''. \tag{6.35}$$

In the formal limit $n \to \infty$, one obtains the path integral representation

$$\langle z'' | U_0(t'', t') | \bar{z}' \rangle = \int \left[\frac{d\bar{z}(t) dz(t)}{2i\pi}\right] e^{\bar{z}(t') z(t')} \exp\left[-\mathcal{S}_0(z, \bar{z})\right], \tag{6.36}$$

$$\mathcal{S}_0(z, \bar{z}) = \int_{t'}^{t''} dt\, \bar{z}(t) \left[-\dot{z}(t) + \omega z(t)\right]$$

with the boundary conditions $z(t'') = z''$, $\bar{z}(t') = \bar{z}'$. The symmetry of the action between initial and final times, is not explicit, but can be verified by an integration by parts of the term $\bar{z}\dot{z}$:

$$\mathcal{S}_0(z, \bar{z}) - \bar{z}(t') z(t') = -\bar{z}(t'') z(t'') + \int_{t'}^{t''} dt\, z(t) \left[\dot{\bar{z}}(t) + \omega \bar{z}(t)\right].$$

Let us point out, however, that the validity of this integration within the path integral assumes that differentiation and expectation value for equal time products commute (see Section 5.3.2) and thus relies on the convention $\mathrm{sgn}(0) = 0$.

Finally, in the case of the holomorphic path integral, the discussion of the existence of a continuum limit is complicated, the nature of the problem being similar to what is encountered in integrals over phase space (see Section 10.2).

6.4.1 General gaussian integral

Expression (6.36) can immediately be generalized to a system coupled linearly to external sources $b(t)$ and $\bar{b}(t)$. To the hamiltonian

$$H(t) = \omega a^\dagger a - b(t)a - \bar{b}(t)a^\dagger \qquad (6.37)$$

are associated the differential operator and the kernel

$$\omega z \partial_z - b(t)\partial_z - \bar{b}(t)z \mapsto \left[\omega z\bar{z} - b(t)\bar{z} - \bar{b}(t)z\right] e^{\bar{z}z}.$$

To order ε, the corresponding statistical operator $U_G(b; t+\varepsilon, t) = \mathbf{1} - \varepsilon H(t)$ and, thus,

$$\begin{aligned}\langle z | U_G(b; t+\varepsilon, t) | \bar{z} \rangle &= e^{\bar{z}z}\left[1 - \varepsilon\left(\omega z\bar{z} - b(t)\bar{z} - \bar{b}(t)z\right)\right] + O(\varepsilon^2) \\ &= \exp\left[\bar{z}z - \varepsilon\left(\omega z\bar{z} - b(t)\bar{z} - \bar{b}(t)z\right)\right) + O(\varepsilon^2)\right].\end{aligned}$$

The corresponding action is

$$\mathcal{S}_G(z, \bar{z}) = \mathcal{S}_0(z, \bar{z}) - \int_{t'}^{t''} dt \left[\bar{z}(t)b(t) + \bar{b}(t)z(t)\right], \qquad (6.38)$$

and the matrix elements of U_G are given by

$$\langle z'' | U_G(b; t'', t') | \bar{z}' \rangle = \int \left[\frac{d\bar{z}(t)dz(t)}{2i\pi}\right] e^{\bar{z}(t')z(t')} \exp\left[-\mathcal{S}_G(z, \bar{z})\right]. \qquad (6.39)$$

Calculation of the path integral. The first step in the explicit calculation of the integral involves solving the classical equations obtained by varying $\bar{z}(t)$ and $z(t)$:

$$\begin{aligned}-\dot{z}(t) + \omega z(t) - b(t) &= 0, \\ \dot{\bar{z}}(t) + \omega \bar{z}(t) - \bar{b}(t) &= 0.\end{aligned}$$

The solutions that satisfy the boundary conditions are

$$\begin{aligned}z(t) &= z'' e^{-\omega(t''-t)} + \int_t^{t''} e^{-\omega(u-t)} b(u)du, \\ \bar{z}(t) &= \bar{z}' e^{-\omega(t-t')} + \int_{t'}^{t} e^{-\omega(t-u)} \bar{b}(u)du.\end{aligned}$$

Shifting integration variables by the solution of the classical equations, one obtains a gaussian integral that does not depend on the external sources and boundary conditions anymore. After integration, the result takes the form

$$\langle z'' | U_G(b; t'', t') | \bar{z}' \rangle = \mathcal{N}\bigl(\omega(t''-t')\bigr) e^{\bar{z}(t')z(t') - \mathcal{S}_G(z, \bar{z})}, \qquad (6.40)$$

where

$$\bar{z}(t')z(t') = \bar{z}'z'' e^{-\omega(t''-t')} + \bar{z}' \int_{t'}^{t''} dt\, e^{-\omega(t-t')} b(t),$$

$$-S_G(z,\bar{z}) = \int_{t'}^{t''} dt\, \bar{b}(t) z(t)$$

$$= \int_{t'}^{t''} dt\, \bar{b}(t)\, e^{-\omega(t''-t)} z'' + \int_{t'}^{t''} du\, dt\, \bar{b}(u)\theta(t-u)\, e^{-\omega(t-u)} b(t),$$

$\theta(t)$ being the step function: $\theta(t) = 1$ for $t > 0$, $\theta(t) = 0$ for $t < 0$.

As in the case of the usual path integral, the normalization $\mathcal{N}$ can be determined entirely only by a comparison with a reference path integral.

6.4.2 Gaussian correlation functions

The trace of expression (6.40) (equation (6.26)) yields, up to a normalization, a generating functional of correlation functions of $z, \bar{z}$ with the gaussian weight $e^{-S_0}/\mathcal{Z}_0$ and, as we will see, periodic boundary conditions. Setting $\beta = t'' - t'$, one finds

$$\mathcal{Z}_G(b, \bar{b}; \beta) = \operatorname{tr} U_G(b; \beta/2, -\beta/2) = \int \frac{dz\, d\bar{z}}{2i\pi}\, e^{-z\bar{z}} \langle z| U_G(b; \beta/2, -\beta/2)|z\rangle$$

$$= \int \frac{d\bar{z}\, dz}{2i\pi}\, \exp\left[-\Sigma(z, \bar{z})\right] \tag{6.41}$$

with

$$\Sigma(z, \bar{z}) = \bar{z}z\left(1 - e^{-\omega\beta}\right) + \int_{-\beta/2}^{\beta/2} dt\, \left[\bar{b}(t)\, e^{-\omega(\beta/2-t)} z + \bar{z}\, e^{-\omega(t+\beta/2)} b(t)\right]$$

$$- \int_{-\beta/2}^{\beta/2} du\, dt\, \bar{b}(u)\theta(t-u)\, e^{-\omega(t-u)} b(t).$$

The integration yields

$$\mathcal{Z}_G(b, \bar{b}; \beta) = \mathcal{Z}_0(\beta) \exp\left[\int_{-\beta/2}^{\beta/2} du\, dt\, \bar{b}(u)\Delta(t-u)b(t)\right], \tag{6.42}$$

where $\mathcal{Z}_0(\beta) = \operatorname{tr} e^{-\beta H_0}$ is the partition function, which is obtained in the form

$$\mathcal{Z}_0(\beta) = \frac{\mathcal{N}(\omega\beta)}{1 - e^{-\omega\beta}} \Rightarrow \mathcal{N}(\omega\beta) = 1,$$

and $\Delta(t)$ is the *propagator*. It can be written as

$$\Delta(t) = \tfrac{1}{2} e^{-\omega t}\left[\operatorname{sgn}(t) + 1/\tanh(\omega\beta/2)\right] = e^{-\omega t}\, \frac{\theta(t)\, e^{\omega\beta/2} + \theta(-t)\, e^{-\omega\beta/2}}{e^{\omega\beta/2} - e^{-\omega\beta/2}}, \tag{6.43}$$

6.4   Path integrals and holomorphic formalism   149

where $\mathrm{sgn}(t) = 1$ for $t > 0$, $\mathrm{sgn}(t) = -1$ for $t < 0$ and $\theta(t) = (1 + \mathrm{sgn}(t))/2$ is the step function. One verifies that it is the solution of the differential equation

$$\dot{\Delta}(t) + \omega\Delta(t) = \delta(t), \tag{6.44}$$

with periodic boundary conditions on the interval $[-\beta/2, \beta/2]$ (we recall that, in the sense of distributions, $\mathrm{d}\,\mathrm{sgn}(t)/\mathrm{d}t = 2\delta(t)$).

In the limit $\beta \to \infty$, the propagator Δ reduces to

$$\Delta(t) = \theta(t)\,\mathrm{e}^{-\omega t}. \tag{6.45}$$

This form leads to a time-ordering or causal property in perturbative calculations.

Notice that, since the propagator $\Delta(t)$ satisfies a differential equation with periodic boundary conditions, the result (6.42) is also obtained from a path integral with periodic boundary conditions:

$$\mathcal{Z}_\mathrm{G}(b, \bar{b}; \beta) = \int \left[\frac{\mathrm{d}\bar{z}(t)\mathrm{d}z(t)}{2i\pi}\right] \exp\left[-\mathcal{S}_\mathrm{G}(z, \bar{z})\right] \tag{6.46}$$

with $z(-\beta/2) = z(\beta/2)$, $\bar{z}(-\beta/2) = \bar{z}(\beta/2)$ and

$$\mathcal{S}_\mathrm{G}(z, \bar{z}) = \int_{-\beta/2}^{\beta/2} \mathrm{d}t\, \left\{\bar{z}(t)\left[-\dot{z}(t) + \omega z(t)\right] - \bar{z}(t)b(t) - \bar{b}(t)z(t)\right\}.$$

With some care, one can also derive these boundary conditions by considering directly the trace of the initial expression (6.36).

The generating functional of correlation functions with periodic boundary conditions, fully determined by the condition $\langle 1 \rangle = 1$, is then

$$\mathcal{Z}_\mathrm{G}(b, \bar{b}; \beta)/\mathcal{Z}_0(\beta) = \exp\left[\int \mathrm{d}u\,\mathrm{d}t\, \bar{b}(u)\Delta(t-u)b(t)\right]. \tag{6.47}$$

The initial expression (6.39) shows that the derivative with respect to b and $\bar{b}$, in the limit $b = \bar{b} = 0$, is the two-point function and, thus,

$$\langle \bar{z}(t)z(u) \rangle = \Delta(t-u). \tag{6.48}$$

6.4.3 Partition function

The derivative of the path integral (6.46) with respect to ω, taken for $b = \bar{b} = 0$, is the derivative of the partition function and, thus,

$$\frac{\mathrm{d}}{\mathrm{d}\omega}\ln\mathcal{Z}_0(\beta) = -\int_{-\beta/2}^{\beta/2}\mathrm{d}t\,\langle\bar{z}(t)z(t)\rangle = -\beta\Delta(0)$$

$$= -\frac{\beta}{2}\left[\mathrm{sgn}(0) + \frac{\cosh(\omega\beta/2)}{\sinh(\omega\beta/2)}\right].$$

Integrating and using the property that the ground state is not degenerate, which determines the normalization, one obtains

$$\mathcal{Z}_0(\beta) = \frac{e^{-\beta\omega(1+\mathrm{sgn}(0))/2}}{1 - e^{-\omega\beta}}. \tag{6.49}$$

Since the result involves sgn(0), it is clearly not defined. Different conventions for the value of sgn(0) correspond to a global shift of the whole spectrum:

$$E_k = E_0 + \omega k, \quad E_0 = \omega(1 + \mathrm{sgn}(0))/2.$$

Let us examine extreme cases. The choice $\mathrm{sgn}(0) = 0$ corresponds to the standard symmetric hamiltonian $H_0 = \frac{1}{2}\omega(aa^\dagger + a^\dagger a)$. A different choice, for instance, $\Delta(0) = \Delta(0_-)$ and, thus, $\mathrm{sgn}(0) = -1$ leads to

$$\mathcal{Z}_0(\beta) = \frac{1}{1 - e^{-\omega\beta}},$$

which corresponds to normal order, $H_0 = \omega a^\dagger a$, and is consistent with the result (6.42). Finally, $\Delta(0) = \Delta(0_+)$ corresponds to a hamiltonian $H_0 = \omega aa^\dagger$.

The ambiguity, which leads to the appearance of the quantity sgn(0), is related to the choice of the order in products of the operators a and $a^\dagger$, which is no longer apparent in the formal continuum limit. Analogous problems have been already encountered in Sections 5.3 and 5.5.2. Different choices of $\Delta(0)$ correspond to different choices of quantization.

However, we have already pointed out in Section 5.1 the merits of the symmetric choice $\mathrm{sgn}(0) = 0$ and we illustrate them again here. Using the explicit form (6.43) of the two-point function, one infers the expectation value

$$\langle [\bar{z}(t+\delta) - \bar{z}(t)][z(t+\delta) - z(t)]\rangle = 2\Delta(0) - \Delta(\delta) - \Delta(-\delta).$$

Only the symmetric choice

$$\Delta(0) = \lim_{\delta \to 0} \tfrac{1}{2}(\Delta(\delta) + \Delta(-\delta)) = \frac{\cosh\omega(\beta/2)}{2\sinh(\omega\beta/2)},$$

(equation (6.43)), and thus $\mathrm{sgn}(0) = 0$, ensures the continuity of this combination when δ goes to zero. Then, the generic values of $[z(t+\delta) - z(t)][\bar{z}(t+\delta) - \bar{z}(t)]$ are of order $|\delta|$, a property analogous to the behaviour of brownian motion.

6.5 Path integral: general hamiltonians

We now construct a path integral representation, based on the holomorphic formalism, of the statistical operator for a general hamiltonian. In Section 10.2, we define an integral over paths in position-momentum phase space. Holomorphic path integrals and phase space integrals (10.27) are related by a simple change of variables of the form (6.17) (the quantum operators being replaced by the corresponding classical variables), but have different boundary conditions and boundary terms.

6.5.1 Path integral

The matrix elements of the statistical operator satisfy the equation

$$\frac{\partial}{\partial t} \langle z| U(t,t') |\bar{z}\rangle = -H(z, \partial/\partial z; t) \langle z| U(t,t') |\bar{z}\rangle \tag{6.50}$$

with

$$\langle z| U(t',t') |\bar{z}\rangle = \mathrm{e}^{\bar{z}z}.$$

A general hamiltonian can be cast into the normal-ordered form

$$H = \sum_{m,n} H_{mn} a^{\dagger m} a^n \equiv \sum_{m,n} H_{mn} z^m \left(\frac{\partial}{\partial z}\right)^n. \tag{6.51}$$

With the definition

$$H(z, \bar{z}; t) = \sum_{m,n} H_{mn} z^m \bar{z}^n, \tag{6.52}$$

the matrix elements of $U(t+\varepsilon, t)$, solution of equation (6.50), are given at order ε by

$$\langle z''| U(t+\varepsilon, t) |\bar{z}'\rangle = \mathrm{e}^{\bar{z}'z''} - \varepsilon\, \mathrm{e}^{\bar{z}'z''} H(z'', \bar{z}'; t) + O(\varepsilon^2) \tag{6.53}$$
$$= \exp\left[\bar{z}'z'' - \varepsilon H(z'', \bar{z}'; t) + O(\varepsilon^2)\right].$$

A simple generalization of the arguments of Section 6.4 shows that the path integral then takes the form

$$\langle z''| U(t'',t') |\bar{z}'\rangle = \int \left[\frac{\mathrm{d}\bar{z}(t)\mathrm{d}z(t)}{2i\pi}\right] \mathrm{e}^{\bar{z}(t')z(t'')} \exp\left[-\mathcal{S}(z,\bar{z})\right] \tag{6.54}$$

with the euclidean action

$$\mathcal{S}(z,\bar{z}) = \int_{t'}^{t''} \mathrm{d}t \left[-\bar{z}(t)\dot{z}(t) + H(z(t), \bar{z}(t); t)\right] \tag{6.55}$$

and the boundary conditions $z(t'') = z''$, $\bar{z}(t') = \bar{z}'$.

Remark. The function $H(z, \bar{z}; t)$, defined by equation (6.52), is related to the kernel representation of the hamiltonian by

$$\langle z| H(t) |\bar{z}\rangle = \mathrm{e}^{z\bar{z}} H(z, \bar{z}; t).$$

In terms of this kernel, equation (6.50) reads

$$\frac{\partial}{\partial t} \langle z''| U(t,t') |\bar{z}'\rangle = -\int \frac{\mathrm{d}\bar{z}\mathrm{d}z}{2i\pi} \langle z''| H(t) |\bar{z}\rangle\, \mathrm{e}^{-z\bar{z}} \langle z| U(t,t') |\bar{z}'\rangle.$$

Perturbative expansion. A perturbative expansion is based on a decomposition of the hamiltonian into the sum of a harmonic oscillator term and the remaining part, called the interaction:
$$H(z,\bar z;t) = \omega z\bar z + H_{\rm I}(z,\bar z;t).$$
The results of Section 6.4 and, in particular Wick's theorem, allows then calculating the path integral as an expansion in powers of $H_{\rm I}$. Formally,
$$\langle z''|U(t'',t')|\bar z'\rangle$$
$$= \exp\left[-\int_{t'}^{t''} dt\, H_{\rm I}\left(\frac{\partial}{\partial \bar b(t)},\frac{\partial}{\partial b(t)};t\right)\right] U_{\rm G}(b;z'',\bar z';t'',t')\bigg|_{b=\bar b=0}. \quad (6.56)$$

Partition function. In the case of a time-independent hamiltonian, one can derive from the the statistical operator the partition function by taking its trace. In terms of its matrix elements, the relation reads:
$$\mathcal{Z}(\beta) = \int \frac{dz d\bar z}{2i\pi}\, e^{-z\bar z}\, \langle z|U(\beta/2,-\beta/2)|\bar z\rangle.$$
Combining the result (6.46) with the perturbative expression (6.56), one verifies that the trace operation leads to periodic boundary conditions in the path integral. One then obtains
$$\mathcal{Z}(\beta) = \int\left[\frac{d\bar z(t) dz(t)}{2i\pi}\right] \exp\left[-\mathcal{S}(z,\bar z)\right] \quad (6.57)$$
with
$$\mathcal{S}(z,\bar z) = \int_{-\beta/2}^{\beta/2} dt\, \left[-\bar z(t)\dot z(t) + H(z(t),\bar z(t))\right]$$
and the boundary conditions $z(-\beta/2) = z(\beta/2)$, $\bar z(-\beta/2) = \bar z(\beta/2)$.

6.5.2 Several complex variables

The generalization of the holomorphic formalism to entire functions of N complex variables z_i is straightforward. The scalar product of two functions is then defined by
$$(g,f) = \int \left(\prod_{i=1}^{N} \frac{dz_i\, d\bar z_i}{2i\pi}\, e^{-z_i\bar z_i}\right) \overline{g(\mathbf{z})} f(\mathbf{z}). \quad (6.58)$$
In terms of the Taylor series expansion
$$f(\mathbf{z}) = \sum_{k,i_1,\ldots,i_k} f_{i_1,\ldots,i_k} z_{i_1}\cdots z_{i_k},$$
where the coefficients $f_{i_1,\ldots,i_k}$ are symmetric in the k indices, the norm of the function is given by
$$\|f\|^2 = (f,f) = \sum_{k,i_1,\ldots,i_k} k!\, |f_{i_1,\ldots,i_k}|^2.$$

The functions with finite norm span a Hilbert space. Such spaces will be encountered in the discussion of boson systems in Section 6.6.

On these functions act hermitian conjugate creation and annihilation operators $z_i, \partial/\partial z_i$, with the commutation relations

$$[\partial/\partial z_i, z_j] = \delta_{ij}.$$

The kernel of the identity is

$$\mathcal{I}(\mathbf{z}, \bar{\mathbf{z}}) = \exp \sum_i z_i \bar{z}_i. \tag{6.59}$$

It is then simple to derive kernels associated to operators written in normal form by acting on $\mathcal{I}$.

A partition function is given by a path integral that generalizes expression (6.57):

$$\mathcal{Z}(\beta) = \int \left[\frac{\mathrm{d}^N z(t) \mathrm{d}^N \bar{z}(t)}{(2i\pi)^N} \right] \exp\left[-\mathcal{S}(\mathbf{z}, \bar{\mathbf{z}})\right] \tag{6.60}$$

with an action of the form

$$\mathcal{S}(\mathbf{z}, \bar{\mathbf{z}}) = \int_{-\beta/2}^{\beta/2} \mathrm{d}t \left[-\sum_i \bar{z}_i(t) \dot{z}_i(t) + H\big(\mathbf{z}(t), \bar{\mathbf{z}}(t)\big) \right] \tag{6.61}$$

and periodic boundary conditions

$$\mathbf{z}(-\beta/2) = \mathbf{z}(\beta/2), \quad \bar{\mathbf{z}}(-\beta/2) = \bar{\mathbf{z}}(\beta/2).$$

6.5.3 Discussion

Operator ordering and ambiguities in the path integral. The difficulties already encountered in Section 6.4.3, induced by the ordering in products of the operators $a, a^\dagger$ that do not commute, clearly increase for more general quantum hamiltonians because, in most interesting examples, interactions contain monomials involving both a and $a^\dagger$. In $H(z, \bar{z})$, a factor $(\bar{z}z)^p$ generates ambiguities of the form of a polynomial in $\bar{z}z$ of degree $p-1$ and thus the number of parameters increases with the degree p. We leave it, as an exercise, to verify that, indeed, the undefined quantity $\Delta(0)$ (equation (6.43)) appears in a perturbative calculation of the path integral, generated by pairings inside the products $(z\bar{z})^p$ which originate from products of operators.

As the calculation of the partition function shows (equation (6.49)), normal order is associated with a time-asymmetric convention $\mathrm{sgn}(0) = -1$. A more time-symmetric construction consists in combining normal and anti-normal order. The hamiltonian (6.51) can also be cast into an anti-normal ordered form

$$H(t) = \tilde{H}(z, \partial/\partial z; t) \equiv \sum_{m,n} \tilde{H}_{mn} \left(\frac{\partial}{\partial z}\right)^m z^n.$$

With this definition, the matrix elements of $U(t+\varepsilon,t)$, solution of equation (6.50), are also given at order ε by

$$\langle z''|U(t+\varepsilon,t)|\bar{z}'\rangle = e^{\bar{z}'z''} - \varepsilon \tilde{H}(z'', \partial/\partial z''; t)\, e^{\bar{z}'z''} + O(\varepsilon^2). \tag{6.62}$$

We then write the identity $U(t+2\varepsilon,t) = U(t+2\varepsilon, t+\varepsilon)U(t+\varepsilon,t)$, in terms of matrix elements, as

$$\langle z_2|U(t+2\varepsilon,t)|\bar{z}_0\rangle = \int \frac{d\bar{z}_1 dz_1}{2i\pi} \langle z_2|U(t+2\varepsilon, t+\varepsilon)|\bar{z}_1\rangle\, e^{-z_1\bar{z}_1} \langle z_1|U(t+\varepsilon,t)|\bar{z}_0\rangle$$

$$= \int \frac{d\bar{z}_1 dz_1}{2i\pi} \left[1 - \varepsilon H(z_2, \bar{z}_1; t)\right] e^{\bar{z}_1 z_2 - \bar{z}_1 z_1}$$
$$\times \left[1 - \varepsilon \tilde{H}(z_1, \partial/\partial z_1; t)\right] e^{\bar{z}_0 z_1} + O(\varepsilon^2),$$

where the forms (6.53) and (6.62) have been successively used. An integration by parts over z_1 leads to the substitution

$$\int dz_1\, e^{-\bar{z}_1 z_1}\left(\frac{\partial}{\partial z_1}\right)^m z_1^n = \int dz_1\, e^{-\bar{z}_1 z_1}\, \bar{z}_1^m z_1^n.$$

One then obtains

$$\langle z_2|U(t+2\varepsilon,t)|\bar{z}_0\rangle = \int \frac{d\bar{z}_1 dz_1}{2i\pi} e^{\bar{z}_1 z_2 - \bar{z}_1 z_1 + \bar{z}_0 z_1}\left[1 - \varepsilon H(z_2, \bar{z}_1; t) - \varepsilon \tilde{H}(z_1, \bar{z}_1; t)\right]$$
$$+ O(\varepsilon^2).$$

From this remark, one concludes that a more time-symmetric formalism involves inserting in the path integral the half-sum of the normal and anti-normal ordered form. This choice is then consistent with the convention $\mathrm{sgn}(0) = 0$. Note, however, that this ansatz solves the problem of the ambiguity of $(z\bar{z})^n$ at order $(z\bar{z})^{n-1}$, but difficulties remain at order $(z\bar{z})^{n-2}$.

Example. In the case of the hamiltonian (6.20), this leads to the insertion of

$$\tfrac{1}{2}\omega[\bar{z}z + (\bar{z}z - 1)] = \omega(\bar{z}z - \tfrac{1}{2}).$$

The result (6.49) is then replaced by

$$\mathcal{Z}_0(\beta) = e^{\beta\omega/2}\,\frac{e^{-\beta\omega(1+\mathrm{sgn}(0))/2}}{1 - e^{-\omega\beta}} = \frac{e^{-\beta\omega\,\mathrm{sgn}(0)/2}}{1 - e^{-\omega\beta}}.$$

The exact result is now indeed recovered for $\mathrm{sgn}(0) = 0$.

Holomorphic formalism and position-momentum phase space. In Section 10.2, we present a formal construction of a path integral in which paths are trajectories in position–momentum phase space. If, in the holomorphic formalism, one introduces real variables with the change

$$z = (q - ip)/\sqrt{2}, \qquad \bar{z} = (q + ip)/\sqrt{2}, \tag{6.63}$$

one obtains the classical euclidean action in the hamiltonian formalism, expressed in terms of phase space variables (equation (10.28)):

$$S(p, q) = \int dt\, [-ip(t)\dot{q}(t) + H(p, q)],$$

up to boundary terms, as well as the Liouville measure of integration. However, the boundary conditions in the integration are different, which is not surprising since the matrix elements are different, except in the case of the partition function.

6.5.4 Harmonic oscillator: real perturbation

As an exercise, we now show how to recover the perturbative expansion of the partition function (2.75), starting from the harmonic oscillator approximation, in the example of a hamiltonian of the form

$$H = \tfrac{1}{2}\hat{p}^2 + \tfrac{1}{2}\omega^2 \hat{q}^2 + V_{\rm I}(\hat{q}).$$

Inverting the relations (6.17), one can express the position operator $\hat{q}$ in terms of the operators $a, a^\dagger$:

$$\hat{q} = \frac{1}{\sqrt{2\omega}}(a + a^\dagger).$$

The action in the corresponding holomorphic path integral is then

$$S(z, \bar{z}) = \int_{t'}^{t''} dt\, \left\{\bar{z}(t)\,[-\dot{z}(t) + \omega z(t)] + V_{\rm I}\!\left((z(t) + \bar{z}(t))/\sqrt{2\omega}\right)\right\}.$$

The perturbative expansion in $V_{\rm I}$ can then be expressed in terms of functional differentiations acting on the path integral of the harmonic oscillator coupled linearly to $z + \bar{z}$, in a form analogous to expression (2.75). This combination corresponds to a gaussian integral (6.38) with a real source: $b(t) = \bar{b}(t)$. Thus, the expressions (6.40) and (6.43) can be symmetrized in time. Changing also $b(t) \mapsto b(t)/\sqrt{2\omega}$, one finds

$$\langle z''|\,U(t'', t')\,|\bar{z}'\rangle = \exp\!\left[-V_{\rm I}\!\left(\frac{\delta}{\delta b(t)}\right)\right] \langle z''|\,U_{\rm G}(b; t'', t')\,|\bar{z}'\rangle\bigg|_{b=0} \tag{6.64}$$

with (see equation (6.40))

$$\langle z''|\,U_{\rm G}(b; t'', t')\,|\bar{z}'\rangle$$
$$= \exp\bigg[-\omega(t''-t')/2 + \bar{z}' z''\, e^{-\omega(t''-t')} + \int_{t'}^{t''} dt\, \left(z''\, e^{-\omega(t''-t)} + \bar{z}'\, e^{\omega(t'-t)}\right)\frac{b(t)}{\sqrt{2\omega}}$$
$$+ \frac{1}{2}\int_{t'}^{t''} dt\, d\tau\, b(t)\frac{e^{-\omega|t-\tau|}}{2\omega} b(\tau)\bigg]. \tag{6.65}$$

In the same way, equation (6.42) can be rewritten as

$$\operatorname{tr} U_G(b; t'', t') = \frac{1}{2\sinh(\omega\beta/2)} \exp\left[\frac{1}{2}\int dt\, d\tau\, b(t) D(t-\tau) b(\tau)\right], \qquad (6.66)$$

where $D(t)$ is related to the function $\Delta(t)$ (equation (6.43)) by

$$D(t) = \frac{1}{4\omega}(\Delta(t) + \Delta(-t)) = \frac{\cosh\omega(\beta/2 - |t|)}{2\omega\sinh(\omega\beta/2)}, \qquad (6.67)$$

a result in agreement with equation (2.52).

6.6 Bosons: second quantization

We now study quantum boson systems, in the so-called second quantization formalism. In a first part, we assume that bosons can occupy only a finite number of quantum states, for example, because only spin degrees of freedom are relevant or because bosons live on a finite space lattice.

6.6.1 Boson states and hamiltonian

One-particle states. A one-boson state is defined by a vector, which we denote by ψ_i, and which belongs to a complex vector space $\mathfrak{H}_1$, which we first assume to be of finite dimension N.

Many-particle states. An n-particle state is described by a vector $\psi_{i_1 i_2 \ldots i_n}$ where the indices i_k take N values. The statistical properties of bosons then imply the invariance of the vector $\psi_{i_1 i_2 \ldots i_n}$ under all permutations P of the indices $\{i_1, \ldots, i_n\}$:

$$\psi_{i_1 i_2 \ldots i_n} = \psi_{i_{P_1} i_{P_2} \ldots i_{P_n}}.$$

The vectors $\psi_{i_1 i_2 \ldots i_n}$ are symmetric tensors with n indices and belong to a complex vector space $\mathfrak{H}_n$ of dimension $\binom{N+n-1}{n}$.

Independent particles: hamiltonian. A one-body or one-particle hamiltonian $\mathbf{H}^{(1)}$ is defined by its action on one-particle states: it is then represented by a hermitian $N \times N$ matrix $H^{(1)}_{ij}$, which thus can be diagonalized. We denote by ω_i its eigenvalues. Then,

$$[\mathbf{H}^{(1)}\psi]_i = \omega_i \psi_i.$$

Its action on an n-particle state is additive:

$$[\mathbf{H}^{(1)}\psi]_{i_1 i_2 \ldots i_n} = \sum_\ell \omega_{i_\ell} \psi_{i_1 i_2 \ldots i_n}.$$

If the complete hamiltonian has this simple form, the bosons do not interact: one then speaks of independent particles.

Two-body or pair interaction. A pair or two-body interaction $\mathbf{H}^{(2)}$ is defined by its action on two-particle states:

$$[\mathbf{H}^{(2)}\psi]_{i_1 i_2} = \sum_{j_1,j_2} H^{(2)}_{i_1 i_2, j_1 j_2} \psi_{j_1 j_2},$$

where $H^{(2)}_{i_1 i_2, j_1 j_2}$ is a hermitian matrix that satisfies

$$H^{(2)}_{i_1 i_2, j_1 j_2} = H^{(2)}_{i_2 i_1, j_2 j_1} = \bar{H}^{(2)}_{j_j i_2, i_1 i_2},$$

and, therefore, is an internal mapping in the vector space $\mathfrak{H}_2$ of symmetric tensors. When $\mathbf{H}^{(2)}$ acts only on symmetric tensors, the matrix can be symmetrized and then satisfies

$$H^{(2)}_{i_1 i_2, j_1 j_2} = H^{(2)}_{i_1 i_2, j_2 j_1} = H^{(2)}_{i_2 i_1, j_1 j_2}.$$

The action of $\mathbf{H}^{(2)}$ on an n-particle state, then, has the form

$$[\mathbf{H}^{(2)}\psi]_{i_1 i_2 \ldots i_n} = \frac{1}{2} \sum_{\ell \neq m} \sum_{j,k} H^{(2)}_{i_\ell i_m, jk} \psi_{i_1 i_2 \ldots i_{\ell-1} j i_{\ell+1} \ldots i_{m-1} k i_{m+1} \ldots i_n}.$$

A simple generalization of this construction allows defining k-particle (k-body) interactions but, in what follows, for simplicity we restrict the discussion to at most two-particle interactions.

6.6.2 State vectors: generating function and hamiltonian

We now consider the set of all state vectors corresponding to an arbitrary number of bosons, that is, which belong to the union of all spaces $\oplus_n \mathfrak{H}_n$, $i = 0, 1, \ldots, \infty$. The space $\mathfrak{H}_0$, which has not been defined yet, corresponds to the zero-particle or empty state (also called the *vacuum*).

We introduce a complex vector $z_i \in \mathbb{C}^N$ and the function of N complex variables

$$\Psi(z) = \sum_{n=0}^{\infty} \frac{1}{n!} \sum_{i_1, i_2, \ldots, i_n} \psi_{i_1 i_2 \ldots i_n} z_{i_1} z_{i_2} \ldots z_{i_n}. \tag{6.68}$$

Because we deal with bosons, the coefficients $\psi_{i_1 i_2 \ldots i_n}$ are symmetric in all indices and, thus, can be recovered by differentiating the function $\Psi(z)$.

The construction that follows will justify the choice of functions $\Psi(z)$ that are normalizable with respect to the scalar product (6.58):

$$\|\Psi\|^2 = \sum_{n=0}^{\infty} \frac{1}{n!} \sum_{i_1, i_2, \ldots, i_n} |\psi_{i_1 i_2 \ldots i_n}|^2,$$

where the factor $1/n!$ cancels the overcounting of states implied by the unrestricted summation over the indices i_k.

These functions belong to a Hilbert space of entire functions of N variables and this explains the name of second quantization given to the formalism.

One then notices that

$$\sum_j \omega_j z_j \frac{\partial \Psi(z)}{\partial z_j} = \sum_{n=1}^{\infty} \frac{1}{(n-1)!} \sum_j z_j \omega_j \sum_{i_1,i_2,\ldots,i_{n-1}} z_{i_1} z_{i_2} \cdots z_{i_{n-1}} \psi_{i_1 i_2 \ldots i_{n-1} j}$$

$$= \sum_{n=1}^{\infty} \frac{1}{n!} \sum_{i_1,i_2,\ldots,i_n} z_{i_1} z_{i_2} \cdots z_{i_n} \sum_{\ell} \omega_{i_\ell} \psi_{i_1 i_2 \ldots i_n}.$$

The operator

$$\mathbf{H}^{(1)} \equiv \sum_i z_i \omega_i \frac{\partial}{\partial z_i}, \tag{6.69}$$

thus, represents the one-particle hamiltonian acting on the vector $\Psi(z)$.

An analogous calculation shows that a pair interaction is represented by the operator

$$\mathbf{H}^{(2)} = \frac{1}{2} \sum_{i_1,i_2,j_1,j_2} z_{i_1} z_{i_2} H^{(2)}_{i_1 i_2, j_1 j_2} \frac{\partial^2}{\partial z_{j_1} \partial z_{j_2}}. \tag{6.70}$$

The total hamiltonian

$$\mathbf{H} = \mathbf{H}^{(1)} + \mathbf{H}^{(2)} \tag{6.71}$$

has exactly the form of the hamiltonians discussed in the framework of the holomorphic representation, generalized to N complex variables. In particular, it is hermitian with respect to the scalar product (6.58).

Particle number operator. The hamiltonian $\mathbf{H}$ conserves the number of particles, a property that implies equality between the number of factors z and $\partial/\partial z$ in each term. In terms of the particle number operator $\mathbf{N}$, which, acting on $\Psi(z)$, has the representation

$$\mathbf{N} = \sum_i z_i \frac{\partial}{\partial z_i}, \tag{6.72}$$

as one verifies immediately, this property is expressed by the commutation relation

$$[\mathbf{N}, \mathbf{H}] = 0.$$

The particle number operator is a special example of the one-particle operator (6.69).

Associated kernels. The kernel associated with the identity is now (expression (6.59))

$$\mathcal{I}(z, \bar{z}) = \exp \sum_i z_i \bar{z}_i. \tag{6.73}$$

The hamiltonian (6.69) and the particle number operator are then represented, respectively, by

$$H^{(1)}(z, \bar{z}) = \mathcal{I}(z, \bar{z}) \sum_i \omega_i z_i \bar{z}_i, \quad N(z, \bar{z}) = \mathcal{I}(z, \bar{z}) \sum_i z_i \bar{z}_i. \tag{6.74}$$

Finally, the pair interaction (6.70) has the representation

$$H^{(2)}(z,\bar{z}) = \frac{1}{2}\mathcal{I}(z,\bar{z}) \sum_{i_1,i_2,j_1,j_2} z_{i_1} z_{i_2} H^{(2)}_{i_1 i_2, j_1 j_2} \bar{z}_{j_1} \bar{z}_{j_2}. \tag{6.75}$$

6.7 Partition function

From the viewpoint of quantum statistical physics, one can use two different strategies to study the thermodynamic limit of a system of particles: one can either work with a fixed number n of particles and then take the limit $n \to \infty$, or consider the direct sum $\oplus \mathfrak{H}_n$, $n = 1, \ldots, \infty$ of Hilbert spaces, and fix the average number of particles by varying the chemical potential (this assumes a weak coupling to a reservoir of particles, the analogue of a thermal bath for the temperature). We study, here, the statistical properties of Bose systems in the latter framework, called the grand canonical formulation of statistical physics, using the formalism presented in Section 6.6.

Chemical potential. In quantum mechanics, conservation laws correspond to operators that commute with the hamiltonian. From the statistical viewpoint, conservation laws lead to a breaking of ergodicity. In such a situation, it is necessary to add to the hamiltonian a term proportional to the corresponding operator, to determine the expectation value of the conserved quantity. In the case of the conservation of the number of particles, one thus replaces the hamiltonian $\mathbf{H}$ by the operator

$$\mathbf{H} - \mu \mathbf{N},$$

which amounts simply to modifying $\mathbf{H}^{(1)}$. The real parameter μ, that is coupled to the particle number operator $\mathbf{N}$, is called the *chemical potential*. It allows varying the average number of particles.

Partition function. To calculate the partition function

$$\mathcal{Z}(\beta,\mu) = \operatorname{tr} e^{-\beta(\mathbf{H}-\mu\mathbf{N})},$$

we now introduce the holomorphic formalism described in the first part of the chapter. The partition function is then given by a path integral of the form (6.60):

$$\mathcal{Z}(\beta,\mu) = \int [\mathrm{d}z(t)\mathrm{d}\bar{z}(t)] \exp\left[-\mathcal{S}(z,\bar{z})\right] \tag{6.76}$$

with an action

$$\mathcal{S}(z,\bar{z}) = \int_{-\beta/2}^{\beta/2} \mathrm{d}t \left\{ -\sum_i \bar{z}_i(t) [\dot{z}_i(t) + \mu z_i(t)] + H(z(t),\bar{z}(t)) \right\} \tag{6.77}$$

and periodic boundary conditions

$$z_i(-\beta/2) = z_i(\beta/2), \quad \bar{z}_i(-\beta/2) = \bar{z}_i(\beta/2).$$

The kernel associated to the hamiltonian (6.71) can be written as

$$\langle z | H | \bar{z} \rangle = \mathcal{I}(z, \bar{z}) H(z, \bar{z}),$$

where $\mathcal{I}$ is defined in (6.73) and $H(z, \bar{z})$ can be inferred from expressions (6.74, 6.75). One then finds

$$H(z, \bar{z}) = \sum_i \omega_i z_i \bar{z}_i + \frac{1}{2} \sum_{i_1, i_2, j_1, j_2} z_{i_1} z_{i_2} H^{(2)}_{i_1 i_2, j_1 j_2} \bar{z}_{j_1} \bar{z}_{j_2} . \qquad (6.78)$$

The conservation of the number of particles leads to a $U(1) \sim SO(2)$ symmetry of the action, corresponding to the rotation

$$z_i \mapsto e^{i\theta} z_i, \quad \bar{z}_i \mapsto e^{-i\theta} \bar{z}_i,$$

since only monomials with equal number of factors z and $\bar{z}$ appear. Of course, the same formalism allows also studying systems where the number of particles is not conserved. This symmetry is then absent and a chemical potential is no longer required.

Equation of state. The equation of state is the relation between average number of particles, temperature and chemical potential. It can be derived from the partition function by differentiating with respect to the chemical potential:

$$\langle \mathbf{N} \rangle = \frac{1}{\beta} \frac{\partial \ln \mathcal{Z}}{\partial \mu}.$$

Using the path integral representation, one obtains

$$\langle \mathbf{N} \rangle = \frac{1}{\beta} \sum_i \int_{-\beta/2}^{\beta/2} dt \, \langle z_i(t) \bar{z}_i(t) \rangle = \sum_i \langle z_i(0) \bar{z}_i(0) \rangle , \qquad (6.79)$$

as a consequence of time-translation invariance.

Operator ordering. In any explicit calculation of the path integral, one is confronted with ambiguities, which are related to the problem of operator ordering. The products that appear in expressions (6.69, 6.70) are naturally written in normal order. Therefore, if one insists using the symmetric convention $\operatorname{sgn}(0) = 0$, for reasons that have been explained in Section 6.5.3, one must modify $\mathbf{H}^{(1)}$ by a term generated by $\mathbf{H}^{(2)}$ and add a constant term to cancel the ground state energy.

6.8 Bose–Einstein condensation

We first discuss the example of a system of independent particles. Of course, in such a case the sophisticated formalism that we have described in Section 6.7 is not really required. But applying it to this simple situation will provide us with a transition to Section 6.9, and allow us to explain Bose–Einstein condensation.

In the case of independent bosons, the partition function factorizes into a product of partition functions of harmonic oscillators corresponding to each energy level, a result that also follows immediately from the path integral representation (6.76).

In the absence of interactions, using the partition function or directly equation (6.79) with the gaussian two-point function (6.43) (with $\text{sgn}(0) = -1$), one obtains the equation of state

$$\langle \mathbf{N} \rangle = \sum_i \langle n_i \rangle, \quad \langle n_i \rangle = \frac{1}{e^{\beta(\omega_i - \mu)} - 1}, \tag{6.80}$$

where $\langle n_i \rangle$ is the average occupation number of the energy level i.

This expression can also be expressed in terms of the one-particle hamiltonian $H^{(1)}$ as

$$\langle \mathbf{N} \rangle = \text{tr}\, \frac{1}{e^{\beta(H^{(1)} - \mu)} - 1}. \tag{6.81}$$

Notice that a boson system is stable only for $\mu < \inf_i \omega_i$.

From finite-dimensional vector space to Hilbert space. Up to now, we have assumed that the one-particle states belong to a finite-dimensional vector space. We now generalize the formalism to the situation where one-particle states belong themselves to a Hilbert space. We discuss this situation more systematically in Section 6.9 but, here, as an introduction, we study the equation of state of a system of independent particles. The formal expression (6.81) is still valid, but $H^{(1)}$ is then a one-particle hamiltonian operator.

6.8.1 Harmonic potential

We first study the example of particles in an isotropic harmonic well in d-dimensional space. The one-particle hamiltonian can be written as

$$H^{(1)}(\hat{\mathbf{p}}, \hat{\mathbf{q}}) = \frac{1}{2m}\hat{\mathbf{p}}^2 + \frac{1}{2}m\omega^2 \hat{\mathbf{q}}^2.$$

A simple extension of the semi-classical formalism of Section 2.10 allows evaluating the average number of particles at high temperature (high compared to the separation $\hbar\omega$ between energy levels and thus for $\beta\hbar\omega \ll 1$). One can, for example, expand expression (6.81) in powers of the statistical operator and replace each term by the semi-classical approximation in the form of equation (2.90). Summing the expansion, one finds

$$\langle \mathbf{N} \rangle \sim \frac{1}{(2\pi\hbar)^d} \int \frac{d^d p\, d^d q}{e^{\beta(H^{(1)}(\mathbf{p}, \mathbf{q}) - \mu)} - 1}. \tag{6.82}$$

The average number of particles is an increasing function of μ. In this semi-classical approximation, μ must be non-positive. We then note that for $d > 1$, the integral has a finite limit for $\mu = 0$, which can be calculated explicitly. Then,

$$\langle \mathbf{N} \rangle = \frac{1}{(\hbar\omega\beta)^d}\zeta(d), \tag{6.83}$$

where ζ is the Riemann function. This result leads to an apparent paradox: if the temperature is decreased at fixed average number of particles, one finds a limiting temperature

$$T_c = \frac{1}{\beta_c} = \hbar\omega \left(\frac{\langle \mathbf{N} \rangle}{\zeta(d)}\right)^{1/d}.$$

At first instance, this phenomenon reflects a limitation of the semi-classical approximation since, for a hamiltonian with a discrete spectrum, $\langle \mathbf{N} \rangle$ diverges when μ tends towards the ground state energy.

Let us, however, examine more closely what happens for energy levels close to the ground state when, at β or $T = T_c$ fixed, $\langle \mathbf{N} \rangle$ increases by a macroscopic amount δN, that is $\delta N = O(\langle \mathbf{N} \rangle) = O((\hbar\omega\beta)^{-d})$. The chemical potential then tends toward the ground state energy $E_0 = d\hbar\omega/2$. For all states other than the ground state, the energy E satisfies $E - \mu \geq E - E_0 \geq \hbar\omega$ and this bounds the individual occupation since

$$n \leq \frac{1}{e^{\beta\hbar\omega}-1} = O(1/\beta\hbar\omega) \ll \delta N = O((\hbar\omega\beta)^{-d}).$$

Therefore, they can absorb only a negligible fraction of the increase, and the sum of all corresponding occupation numbers is still given by equation (6.83). In contrast, for the ground state the equation

$$\delta N \sim \frac{1}{e^{\beta(E_0-\mu)}-1} \sim \frac{1}{\beta(E_0-\mu)}$$

has the solution

$$\beta(E_0 - \mu) = \frac{1}{\delta N} = O(\hbar\omega\beta)^d \ll \hbar\omega\beta.$$

For space dimensions larger than one, one thus observes a remarkable phenomenon characteristic of bosons: at fixed average number of particles, below T_c a macroscopic fraction of the gas occupies only one quantum state, the ground state of the one-particle hamiltonian. This is the essence of the physical phenomenon called *Bose–Einstein condensation* and T_c is the condensation temperature.

6.8.2 Free boson gas in a box

We now consider the hamiltonian of a free particle in the limit of a box of equal size L in all dimensions and, thus, of volume L^d in dimension d. The one-particle quantum hamiltonian is simply

$$H^{(1)} = \hat{\mathbf{p}}^2/2m.$$

In a box, momenta are quantized with the precise form depending on the boundary conditions. Assuming periodic boundary conditions for convenience, but this plays no role in the discussion, one finds

$$\mathbf{p} = 2\pi\hbar\,\mathbf{n}/L\,,\ \mathbf{n} \in \mathbb{Z}^d.$$

In the infinite volume limit $L \to \infty$, one always ends up here in a high temperature situation, since the splittings between neighbouring energies decrease as $\hbar^2/2mL^2$.

The equation of state, limit of equation (6.80) where sums are replaced by integrals, in d space dimensions takes a form analogous to equation (6.82). The integration over $\mathbf{q}$ yields a factor L^d and, thus, the density is given by

$$\rho(\beta,\mu) = \frac{\langle \mathbf{N} \rangle}{L^d} \underset{L \to \infty}{\sim} \frac{1}{(2\pi\hbar)^d} \int \frac{\mathrm{d}^d p}{\mathrm{e}^{\beta(\mathbf{p}^2/2m-\mu)}-1}\,. \qquad (6.84)$$

One notes that μ must be negative, otherwise the boson system is unstable, and that, for $d > 2$, ρ, which is an increasing function of μ, is bounded by

$$\rho_c = \frac{1}{(2\pi\hbar)^d} \int \frac{\mathrm{d}^d p}{\mathrm{e}^{\beta \mathbf{p}^2/2m}-1} = \zeta(d/2)/\lambda^d,$$

where $\zeta(z)$ is the Riemann function, and λ the thermal wave length:

$$\lambda = 2\pi\hbar/\sqrt{mT}\,.$$

Alternatively, at ρ fixed, the equation of state as no solution for temperatures $T < T_c(\rho) \propto (\hbar^2/m)\rho^{2/d}$. Returning to a box of finite size, in which momenta are quantized and energy levels discrete (as in equation (6.80)), one verifies that the remaining particles accumulate in the ground state, here the state with zero momentum. This provides another example of Bose–Einstein condensation.

Finally, we recall that the chemical potential is not a physical observable and is generally eliminated in favour of the gas pressure $P = \ln \mathcal{Z}/\beta L^d$.

Remark. Thrice, we have used implicitly the identity

$$\int_0^\infty \frac{\mathrm{d}s\, s^{\alpha-1}}{\mathrm{e}^s - 1} = \Gamma(\alpha)\zeta(\alpha)$$

which can, for example, be proved by expanding

$$\frac{1}{\mathrm{e}^s - 1} = \sum_{n=1}^\infty \mathrm{e}^{-ns},$$

integrating each term and summing.

6.9 Generalized path integrals: the quantum Bose gas

In this section, we show how a natural generalization of the path integral formalism presented in Sections 6.6 and 6.7 allows deriving a field or functional integral representation (one integrates over classical fields) of the partition function for non-relativistic boson systems.

We again consider the thermodynamic properties of a system of particles obeying the Bose statistics, in the grand canonical formulation.

6.9.1 Hamiltonian in Fock space

We consider the hamiltonian of a quantum Bose gas, in d space dimensions, of the form

$$\mathbf{H} = \mathbf{T} + \mathbf{V}, \tag{6.85}$$

where $\mathbf{T}$ is the kinetic term, which in the sub-space of n-particle wave functions is represented by

$$T_n = -\frac{\hbar^2}{2m} \sum_{i=1}^{n} \nabla_{x_i}^2 \tag{6.86}$$

and $\mathbf{V}$ is a pair interaction represented by

$$V_n = \sum_{i<j\leq n} V(x_i - x_j) \quad \text{with} \quad V(x) = V(-x). \tag{6.87}$$

For simplicity, we do not introduce here a one-particle potential.

We then use directly the formalism presented in Sections 6.6 and 6.7.

Generating functionals and scalar product. To show how functional methods can be used in this context, we proceed in several steps.

We denote by $\psi_n(x_1, ..., x_n)$ the n-boson wave function, a function invariant under permutation of its n arguments. We then introduce a complex function (a field) $\varphi(x)$ (which generalizes the complex vector z_i of Section 6.6) and the functional

$$\Psi(\varphi) = \sum_{n=0}^{\infty} \frac{1}{n!} \left(\int \prod_i \mathrm{d}^d x_i \, \varphi(x_i) \right) \psi_n(x_1, \ldots, x_n). \tag{6.88}$$

Because the wave functions are symmetric, $\Psi(\varphi)$ is a generating functional of these wave functions (see Section 2.5.2). The vector space of generating functionals is then endowed with a scalar product, which takes the form of a generalized path integral, a *field* or *functional* integral, because it involves integrating over complex functions $\varphi(x), \bar\varphi(x)$:

$$(\Psi_1, \Psi_2) = \int [\mathrm{d}\varphi \mathrm{d}\bar\varphi] \, \bar\Psi_1(\varphi) \Psi_2(\varphi) \exp\left[-\int \mathrm{d}^d x \, \bar\varphi(x)\varphi(x)\right], \tag{6.89}$$

6.9 Path integrals and holomorphic formalism

where the integral is implicitly normalized by the condition

$$(1,1) = 1 = \int [\mathrm{d}\varphi \mathrm{d}\bar\varphi] \exp\left[-\int \mathrm{d}^d x\, \bar\varphi(x)\varphi(x)\right].$$

This normed complex vector space is called a Fock space.

Since the scalar product is given by a gaussian integral, to calculate scalar products one only needs the two-point function. It can be derived from the general integral

$$\mathcal{J}(J,\bar J) = \int [\mathrm{d}\varphi \mathrm{d}\bar\varphi] \exp\left\{\int \mathrm{d}^d x\, \left[-\bar\varphi(x)\varphi(x) + J(x)\bar\varphi(x) + \bar J(x)\varphi(x)\right]\right\}.$$

Translating the field and using the normalization condition, one obtains

$$\mathcal{J}(J,\bar J) = \exp\left[\int \mathrm{d}^d x\, \bar J(x)J(x)\right].$$

Functional differentiation then yields the two-point function

$$\int [\mathrm{d}\varphi \mathrm{d}\bar\varphi]\, \bar\varphi(x_1)\varphi(x_2) \exp\left[-\int \mathrm{d}^d x\, \bar\varphi(x)\varphi(x)\right] = \delta^{(d)}(x_1 - x_2),$$

where $\delta^{(d)}$ is the d-dimensional Dirac function.

As a consequence, the norm of the functional (6.88) is given by

$$|\Psi|^2 = (\Psi,\Psi) = \sum_{n=0}^\infty \frac{1}{n!} \left(\int \prod_i \mathrm{d}^d x_i\right) |\psi_n(x_1,\ldots,x_n)|^2.$$

Hamiltonian representation. One then represents the kinetic term and the potential as operators acting on Ψ.

A short calculation allows deriving the representation of the kinetic term $\mathbf{T}$. One considers

$$\int \mathrm{d}^d x\, \varphi(x) \nabla_x^2 \frac{\delta}{\delta\varphi(x)} \Psi(\varphi)$$

$$= \int \mathrm{d}^d x\, \varphi(x) \nabla_x^2 \sum_n \frac{1}{(n-1)!} \int \left(\prod_{i<n} \mathrm{d}^d x_i\, \varphi(x_i)\right) \psi_n(x_1,\ldots,x_{n-1},x).$$

In the r.h.s. the argument x can be renamed x_n, and the coefficient of $\prod_{i\le n} \varphi(x_i)$ can then be symmetrized. This generates a factor $1/n$ and yields the sum of all second derivatives, which is proportional to the kinetic term (6.86). Thus,

$$[\mathbf{T}\Psi](\varphi) = -\frac{\hbar^2}{2m} \int \mathrm{d}^d x\, \varphi(x) \nabla_x^2 \frac{\delta}{\delta\varphi(x)} \Psi(\varphi). \tag{6.90}$$

The two-body potential $\mathbf{V}$, which in the n-particle subspace is given by (6.87), can then be generated by differentiating twice with respect to φ with two different arguments. One finds

$$[\mathbf{V}\Psi](\varphi) = \frac{1}{2}\int d^d x\, d^d y\, \varphi(x)\varphi(y) V(x-y) \frac{\delta^2}{\delta\varphi(x)\delta\varphi(y)} \Psi(\varphi). \tag{6.91}$$

We have now found a representation of the total hamiltonian $\mathbf{H} = \mathbf{T} + \mathbf{V}$ when acting on generating functionals. Finally, the representation of the particle number operator is simply

$$\mathbf{N} = \int d^d x\, \varphi(x) \frac{\delta}{\delta\varphi(x)} \quad \text{and} \quad [\mathbf{N}, \mathbf{H}] = 0. \tag{6.92}$$

For reasons already explained in Section 6.7, one considers, in what follows, the modified hamiltonian $\mathbf{H} - \mu\mathbf{N}$, where the parameter μ is the chemical potential that allows varying the expectation value of $\mathbf{N}$.

6.9.2 Functional or field integrals

The representation of matrix elements of the statistical operator in the form of a field integral can then be inferred from the results already obtained in quantum mechanics in Sections 6.6 and 6.7. The complex variables z_i of the holomorphic representation (Section 6.2) are replaced by $\varphi(x)$ where the continuum coordinate x plays the role of the indices i. We denote by $\bar\varphi(x)$ the field conjugate to $\varphi(x)$. One verifies that in the mixed representation $\bar\varphi, \varphi$, the identity associated with the scalar product (6.89) reads (see also equation (6.21))

$$\mathcal{I}(\varphi, \bar\varphi) = \exp\left[\int d^d x\, \bar\varphi(x)\varphi(x)\right].$$

The kernel representation of the hamiltonian is then

$$\langle \varphi|\mathbf{H}|\bar\varphi\rangle = \left[-\frac{\hbar^2}{2m}\int d^d x\, \varphi(x)\nabla_x^2 \bar\varphi(x)\right.$$
$$\left. + \frac{1}{2}\int d^d x\, d^d y\, \varphi(x)\varphi(y) V(x-y)\bar\varphi(x)\bar\varphi(y)\right]\mathcal{I}(\varphi,\bar\varphi). \tag{6.93}$$

The particle number operator is represented by $\int d^d x\, \bar\varphi(x)\varphi(x)\mathcal{I}(\varphi,\bar\varphi)$.

Adapting the expressions of Section 6.7, in particular equation (6.76), one obtains a representation of the partition function as a *field or functional integral*:

$$\mathcal{Z}(\tau/\hbar) = \operatorname{tr} \mathbf{U}(\tau/2, -\tau/2) = \int [d\varphi(t,x) d\bar\varphi(t,x)] \exp[-\mathcal{S}(\bar\varphi,\varphi)/\hbar] \tag{6.94}$$

with the periodic boundary conditions

$$\varphi(\tau/2, x) = \varphi(-\tau/2, x), \quad \bar\varphi(\tau/2, x) = \bar\varphi(-\tau/2, x),$$

and the euclidean action

$$S(\bar\varphi, \varphi) = -\int dt\, d^d x\, \bar\varphi(t,x)\left(\hbar\frac{\partial}{\partial t} + \frac{\hbar^2}{2m}\nabla_x^2 + \mu\right)\varphi(t,x)$$
$$+ \frac{1}{2}\int dt\, d^d x\, d^d y\, \bar\varphi(t,x)\varphi(t,x)V(x-y)\bar\varphi(t,y)\varphi(t,y). \quad (6.95)$$

The addition of a one-particle potential $V_1(x)$ leads simply to the substitution $\mu \mapsto \mu - V_1(x)$. Moreover, we have given here the representation of $e^{-\tau(\mathbf{H}-\mu\mathbf{N})/\hbar}$. The statistical operator is obtained by setting $\tau = \hbar\beta$.

Again, particle number conservation leads to a $U(1)$ symmetry of the action, corresponding to the transformations

$$\varphi(x) \mapsto e^{i\theta}\varphi(x),\quad \bar\varphi(x) \mapsto e^{-i\theta}\bar\varphi(x).$$

Remarks.
(i) Let us point out, however, that in explicit calculations of field integrals new divergences in general appear, which require operations like *regularizations* and *renormalizations*, but this goes somewhat beyond the scope of the present work.

(ii) We have constructed here a non-relativistic quantum field theory. The transition to a relativistic quantum field theory now is essentially a kinematic problem.

The free theory. The action of the free theory reduces to

$$S(\bar\varphi, \varphi) = -\int dt\, d^d x\, \bar\varphi(t,x)\left(\hbar\frac{\partial}{\partial t} + \frac{\hbar^2}{2m}\nabla_x^2 + \mu\right)\varphi(t,x).$$

A convenient representation is obtained by introducing the Fourier representation of the field:

$$\varphi(t,x) = \int d^d p\, e^{ipx/\hbar}\, \tilde\varphi(t,p),\quad \bar\varphi(t,x) = \int d^d p\, e^{-ipx/\hbar}\, \tilde\varphi^*(t,p).$$

The quadratic form in the action is then diagonalized and the action becomes

$$S(\bar\varphi, \varphi) = (2\pi\hbar)^d \int dt\, d^d p\, \tilde\varphi^*(t,p)\left(-\hbar\frac{\partial}{\partial t} + \frac{p^2}{2m} - \mu\right)\tilde\varphi(t,p).$$

In a free (gaussian) theory, all quantities can be expressed in terms of the two-point function. Using the result (6.43), one obtains

$$\langle \tilde\varphi^*(t,p)\tilde\varphi(t',p')\rangle = \frac{1}{(2\pi\hbar)^d}\Delta(t-t',p)\delta^d(p-p')$$

with

$$\Delta(t,p) = \frac{1}{2}e^{-\omega(p)t/\hbar}\left\{\mathrm{sgn}(t) + \frac{\cosh[\omega(p)\tau/2\hbar]}{\sinh[\omega(p)\tau/2\hbar]}\right\},$$

where $\omega(p) = p^2/2m - \mu$. The equation of state is obtained by differentiating the partition function (6.94). Assuming a periodic box of linear size L, one finds the density

$$\rho(\beta,\mu) = \frac{1}{\beta L^d} \frac{\partial \ln \mathcal{Z}}{\partial \mu} = \frac{1}{\beta L^d \hbar} \int \mathrm{d}t\, \mathrm{d}^d x\, \langle \bar{\varphi}(t,x)\varphi(t,x) \rangle = \langle \bar{\varphi}(0,0)\varphi(0,0)\rangle,$$

where translation invariance in space and time has been used. Introducing the field Fourier representation, one obtains

$$\rho(\beta,\mu) = \int \mathrm{d}^d p\, \mathrm{d}^d p'\, \langle \tilde{\varphi}^*(0,p)\tilde{\varphi}(0,p')\rangle = \frac{1}{(2\pi\hbar)^d} \int \mathrm{d}^d p\, \Delta(0,p)$$

$$= \frac{1}{(2\pi\hbar)^d} \int \mathrm{d}^d p\, \frac{1}{2} \left\{ \mathrm{sgn}(0) + \frac{\cosh[\beta\omega(p)/2]}{\sinh[\beta\omega(p)/2]} \right\}.$$

This expression coincides with the result (6.84) obtained directly when $\mathrm{sgn}(0)$ is chosen to be -1. Another choice like $\mathrm{sgn}(0) = 0$ leads to a divergent result, but in the latter case, a term proportional to μ must be added to the action to remove this additional contribution to ρ.

Interactions. In presence of weak repulsive local interactions, this formalism allows, for example, studying the crossover between Bose–Einstein condensation and superfluid Helium phase transition. The potential can then be simulated by a Dirac δ function (but regularized at short distance in dimensions $d > 1$)

$$V(x) = g\, \delta^d(x), \quad g > 0.$$

and the action becomes local in space and time, in the sense that it becomes the integral of a lagrangian density depending only on the field and its derivatives:

$$S(\bar{\varphi},\varphi) = -\int \mathrm{d}t\, \mathrm{d}^d x\, \left[\bar{\varphi}(t,x) \left(\hbar\frac{\partial}{\partial t} + \frac{\hbar^2}{2m} \nabla_x^2 + \mu \right) \varphi(t,x) \right.$$

$$\left. + \frac{1}{2} g \big(\bar{\varphi}(t,x)\varphi(t,x)\big)^2 \right]. \qquad (6.96)$$

$$S(\bar{\varphi},\varphi) = -\int \mathrm{d}t\, \mathrm{d}^d x\, \left[\bar{\varphi}(t,x) \left(\hbar\frac{\partial}{\partial t} + \frac{\hbar^2}{2m} \nabla_x^2 + \mu \right) \varphi(t,x) + \frac{1}{2} g\big(\bar{\varphi}(t,x)\varphi(t,x)\big)^2 \right]. \qquad (6.97)$$

The field integral can be calculated by the steepest descent method. General arguments indicate that saddle points correspond to constant fields. The saddle point equations then are

$$-\mu\varphi + g\varphi^2 \bar{\varphi} = -\mu\bar{\varphi} + g\bar{\varphi}^2 \varphi = 0.$$

For $\mu < 0$, the equations have only the trivial solution $\varphi = \bar{\varphi} = 0$. In contrast, for $\mu > 0$, they have other solutions: $\bar{\varphi}\varphi = \mu/g$ and one verifies that these are the leading saddle points since the corresponding action in a large volume L^d is

$$S = -\beta L^d \mu^2/2g \;\Rightarrow\; \rho = \mu/g.$$

For $\mu < 0$, the solution is $U(1)$ invariant and the $U(1)$ symmetry is not broken. For $\mu > 0$, the leading saddle points are not $U(1)$ invariant and this corresponds to a spontaneous breaking of the $U(1)$ symmetry. In presence of the interaction, the value of the chemical potential $\mu = 0$ is a transition point of the class of the He^4 phase transition. Note, however, that beyond leading order, the first correction to ρ has the form (6.84) and is finite for $\mu = 0$ only in the thermodynamic limit and for dimensions $d > 2$, as for the Bose–Einstein condensation. The marginal dimension $d = 2$ is special and corresponds to a phase transition without symmetry breaking, the Kosterlitz–Thouless phase transition.

Exercises

Exercise 6.1

Use the result (6.7) to give a simple proof of an identity about determinants: one writes a matrix $\mathbf{A}$ in the form

$$\mathbf{A} = \begin{pmatrix} \mathbf{A}_{11} & \mathbf{A}_{12} \\ \mathbf{A}_{21} & \mathbf{A}_{22} \end{pmatrix},$$

where $\mathbf{A}_{11}, \mathbf{A}_{12}, \mathbf{A}_{21}$ and $\mathbf{A}_{22}$ are four $p \times p$, $p \times n - p$, $n - p \times p$ and $n - p \times n - p$ matrices, respectively. Then,

$$\det \mathbf{A} = \det \mathbf{A}_{11} \det \left(\mathbf{A}_{22} - \mathbf{A}_{21} \mathbf{A}_{11}^{-1} \mathbf{A}_{12} \right).$$

Solution. One writes the determinant of $\mathbf{A}$ as a complex gaussian integral. In the gaussian integral, one integrates over the p first variables and then over the remaining $n - p$. One verifies that this leads to the identity.

Exercise 6.2

Calculate the expectation value

$$\langle \bar{z}_1 z_1 \bar{z}_2 z_2 \rangle,$$

where $\bar{z}_1, z_1, \bar{z}_2, z_2$ are complex variables, with the measure

$$\frac{d\bar{z}_1 dz_1}{2i\pi} \frac{d\bar{z}_2 dz_2}{2i\pi} \exp\left[-2\bar{z}_1 z_1 - \bar{z}_1 z_2 - \bar{z}_2 z_1 - \bar{z}_2 z_2 \right].$$

Solution. The matrix associated to the quadratic form is

$$M = \begin{pmatrix} 2 & 1 \\ 1 & 1 \end{pmatrix} \Rightarrow \det M = 1, \ M^{-1} = \begin{pmatrix} 1 & -1 \\ -1 & 2 \end{pmatrix},$$

and thus

$$\langle \bar{z}_1 z_1 \bar{z}_2 z_2 \rangle = 3.$$

Exercise 6.3

Calculate the expectation values

$$\langle \bar{z}_1 z_1 \rangle,\ \langle \bar{z}_2 z_1 \rangle,\ \langle \bar{z}_1 z_2 \rangle,\ \langle \bar{z}_2 z_2 \rangle,\ \langle (\bar{z}_1)^2 z_1 z_2 \rangle,\ \langle \bar{z}_1 z_1 \bar{z}_2 z_2 \rangle,$$

where $\bar{z}_1, z_1, \bar{z}_2, z_2$ are complex variables, with the measure

$$\frac{d\bar{z}_1 dz_1}{2i\pi} \frac{d\bar{z}_2 dz_2}{2i\pi} \exp\left[-2\bar{z}_1 z_1 - 3\bar{z}_1 z_2 - 3\bar{z}_2 z_1 - 5\bar{z}_2 z_2\right].$$

Solution.

$\langle \bar{z}_1 z_1 \rangle = 5$, $\langle \bar{z}_2 z_1 \rangle = -3$, $\langle \bar{z}_1 z_2 \rangle = -3$, $\langle \bar{z}_2 z_2 \rangle = 2$, $\langle (\bar{z}_1)^2 z_1 z_2 \rangle = -30$, $\langle \bar{z}_1 z_1 \bar{z}_2 z_2 \rangle = 19$.

Exercise 6.4

Calculate the square of the norm of the holomorphic function

$$f = 1 + 2z_1 + 5z_1^3 + 3z_1^2 z_2^2,$$

where $\bar{z}_1, z_1, \bar{z}_2, z_2$ are complex variables, with the measure

$$\frac{1}{2} \frac{d\bar{z}_1 dz_1}{2i\pi} \frac{d\bar{z}_2 dz_2}{2i\pi} \exp\left[-\bar{z}_1 z_1 - \bar{z}_2 z_2/2\right].$$

Solution. 299.

Exercise 6.5

In the holomorphic formulation of quantum mechanics, one represents the position and momentum operators by

$$\hat{q} \mapsto (2\omega)^{-1/2}\left(z + \frac{d}{dz}\right),\ \hat{p} \mapsto i(\omega/2)^{1/2}\left(z - \frac{d}{dz}\right). \qquad (6.98)$$

One then considers the state corresponding to the entire function $\psi(z) = e^{\alpha z}$, where α is an arbitrary complex number.

Calculate the expectation values of the position and momentum operators $\hat{q}, \hat{p}$ and the mean squared deviation $\Delta q = (\langle \hat{q}^2 \rangle - \langle \hat{q} \rangle^2)^{1/2}$, $\Delta p = (\langle \hat{p}^2 \rangle - \langle \hat{p} \rangle^2)^{1/2}$ (setting $\hbar = 1$). Show that the states ψ are minimum dispersion states: $\Delta q \Delta p = \frac{1}{2}$ (this is the minimal value allowed by Heisenberg's uncertainty principle).

Solution. One first calculates the norm of $\psi(z)$, and one finds $e^{\alpha \bar{\alpha}}$. A simple calculation then yields

$$\langle \hat{q} \rangle = (2/\omega)^{1/2} \operatorname{Re} \alpha,\ \langle \hat{q} \rangle = (2\omega)^{1/2} \operatorname{Im} \alpha,\ (\Delta q)^2 = 1/2\omega,\ (\Delta p)^2 = \omega/2,$$

which shows that ψ is a minimum dispersion state.

Exercise 6.6

Calculate the square of the norm of the function $f(z) = e^{-\alpha z^2/2}$, with α real, using the definition (6.11). Comment. Generalize to α complex.

Solution. One can return to 'real' variables (6.1) to integrate. One finds

$$(f,f) = \frac{1}{\sqrt{1-\alpha^2}}.$$

In agreement with the general bound, the function is normalizable only for $|\alpha| < 1$.

The case of complex α reduces to the real case by a change of variables $z \mapsto e^{i\theta} z$, $\bar{z} \mapsto e^{-i\theta} \bar{z}$ with $\theta = -\mathrm{Arg}(\alpha)/2$. Then, α^2 is replaced by $|\alpha|^2$.

Exercise 6.7

Space and momentum translation operators. In this exercise, one still uses the representation (6.98) for the operators $\hat{p}, \hat{q}$, but one sets $\omega = 1$.

1. Find the eigenvectors of the position operator $\hat{q}$ in the space of entire functions. Calculate the scalar product of two eigenvectors (it may be useful to return to real integration).

Solution. The eigenvector $f_q(z)$, corresponding to the real eigenvalue q, is solution of

$$\frac{1}{\sqrt{2}}\left(z + \frac{\mathrm{d}}{\mathrm{d}z}\right) f_q(z) = q f_q(z).$$

The solution can be written as

$$f_q(z) = e^{-(z-q\sqrt{2})^2/2}.$$

Then,

$$(f_{q'}, f_q) = \sqrt{\pi}\, e^{-q^2} \delta(q - q').$$

2. One considers the operator $T(\alpha)$ as defined by its action on holomorphic vectors:

$$[T(\alpha)f](z) \equiv f(z,\alpha) = e^{-\alpha z/\sqrt{2} - \alpha^2/4} f(z + \alpha/\sqrt{2}),$$

with α real. Verify the abelian multiplication law

$$T(\alpha)T(\beta) = T(\alpha + \beta),$$

by acting on an arbitrary vector f.

Verify explicitly that the operator $T(\alpha)$ preserves the scalar product (and, thus, that it is unitary).

Show by acting with $T(\alpha)$ on the eigenvectors f_q of $\hat{q}$:

$$[T(\alpha)f_q](z) = f_{q+\alpha}(z).$$

Show that $f(z, \alpha)$ satisfies the partial derivative equation
$$\frac{\partial f(z, \alpha)}{\partial \alpha} = \frac{1}{\sqrt{2}} \left(\frac{\partial}{\partial z} - z \right) f(z, \alpha). \tag{6.99}$$

Infer that $T(\alpha)$ is the translation operator: $T(\alpha) = e^{i\alpha \hat{p}}$.

3. Determine the kernel associated with $T(\alpha)$ and directly verify unitarity.

Solution. One acts with $T(\alpha)$ on the identity kernel $\mathcal{I}$:
$$T(\alpha, z, \bar{z}) \equiv [T(\alpha)\mathcal{I}](z, \bar{z}) = e^{-\alpha z/\sqrt{2} - \alpha^2/4} \mathcal{I}(z + \alpha/\sqrt{2}, \bar{z})$$
$$= e^{z\bar{z} + \alpha(\bar{z}-z)/\sqrt{2} - \alpha^2/4}.$$

The proof of unitarity then relies on the verification
$$\int \frac{dv d\bar{v}}{2i\pi} e^{-v\bar{v}} \overline{T}(\alpha, z, \bar{v}) T(\alpha, v, \bar{z}) = e^{z\bar{z}}.$$

4. One now defines the operator
$$[V(\beta) f](z) = e^{-i\beta z/\sqrt{2} - \beta^2/4} f(z - i\beta/\sqrt{2}),$$
with β real. Same questions as for $T(\alpha)$, equation (6.99) being replaced by
$$\frac{\partial [V(\beta) f]}{\partial \beta} = -\frac{i}{\sqrt{2}} \left(\frac{\partial}{\partial z} + z \right) [V(\beta) f].$$

Infer that $V(\beta)$ is the momentum translation operator $V(\alpha) = e^{-i\alpha \hat{q}}$.

5. Acting on holomorphic vectors, derive the commutation relation
$$V(\beta) T(\alpha) = e^{i\alpha\beta} T(\alpha) V(\beta).$$

Exercise 6.8

Boson systems in the holomorphic representation.

One considers the Hilbert space of analytic functions $f(z)$ endowed with the scalar product (6.11).

The unperturbed hamiltonian, in the holomorphic representation,
$$H_0 = z \frac{d}{dz},$$
describes a system of bosons that can occupy only one energy state. One then adds an interaction between the bosons and a medium that can absorb and emit boson pairs with equal probability. This corresponds to adding a potential
$$V = \frac{\alpha}{1 + \alpha^2} \left[\left(\frac{d}{dz} \right)^2 + z^2 \right]$$

to H_0, where α is chosen real and the parametrization is convenient for what follows. With these assumptions, the hamiltonian $H = H_0 + V$ remains hermitian.

1. Introducing the operator
$$B = \frac{d}{dz} + \alpha z,$$
express H in terms of $B^\dagger B$.

Solution.
$$B^\dagger = \alpha \frac{d}{dz} + z, \quad H = \frac{1}{1+\alpha^2}\left(B^\dagger B - \alpha^2\right).$$

2. Determine the holomorphic eigenvectors $f_\pm(z)$ with $f_\pm(0) = 1$, such that
$$B f_+ = 0, \quad B^\dagger f_- = 0,$$
and calculate their norm. Under which conditions are the norms finite?

Solution. $f_+(z) = e^{-\alpha z^2/2}$ and, thus, following the result of exercise 6.6,
$$(f_+, f_+) = \frac{1}{\sqrt{1-\alpha^2}}.$$

The norm is finite if $|\alpha| < 1$.

Moreover,
$$[B^\dagger f_-](z) = \alpha f'(z) + z$$
and thus $f_-(z) = e^{-z^2/2\alpha}$ and
$$(f_-, f_-) = \frac{|\alpha|}{\sqrt{\alpha^2-1}}.$$

One finds the condition $\alpha^2 > 1$. Both vectors cannot be normalized simultaneously and for $\alpha^2 = 1$, neither is normalizable.

3. Calculate the commutator $[B, B^\dagger]$ and relate B and $B^\dagger$ to creation and annihilation operators of a harmonic oscillator (distinguishing the two cases $|\alpha| < 1$ and > 1). Infer the spectrum of H. Show that for $|\alpha| \neq 1$, it is a spectrum of independent particles, which one can call quasi-bosons.

Solution. $[B, B^\dagger] = 1 - \alpha^2$.

For $|\alpha| < 1$, one can set $B = A\sqrt{1-\alpha^2}$, where A is the annihilation operator with the standard normalization $[A, A^\dagger] = 1$ and, thus,
$$H = \frac{(1-\alpha^2)A^\dagger A - \alpha^2}{1+\alpha^2} \Rightarrow E_N = \frac{(1-\alpha^2)N - \alpha^2}{1+\alpha^2}, \quad N \geq 0.$$

For $|\alpha| > 1$, one can set $B^\dagger = A\sqrt{\alpha^2 - 1}$ and the spectrum of H is then given by
$$E_N = \frac{(\alpha^2-1)N - 1}{1+\alpha^2}, \quad N \geq 0.$$

These results are consistent with the normalizability conditions of $f_\pm$. One verifies that, like the hamiltonian, the spectrum is invariant by $\alpha \mapsto 1/\alpha$. The change $z \mapsto iz$ shows also directly that $H(\alpha)$ has the same spectrum as $H(-\alpha)$, which is consistent with the explicit result.

For $|\alpha| \neq 1$, one thus finds a spectrum of independent particles, of quasi-bosons in the sense that these boson states consist in a superposition of states with 1, 3, ..., initial bosons.

4. Determine the kernel $\langle z| H |\bar z \rangle$ corresponding to the hamiltonian H. Infer, then, the path integral representation of the kernel $\langle z| U(\beta, 0) |\bar z \rangle$ associated with the statistical operator $U(\beta, 0) = e^{-\beta H}$.

Solution.

$$\langle z| H |\bar z \rangle = e^{\bar z z} H(z, \bar z), \quad H(z, \bar z) = z\bar z + \frac{\alpha}{1+\alpha^2}\left(z^2 + \bar z^2\right).$$

One infers the path integral representation

$$\langle z| U(\beta, 0) |\bar z \rangle = \int \left[\frac{d\bar z(t) dz(t)}{2i\pi}\right] e^{\bar z(0) z(0)} \exp\left[-\mathcal{S}(z, \bar z)\right]$$

with the euclidean action

$$\mathcal{S}(z, \bar z) = \int_0^\beta dt \left[-\bar z(t)\dot z(t) + \bar z(t) z(t) + \frac{\alpha}{1+\alpha^2}\left(z^2(t) + \bar z^2(t)\right)\right] \tag{6.100}$$

and the boundary conditions $z(\beta) = z$, $\bar z(0) = \bar z$.

5. From now on, one restricts the discussion to $0 < \alpha < 1$ and sets

$$\beta(1-\alpha^2)/(1+\alpha^2) = \lambda, \quad \alpha = e^{-\mu}.$$

Infer from the calculation of the path integral, the dependence of $U(z, \bar z; \beta, 0)$ on $z, \bar z$.

One may use linear combinations of the classical equations of motion to cast the action into the form

$$\mathcal{S}(z_c, \bar z_c) = -\tfrac{1}{2}\left[z_c(\beta)\bar z_c(\beta) - z_c(0)\bar z_c(0)\right],$$

where $z_c(t), \bar z_c(t)$ are solutions. Then, to solve the equations, one may introduce the two linear combinations $z(t) \pm \bar z(t)$.

Solution. The integral is gaussian and can thus be evaluated. One first solves the equations of motion to eliminate the boundary conditions:

$$-\dot z(t) + z(t) + \frac{2\alpha}{1+\alpha^2}\bar z(t) = 0, \tag{6.101}$$

$$\dot{\bar z}(t) + \bar z(t) + \frac{2\alpha}{1+\alpha^2} z(t) = 0. \tag{6.102}$$

Adding $\bar z(t)$ times equation (6.101) to $z(t)$ times equation (6.102), one finds

$$z(t)\bar z(t) + \frac{\alpha}{1+\alpha^2}\left(z^2(t)+\bar z^2(t)\right) = \frac{1}{2}\left(\bar z(t)\dot z(t) - \dot{\bar z}(t)z(t)\right).$$

Substituting this identity into the action (6.100), one obtains a total derivative, which leads to the first result:

$$\mathcal{S}(z,\bar z) = -\tfrac{1}{2}\left[z(\beta)\bar z(\beta) - z(0)\bar z(0)\right].$$

Since $z(\beta)$ and $\bar z(0)$ are fixed by the boundary conditions, it suffices to calculate $z(0)$ and $\bar z(\beta)$. The solutions of the equations can be written as

$$z(t) = a\,\mathrm{e}^{\omega t} - \alpha b\,\mathrm{e}^{-\omega t}$$
$$\bar z(t) = -\alpha a\,\mathrm{e}^{\omega t} + b\,\mathrm{e}^{-\omega t},$$

with $\omega = (1-\alpha^2)/(1+\alpha^2)$ and

$$a = \frac{z\,\mathrm{e}^\mu + \bar z\,\mathrm{e}^{-\lambda}}{2\sinh(\lambda+\mu)}, \qquad b = \frac{z+\bar z\,\mathrm{e}^{\lambda+\mu}}{2\sinh(\lambda+\mu)}.$$

Thus,

$$z(0) = \frac{z\sinh\mu - \bar z\sinh\lambda}{\sinh(\lambda+\mu)}, \qquad \bar z(\beta) = \frac{\bar z\sinh\mu - z\sinh\lambda}{\sinh(\lambda+\mu)},$$

and, finally,

$$\mathcal{S}(z,\bar z) - \bar z(0)z(0) = \frac{(z^2+\bar z^2)\sinh\lambda - 2z\bar z\sinh\mu}{2\sinh(\lambda+\mu)}.$$

6. Infer the normalization from the calculation of $\mathrm{tr}\,U(\beta,0)$ by comparing the result, for example, with the expression obtained directly by using the spectrum determined previously.

Solution. The gaussian integration yields a normalization $\mathcal{N}$ and

$$\langle z|U(\beta,0)|\bar z\rangle = \mathcal{N}\,\mathrm{e}^{\bar z(0)z(0)-\mathcal{S}(z,\bar z)}.$$

Calculating the trace, one obtains the partition function

$$\mathcal{Z}(\beta) = \mathcal{N}\int\frac{dz d\bar z}{2i\pi}\,\mathrm{e}^{-z\bar z}\,\mathrm{e}^{\bar z(0)z(0)-\mathcal{S}(z,\bar z)} = \frac{\mathcal{N}}{2\sinh(\lambda/2)}\sqrt{\frac{\sinh(\lambda+\mu)}{\sinh\mu}}.$$

The direct calculation based on the already determined spectrum yields

$$\mathcal{Z}(\beta) = \frac{\mathrm{e}^{\beta/2}}{2\sinh(\lambda/2)} \;\Rightarrow\; \mathcal{N} = \mathrm{e}^{\beta/2}\sqrt{\frac{\sinh\mu}{\sinh(\lambda+\mu)}}.$$

Exercise 6.9

One considers the partition function (6.57) with

$$\mathcal{S}(z,\bar{z}) = \int_{-\beta/2}^{\beta/2} dt \left\{ \bar{z}(t)\left[-\dot{z}(t) + \omega z(t)\right] + \tfrac{1}{4}\lambda\left[\bar{z}(t)z(t)\right]^2 \right\},$$

where ω, λ are two positive constants. Calculate the ground state energy E_0, up to order λ^2 with the convention $\theta(0) = \tfrac{1}{2}$, that is $\text{sgn}(0) = 0$. Determine then the whole spectrum to order λ. More generally, using the remark (3.1), determine the exact spectrum.

Solution. Help can be found in the calculations presented in Section 6.1.3. First, using Wick's theorem, one obtains

$$\text{tr}\, e^{-\beta H} = \frac{1}{2\sinh(\omega\beta/2)} \left[1 - \tfrac{1}{2}\lambda \int dt\, \langle \bar{z}(t)z(t)\rangle^2 + O\left(\lambda^2\right)\right],$$

where $\langle \bar{z}(t)z(t)\rangle$ is given by the propagator (6.43) at time 0:

$$\langle \bar{z}(t)z(t)\rangle \equiv \Delta(0) = \tfrac{1}{2}\left(\text{sgn}(0) + 1/\tanh(\omega\beta/2)\right).$$

Comparing with expression (3.2), one obtains the spectrum to order λ:

$$E_k = (k + \tfrac{1}{2})\omega + \tfrac{1}{4}\lambda(k^2 + k + \tfrac{1}{2}).$$

This is also the exact spectrum. In the limit $\beta \to \infty$, the gaussian two-point function (6.45) implies time ordering. Therefore, all connected diagrams, except the order λ that is ambiguous, vanish and this proves the property for the ground state energy. More generally, one verifies, before integrating over time, that all perturbative contributions are time-independent. Time integration then yields a factor β^n at order λ^n. The expansion (3.1) then implies that contributions of order higher than one no longer modify the spectrum.

Exercise 6.10

Calculate the two-point function $\langle \bar{z}(u)z(v)\rangle$ corresponding to the weight $e^{-\mathcal{S}}/\mathcal{Z}$ to order λ^2, for $\beta \to \infty$. Infer the energy $\Omega = E_1 - E_0$, where E_1 corresponds to the first excited state (the one-particle state).

Solution. Expanding in powers of λ^2 and using Wick's theorem, one finds (again the causal property of the propagator much simplifies the calculation)

$$\langle \bar{z}(u)z(v)\rangle = \theta(u-v)\, e^{-\omega(u-v)} \left(1 - \tfrac{1}{2}\lambda(u-v) + \tfrac{1}{8}\lambda^2(u-v)^2\right) + O(\lambda^3)$$
$$= \theta(u-v)\, e^{-(\omega+\lambda/2)(u-v)} + O(\lambda^3).$$

Thus, $\Omega = \omega + \tfrac{1}{2}\lambda + O(\lambda^3)$, in agreement with the exact spectrum.

Exercise 6.11

Explain these results in terms of the corresponding hamiltonian expressed in terms of creation and annihilation operators.

Solution. The path integral defines the hamiltonian up to the problem of operator ordering. Therefore, it has the form

$$H = \omega(a^\dagger a + \tfrac{1}{2}) + \tfrac{1}{4}\lambda((a^\dagger a)^2 + \alpha a^\dagger a + \beta),$$

where α, β are two numerical constants. The spectrum is then

$$E_k = \omega(k + \tfrac{1}{2}) + \lambda(k^2 + \alpha k + \beta).$$

Comparing with the result of the calculation, one concludes

$$\alpha = 1, \quad \beta = \tfrac{1}{2}.$$

Finally, since the eigenvectors of the hamiltonian are those of the harmonic oscillator, the two-point function is given by

$$\begin{aligned}\langle \bar{z}(u) z(v) \rangle &= \theta(u - v) \langle 0 | \, e^{-(H-E_0)(u-v)} \, | 0 \rangle \\ &= \sum_{k=0} \langle 0 | \, a \, | k \rangle \, e^{-(E_k - E_0)(u-v)} \langle k | \, a^\dagger \, | 0 \rangle \\ &= \theta(u - v) \, e^{-(E_1 - E_0)(u-v)}.\end{aligned}$$

Exercise 6.12

Use the holomorphic path integral to calculate, by two different methods, the correction of order λ to the energy eigenvalues of the hamiltonian

$$H = \tfrac{1}{2}\hat{p}^2 + \tfrac{1}{2}\hat{q}^2 + \lambda \hat{q}^4,$$

and compare with the result with expression (3.2).

In the first method, one expresses the quartic interaction in terms of creation and annihilation operators, one normal-orders the products and one then replaces them by the classical variables $z, \bar{z}$, following the method described in Section 6.5. To obtain the normal-order form, one may first prove the identity

$$e^{x(a+a^\dagger)} = e^{xa^\dagger} e^{xa} e^{x^2/2},$$

either by differentiating or by using Baker–Hausdorf formula.

In the second method, one replaces directly $\hat{q}$ by the classical variable $(z + \bar{z})/\sqrt{2}$ in the quartic term. Finally, compare this expression with the half-sum of the normal and anti-normal order form. One may use the second identity

$$e^{x(a+a^\dagger)} = e^{xa} e^{xa^\dagger} e^{-x^2/2}.$$

Solution. One finds

$$(a+a^\dagger)^4 = a^{\dagger 4} + 4a^\dagger a^3 + 6a^{\dagger 2}a^2 + 4a^{\dagger 3}a + a^4 + 6a^{\dagger 2} + 12a^\dagger a + 6a^2 + 3\,.$$

One then replaces the quantum operators $a, a^\dagger$ by the classical variables $z, \bar{z}$. At this order, only the terms with an equal number of z and $\bar{z}$ factors contribute. Using Wick's theorem, one thus obtains

$$\operatorname{tr} e^{-\beta H} = \frac{1}{2\sinh(\omega\beta/2)}\left[1 - \tfrac{3}{4}\lambda\beta\left(1 + 4\Delta(0) + 4\Delta^2(0)\right) + O\left(\lambda^2\right)\right],$$

where $\langle \bar{z}(t)z(t)\rangle$ is given by the propagator (6.43) at time 0. With the convention $\operatorname{sgn}(0) = -1$, one finds

$$\operatorname{tr} e^{-\beta H} = \frac{1}{2\sinh(\omega\beta/2)}\left[1 - \tfrac{3}{4}\lambda\beta\cotanh^2(\omega\beta/2) + O\left(\lambda^2\right)\right],$$

an expression that is consistent with the result (3.2).

In the second case, the quartic term leads to a contribution

$$\tfrac{1}{4}\lambda \int \mathrm{d}t\,\left(\bar{z}(t) + z(t)\right)^4,$$

to the action. At order λ, only the term proportional to $(\bar{z}z)^2$ contributes. Using Wick's theorem, one finds

$$\operatorname{tr} e^{-\beta H} = \frac{1}{2\sinh(\omega\beta/2)}\left[1 - 3\lambda\beta\Delta^2(0) + O\left(\lambda^2\right)\right],$$

With the convention $\operatorname{sgn}(0) = 0$, one recovers the exact result.

Finally, the half-sum of the normal and anti-normal order expression leads to $(\bar{z} + z)^4 + 3$, and differs from the exact result by a global shift.

7 PATH INTEGRALS: FERMIONS

The methods that we have described so far, allow studying general quantum Bose systems. They are based in a direct way on the introduction of a generating function of symmetric wave functions of bosons, as we have explained in Section 6.6.

In contrast, in the case of fermion systems, one faces the problem that fermion wave functions, or fermion correlation functions (or Green functions) are antisymmetric with respect to the exchange of a fermion pair. The construction of generating functions, thus, requires the introduction of an antisymmetric or Grassmann algebra of 'classical functions'.

Remarkably enough, it is then possible to generalize to Grassmann algebras the notions of derivatives and integrals. This allows constructing quite parallel formalisms for bosons and fermions, in particular, to define a path integral for fermion systems, analogous to the holomorphic path integral for bosons.

In the limit of an infinite number of available fermion states, the formalism allows expressing the partition function of the Fermi gas as an integral over Grassmann fields with anti-periodic boundary conditions.

7.1 Grassmann algebras

A Grassmann algebra $\mathfrak{A}$ on $\mathbb{R}$ or $\mathbb{C}$ (real or complex numbers) is an associative algebra generated by a unit (denoted by 1 in what follows) and a set of generators $\{\theta_i\}$ that satisfy the anti-commutation relations

$$\theta_i \theta_j + \theta_j \theta_i = 0 \qquad \forall i, j. \tag{7.1}$$

(In what follows, unless stated otherwise, we consider only complex algebras and when we speak of generators, we omit the unit that plays a special role.)

As a consequence:

(i) If the number n of generators is finite, the elements of the algebra form a vector space of finite dimension 2^n over $\mathbb{R}$ or $\mathbb{C}$. All elements can be written as linear combinations of the elements

$$1 \text{ and } \{\theta_{i_1} \theta_{i_2} \ldots \theta_{i_p}\} \text{ with } i_1 < i_2 < \cdots < i_p, \ 1 \le p \le n. \tag{7.2}$$

(ii) $\mathfrak{A}$ is a graded algebra, in the sense that to each monomial $\theta_{i_1} \theta_{i_2} \cdots \theta_{i_p}$, one can associate an integer p that counts the number of generators in a product.

(iii) Elements of $\mathfrak{A}$ are invertible if and only if in the expansion on the basis (7.2) the term of degree zero does not vanish. For example, the element $1 + \theta$ is invertible and its inverse is $1 - \theta$; in contrast θ is not invertible. The inverse can be calculated by expanding in a formal power series starting from the inverse of the term of degree zero.

(vi) All elements in a Grassmann algebra, considered as functions of a generator θ_i, are first degree polynomials, that is, affine functions.

Reflection. In the algebra $\mathfrak{A}$, one can define a simple automorphism P, that is,

$$P(A+B) = P(A) + P(B), \quad P(AB) = P(A)P(B),$$
$$P(\lambda A) = \lambda P(A) \quad \forall A, B \in \mathfrak{A}, \quad \lambda \in \mathbb{C},$$

which has the nature of a reflection:

$$P(\theta_i) = -\theta_i \Rightarrow P^2 = \mathbf{1}. \tag{7.3}$$

Acting on a monomial of degree p, it yields

$$P(\theta_{i_1} \cdots \theta_{i_p}) = (-1)^p \theta_{i_1} \cdots \theta_{i_p}. \tag{7.4}$$

The vector space generated by the elements of $\mathfrak{A}$ can be divided into the sum of two vector spaces $\mathfrak{A}^{\pm}$ containing the even and odd elements:

$$P(\mathfrak{A}^{\pm}) = \pm \mathfrak{A}^{\pm}. \tag{7.5}$$

Note also the property

$$A\theta_i = \theta_i P(A). \tag{7.6}$$

When A_+ belongs to $\mathfrak{A}^+$, it commutes with all elements of $\mathfrak{A}$:

$$A_+ \in \mathfrak{A}^+ \Rightarrow A_+ B = B A_+, \quad \forall B.$$

In particular, $\mathfrak{A}^+$ can be identified with the maximal commutative sub-algebra of $\mathfrak{A}$.

On the other hand, if A_-, B_- both belong to $\mathfrak{A}^-$, they anti-commute:

$$A_- \text{ and } B_- \in \mathfrak{A}^- \Rightarrow A_- B_- + B_- A_- = 0.$$

As a consequence, all elements of $\mathfrak{A}^-$ are nilpotent with a vanishing square.

Complex conjugation. In quantum mechanics, one needs mainly Grassmann algebras $\mathfrak{A}$ with an even number of generators, which can be divided into two subsets $\{\theta_i\}$ and $\{\bar\theta_i\}$, $i = 1, \ldots, n$. One can then define in the algebra generated by $\{\theta_i, \bar\theta_i\}$ the analogue of usual complex conjugation. Actually, it has the properties of hermitian conjugation for matrices or operators (and we use the same notation):

$$\begin{cases} \theta^\dagger = \bar\theta, \quad \bar\theta^\dagger = \theta, \\ (\lambda A_1 + \mu A_2)^\dagger = \bar\lambda A_1^\dagger + \bar\mu A_2^\dagger, \\ (A_1 A_2)^\dagger = A_2^\dagger A_1^\dagger, \quad \forall A_1, A_2 \in \mathfrak{A} \text{ and } \lambda, \mu \in \mathbb{C}. \end{cases} \tag{7.7}$$

Note that, as a consequence, $(\vartheta\bar\vartheta)^\dagger = \vartheta\bar\vartheta$.

7.2 Differentiation in Grassmann algebras

In Grassmann algebras, it is possible to define a generalized derivative. A too naive definition, however, would be inconsistent with the non-commutative character of the algebra.

Considered as functions of a given generator θ_i, all elements A of $\mathfrak{A}$ can be written as (in general, after some commutations)
$$A = A_1 + \theta_i A_2 \,,$$
where A_1 and A_2 do not depend on θ_i. One then defines the derivative with respect to θ_i by
$$\frac{\partial A}{\partial \theta_i} = A_2 \,. \tag{7.8}$$
As the usual differentiation, it is a linear operation. However, the operator $\partial/\partial\theta_i$ is *nilpotent* with vanishing square: $(\partial/\partial\theta_i)^2 = 0$.

Remark. Equation (7.8) defines a left derivative in the sense that the action of $\partial/\partial\theta_i$ consists in commuting θ_i to the left in all monomials before suppressing it. In a similar way, one could define a right derivative by commuting θ_i to the right before suppressing it.

Derivative of a product. It follows from the definition (7.8) that the derivative of a product does not satisfy the usual Leibnitz rule $D(AB) = AD(B)+D(A)B$. Using the remark (7.6), one verifies that this rule is replaced by
$$\frac{\partial(AB)}{\partial \theta_i} = P(A)\frac{\partial B}{\partial \theta_i} + \frac{\partial A}{\partial \theta_i}B \,, \tag{7.9}$$
a rule that is consistent for any associative algebra and homomorphism P.

Chain rule. The Grassmann derivative implies a special form of chain rule. If $\sigma(\theta)$ belongs to $\mathfrak{A}^-$ and $x(\theta)$ belongs to $\mathfrak{A}^+$, one finds
$$\frac{\partial}{\partial \theta} f(\sigma, x) = \frac{\partial \sigma}{\partial \theta}\frac{\partial f}{\partial \sigma} + \frac{\partial x}{\partial \theta}\frac{\partial f}{\partial x} \,. \tag{7.10}$$
The verification is simple since F is necessarily an affine function of σ, and σ and x are affine functions of θ.

Note that in the second term on the r.h.s., the ordering of factors is important.

Clifford algebra. The nilpotent differentiation operators $\partial/\partial\theta_i$, combined with the generators θ_i considered as operators acting on $\mathfrak{A}$ by left multiplication, generate an operator algebra whose generators satisfy the anti-commutation relations
$$\theta_i\theta_j + \theta_j\theta_i = 0\,, \quad \frac{\partial}{\partial \theta_i}\frac{\partial}{\partial \theta_j} + \frac{\partial}{\partial \theta_j}\frac{\partial}{\partial \theta_i} = 0,\ \theta_i\frac{\partial}{\partial \theta_j} + \frac{\partial}{\partial \theta_j}\theta_i = \delta_{ij}\,. \tag{7.11}$$
The operators
$$D_i^\pm = \frac{\partial}{\partial \theta_i} \pm \theta_i$$
then satisfy the anti-commutation relations
$$\{D_i^\pm, D_j^\pm\} = \pm 2\delta_{ij}\,, \quad \{D_i^+, D_j^-\} = 0\,.$$
This representation shows that the algebra is the direct sum of two Clifford algebras.

7.3 Integration in Grassmann algebras

The integration over Grassmann variables, which we denote by the integration symbol, is defined to be an operation identical to differentiation:

$$\int \mathrm{d}\theta_i\, A \equiv \frac{\partial}{\partial \theta_i} A, \quad \forall A \in \mathfrak{A}. \tag{7.12}$$

One may, thus, wonder whether it is really useful to introduce two symbols, integral and derivative, for one unique operation. One first verifies that this operation has also the formal properties that one expects from integration in the case of *definite integrals*:

(i) The integral of a total derivative vanishes, a property that legitimates integration by parts.

(ii) After integration over a variable, an expression does not depend on this variable anymore.

(iii) A factor in a product that does not depend on the integration variable can be factorized in front of the integration sign.

Then, the choice of using the integration or differentiation symbol depends on the context, and allows constructing for fermions a formalism quite parallel to the formalism for bosons described in Chapter 6, as we will show.

Change of variables. We consider the integral

$$\int \mathrm{d}\theta\, f(\theta), \tag{7.13}$$

and change variables, setting (the change of variables is necessarily affine)

$$\theta = \theta' A + B. \tag{7.14}$$

We demand that parity, in the sense of the reflection (7.3), is conserved: since $\mathrm{P}(\theta) = -\theta$,

$$\mathrm{P}(\theta') = -\theta' \Leftrightarrow \theta' \in \mathfrak{A}^- \Rightarrow A \in \mathfrak{A}^+,\ B \in \mathfrak{A}^-.$$

Moreover, the element A must be invertible and, thus, its term of degree zero in the Grassmann variables must be non-vanishing. These conditions, in fact, imply that θ and θ' are two equivalent generators in the algebra.

Then, using the definition (7.12), one finds

$$\int \mathrm{d}\theta\, f(\theta) = A^{-1} \int \mathrm{d}\theta'\, f(\theta' A + B) = \left(\frac{\partial \theta}{\partial \theta'}\right)^{-1} \int \mathrm{d}\theta'\, f\bigl(\theta(\theta')\bigr),$$

where the latter form is independent of the special parametrization (7.14). This is a very important property of Grassmann integrals: the jacobian is $(\partial \theta / \partial \theta')^{-1}$, instead of $\partial \theta / \partial \theta'$ in the case of real or complex variables. This difference is also a reflection of the identity between differentiation and integration in Grassmann algebras.

Generalization. More generally, we now show that the change of variables

$$\theta_i = \theta_i(\theta'),\ \theta'_i \in \mathfrak{A}^-,$$

where the matrix $\partial \theta_i / \partial \theta'_j$ is invertible (which is equivalent to the invertibility of the matrix of the terms of degree zero), generates a jacobian that is the *inverse* of the determinant of $\partial \theta_i / \partial \theta'_j$:

$$d\theta_1 \ldots d\theta_n = d\theta'_1 \ldots d\theta'_n J(\theta') \tag{7.15}$$

with

$$J^{-1} = \det \frac{\partial \theta_i}{\partial \theta'_j}. \tag{7.16}$$

Notice that the determinant is defined because all elements of the matrix $\partial \theta_i / \partial \theta'_j$ belong to the commutative sub-algebra $\mathfrak{A}^+$.

The result can be derived by exploiting the identity between differentiation and integration in the form

$$\int d\theta_1 \ldots d\theta_n f(\boldsymbol{\theta}) \equiv \prod_i \frac{\partial}{\partial \theta_i} f(\boldsymbol{\theta}),$$

where the product in the l.h.s. is ordered. We then consider f as a function of the variables θ'_i and, thus, using chain rule (see equation (7.10))

$$\prod_i \frac{\partial}{\partial \theta_i} f(\boldsymbol{\theta}) = \prod_i \frac{\partial \theta'_{j_i}}{\partial \theta_i} \frac{\partial}{\partial \theta'_{j_i}} f(\boldsymbol{\theta}).$$

The elements $\partial \theta'_{j_i} / \partial \theta_i$ commute and can be factorized. The differentiations $\partial / \partial \theta'_{j_i}$ anti-commute (see equations (7.11)) and, thus, all products are proportional to the product ordered from 1 to n. A sign is generated, which is the signature of the permutation $j_1, j_2, \ldots, j_n$. One then recognizes the determinant of the matrix $\partial \theta'_j / \partial \theta_i$:

$$\prod_i \frac{\partial}{\partial \theta_i} = \det \frac{\partial \theta'_j}{\partial \theta_i} \prod_k \frac{\partial}{\partial \theta'_k}.$$

The identity between differentiation and integration then leads immediately to equations (7.15, 7.16).

Remark. The following example allows a direct verification of equation (7.15):

$$1 = \int d\theta_1 \ldots d\theta_n \, \theta_n \ldots \theta_1.$$

After the linear change of variables

$$\theta_i = \sum_j a_{ij} \theta'_j,$$

the result then relies on the identity

$$\theta_n \ldots \theta_1 = \theta'_n \ldots \theta'_1 \det \mathbf{a}.$$

7.3.1 Integration and complex conjugation

In what follows, we consider algebras with a double family of generators $\{\theta_i, \bar{\theta}_i\}$, $1 = 1, \ldots, n$, related by the complex conjugation defined in (7.7). In these algebras, one considers integrals of the form

$$I = \int d\theta d\bar{\theta}\, f(\theta, \bar{\theta}),$$

where the pair $\theta, \bar{\theta}$ stands for any pair of conjugate generators, which are the direct analogues of the complex integrals of Chapter 6.

If we render the dependence of the function f on θ and $\bar{\theta}$ explicit:

$$f = a_0 + \theta a_1 + \bar{\theta} b_1 + \bar{\theta}\theta a_2,$$

where the coefficients belong to the algebra, we can integrate and find $I = a_2$. We now integrate the complex conjugate function:

$$J = \int d\theta d\bar{\theta}\, f^\dagger(\theta, \bar{\theta}).$$

With the same parametrization

$$f^\dagger(\theta, \bar{\theta}) = a_0^\dagger + a_1^\dagger \bar{\theta} + b_1^\dagger \theta + a_2^\dagger \bar{\theta}\theta,$$

and, thus, since $\bar{\theta}\theta$ commutes with $a_2^\dagger$, $J = a_2^\dagger = I^\dagger$. The integral of the conjugate function is the conjugate of the integral of the function. One can thus consider the measure $d\theta_i d\bar{\theta}_i$ as being invariant under complex conjugation.

In particular, if a function is self-conjugate, the same applies to the integral.

In what follows, we also meet integrals of the form

$$I = \int d\theta d\bar{\theta}\, e^{\bar{\theta}\theta}\, f(\theta, \bar{\theta}) = a_0 + a_2.$$

Again, the substitution $f \mapsto f^\dagger$ leads to the conjugate result.

7.4 Gaussian integrals and perturbative expansion

We now define gaussian integrals with an integration over two families of generators $\{\theta_i, \bar{\theta}_i\}$, $1 = 1, \ldots, n$, analogues of the complex gaussian integrals of Section 6.1.

7.4.1 Gaussian integrals

As in the case of complex variables, we first calculate gaussian integrals, and for the same reason: one often tries to reduce any integral to a formal sum of a finite or an infinite number of gaussian integrals.

7.4 Path integrals: fermions

We first consider the integral

$$\mathcal{Z}(\mathbf{M}) = \int \mathrm{d}\theta_1 \mathrm{d}\bar\theta_1 \mathrm{d}\theta_2 \mathrm{d}\bar\theta_2 \ldots \mathrm{d}\theta_n \mathrm{d}\bar\theta_n \, \exp\left(\sum_{i,j=1}^n \bar\theta_i M_{ij} \theta_j\right). \tag{7.17}$$

According to the rules of Grassmann integration, the result is simply the coefficient of the product $\bar\theta_n \theta_n \ldots \bar\theta_1 \theta_1$ in the expansion of the integrand. The argument of the exponential function contains only terms belonging to $\mathfrak{A}^+$, which commute. The integrand can thus be written as

$$\exp\left(\sum_{i,j=1}^n \bar\theta_i M_{ij} \theta_j\right) = \prod_{i=1}^n \exp\left(\bar\theta_i \sum_{j=1}^n M_{ij}\theta_j\right)$$

$$= \prod_{i=1}^n \left(1 + \bar\theta_i \sum_{j_i=1}^n M_{ij_i}\theta_{j_i}\right).$$

Expanding the product, one observes that in each factor only the term proportional to $\bar\theta_i$ contributes to the integral. It thus remains to integrate

$$\prod_{i=1}^n \bar\theta_i \left(\sum_{j_i=1}^n M_{ij_i}\theta_{j_i}\right).$$

The terms that give a non-vanishing contribution to the integral, are those that contain the product $\theta_n \ldots \theta_2 \theta_1$ up to a permutation of the factors θ_j. They have the form

$$\sum_{\substack{\text{permutations} \\ \{j_1 \ldots j_n\}}} M_{n j_n} M_{n-1\, j_{n-1}} \ldots M_{1 j_1} \bar\theta_n \theta_{j_n} \ldots \bar\theta_1 \theta_{j_1}.$$

A commutation of the generators to cast all products into some standard order, for example $\bar\theta_n \theta_n \ldots \bar\theta_1 \theta_1$, yields a sign, the signature of the permutation, and one then recognizes in the coefficient the determinant of the matrix $\mathbf{M}$. Thus,

$$\mathcal{Z}(\mathbf{M}) = \det \mathbf{M}. \tag{7.18}$$

This result is the inverse of the result (6.4), obtained by an integration over complex variables.

This calculation is mainly a verification since, for $\det \mathbf{M} \ne 0$, one can also change variables,

$$\theta_i \mapsto \theta'_i = \sum_j M_{ij}\theta_j \tag{7.19}$$

and use the form (7.15) of the jacobian. One verifies

$$\mathcal{Z}(\mathbf{M}) = \det \mathbf{M} \int d\theta'_1 d\bar{\theta}_1 \ldots d\theta'_n d\bar{\theta}_n \exp\left(\sum_{i=1}^{n} \bar{\theta}_i \theta'_i\right)$$

$$= \det \mathbf{M} \int \prod_{i=1}^{n} d\theta'_i d\bar{\theta}_i \left(1 + \bar{\theta}_i \theta'_i\right) = \det \mathbf{M}.$$

Hermitian quadratic form. With the definition (7.7) of complex conjugation, θ_i and $\bar{\theta}_i$ are conjugate. Moreover, the conjugate of a quadratic form is

$$\sum_{i,j=1}^{n} (\bar{\theta}_i M_{ij} \theta_j)^\dagger = \sum_{i,j=1}^{n} \bar{\theta}_j \bar{M}_{ij} \theta_i = \sum_{i,j=1}^{n} \bar{\theta}_i M^\dagger_{ij} \theta_j.$$

If the matrix $\mathbf{M}$ is hermitian, the quadratic form is invariant under complex conjugation. Then, the result of the integral is real since

$$\det \mathbf{M} = \det \mathbf{M}^\dagger = \overline{\det \mathbf{M}},$$

in agreement with the discussion of Section 7.3.1.

7.4.2 General gaussian integrals

We now introduce another copy of the Grassmann algebra $\mathfrak{A}$ whose generators will be denoted by η_i and $\bar{\eta}_i$, and consider the Grassmann algebra generated by the set $\{\theta, \bar{\theta}, \eta, \bar{\eta}\}$. Adapting the strategy of Section 6.1, we first evaluate the integral

$$\mathcal{Z}_{\mathrm{G}}(\eta, \bar{\eta}) = \int \prod_i d\theta_i d\bar{\theta}_i \exp E_{\mathrm{G}}(\theta, \bar{\theta}, \eta, \bar{\eta}) \qquad (7.20)$$

with

$$E_{\mathrm{G}}(\theta, \bar{\theta}, \eta, \bar{\eta}) = \sum_{i,j=1}^{n} \bar{\theta}_i M_{ij} \theta_j + \sum_{i=1}^{n} (\bar{\eta}_i \theta_i + \bar{\theta}_i \eta_i), \qquad (7.21)$$

where E_{G}, thus, is an element of the direct sum of the two copies of the initial Grassmann algebra. Moreover, we assume $\det \mathbf{M} \neq 0$.

To eliminate the terms linear in θ and $\bar{\theta}$, we solve the equations

$$\frac{\partial E_{\mathrm{G}}}{\partial \theta_i} = 0, \qquad \frac{\partial E_{\mathrm{G}}}{\partial \bar{\theta}_i} = 0.$$

Introducing the inverse matrix

$$\boldsymbol{\Delta} = \mathbf{M}^{-1},$$

one can write the solutions $\theta^{\mathrm{s}}, \bar{\theta}^{\mathrm{s}}$ as

$$\theta^{\mathrm{s}}_i = -\sum_j \Delta_{ij} \eta_j, \quad \bar{\theta}^{\mathrm{s}}_i = -\sum_j \bar{\eta}_j \Delta_{ji}.$$

As in the case of usual integrals, one then change variables, $\{\theta_i\} \mapsto \{\theta'_i\}$, setting

$$\theta_i = \theta'_i - \sum_j \Delta_{ij}\eta_j, \quad \bar{\theta}_i = \bar{\theta}'_i - \sum_j \bar{\eta}_j \Delta_{ji}. \tag{7.22}$$

After this translation, the integral over $\theta', \bar{\theta}'$ takes the form (7.17) calculated before (equation (7.18)). The result is

$$\mathcal{Z}_G(\eta, \bar{\eta}) = \det \mathbf{M} \, \exp\left(-\sum_{i,j=1}^n \bar{\eta}_i \Delta_{ij} \eta_j\right). \tag{7.23}$$

7.4.3 Gaussian expectation values, Wick's theorem

We denote by $\langle \bullet \rangle_\eta$ expectation values with respect to the gaussian weight corresponding to the integrand (7.20). From the definitions (7.20, 7.21), it follows that

$$\langle \theta_i \rangle_\eta = \mathcal{Z}_G^{-1} \frac{\partial}{\partial \bar{\eta}_i} \mathcal{Z}_G, \tag{7.24}$$

$$\langle \bar{\theta}_i \rangle_\eta = -\mathcal{Z}_G^{-1} \frac{\partial}{\partial \eta_i} \mathcal{Z}_G. \tag{7.25}$$

Note the sign in equation (7.25). Another useful identity is obtained by differentiating twice (note the order):

$$\langle \bar{\theta}_i \theta_j \rangle_\eta = \mathcal{Z}_G^{-1} \frac{\partial}{\partial \bar{\eta}_j} \frac{\partial}{\partial \eta_i} \mathcal{Z}_G - \left(\mathcal{Z}_G^{-1} \frac{\partial}{\partial \bar{\eta}_j} \mathcal{Z}_G\right)\left(\mathcal{Z}_G^{-1} \frac{\partial}{\partial \eta_i} \mathcal{Z}_G\right). \tag{7.26}$$

More generally, differentiating with respect to θ and $\bar{\theta}$, one proves a version of Wick' theorem for Grassmann variables.

Wick' theorem. We now define

$$\det \mathbf{M} \, \langle \bar{\theta}_{i_1} \theta_{j_1} \bar{\theta}_{i_2} \theta_{j_2} \ldots \bar{\theta}_{i_p} \theta_{j_p} \rangle$$
$$= \int \left(\prod_i d\theta_i d\bar{\theta}_i\right) \bar{\theta}_{i_1} \theta_{j_1} \ldots \bar{\theta}_{i_p} \theta_{j_p} \exp\left(\sum_{i,j=1}^n M_{ij} \bar{\theta}_i \theta_j\right) \tag{7.27}$$

with $p \leq n$. We can restrict the calculation to integrals with an equal number of factors θ and $\bar{\theta}$ since the other integrals vanish trivially.

First, equation (7.26) in the limit $\eta = \bar{\eta} = 0$ yields

$$\langle \bar{\theta}_i \theta_j \rangle = \Delta_{ji}.$$

Then, repeatedly differentiating the integral (7.20) with respect to η and $\bar{\eta}$ and using the remarks (7.24-7.26), one obtains the identity

$$\det \mathbf{M} \, \langle \bar{\theta}_{i_1} \theta_{j_1} \bar{\theta}_{i_2} \theta_{j_2} \ldots \bar{\theta}_{i_p} \theta_{j_p} \rangle = \left[\frac{\partial}{\partial \bar{\eta}_{j_1}} \frac{\partial}{\partial \eta_{i_1}} \cdots \frac{\partial}{\partial \bar{\eta}_{j_p}} \frac{\partial}{\partial \eta_{i_p}} \mathcal{Z}_G(\eta, \bar{\eta})\right]\bigg|_{\eta=\bar{\eta}=0}. \tag{7.28}$$

Replacing $\mathcal{Z}_G$ by the explicit result (7.23), one finds

$$\langle \bar\theta_{i_1}\theta_{j_1}\bar\theta_{i_2}\theta_{j_2}\ldots\bar\theta_{i_p}\theta_{j_p}\rangle$$
$$= \left\{ \frac{\partial}{\partial\bar\eta_{j_1}}\frac{\partial}{\partial\eta_{i_1}}\cdots\frac{\partial}{\partial\bar\eta_{j_p}}\frac{\partial}{\partial\eta_{i_p}} \exp\left[-\sum_{i,j=1}^n \bar\eta_j \Delta_{ji}\eta_i\right]\right\}\bigg|_{\eta=\bar\eta=0}. \quad (7.29)$$

All variables η and $\bar\eta$ which are not differentiated can immediately be suppressed. The matrix $\boldsymbol{\Delta}$ is then reduced to a $p\times p$ matrix with elements $\Delta_{j_l i_k}$. The identity between differentiation and integration then allows reducing the explicit calculation to a gaussian integration. One concludes

$$\langle\bar\theta_{i_1}\theta_{j_1}\ldots\bar\theta_{i_p}\theta_{j_p}\rangle = \det \Delta_{j_l i_k}$$
$$= \sum_{\substack{\text{permutations}\\ P\text{ of }\{j_1\ldots j_p\}}} \epsilon(P)\Delta_{j_{P_1}i_1}\Delta_{j_{P_2}i_2}\cdots\Delta_{j_{P_p}i_p},$$
$$= \sum_{\substack{\text{permutations}\\ P\text{ of }\{j_1\ldots j_p\}}} \epsilon(P)\langle\bar\theta_{i_1}\theta_{j_{P_1}}\rangle\langle\bar\theta_{i_2}\theta_{j_{P_2}}\rangle\cdots\langle\bar\theta_{i_p}\theta_{j_{P_p}}\rangle, \quad (7.30)$$

where $\epsilon(P)=\pm 1$ is the signature of the permutation P.

This result, which is the form Wick's theorem assumes in the case of 'complex' Grassmann variables, differs from expression (6.9a), obtained in the case of usual complex variables, only by *the signature*.

7.4.4 Perturbative expansion

To calculate expectation values with the weight $e^{E(\theta,\bar\theta)}/\mathcal{Z}$, where

$$E(\theta,\bar\theta) = \sum_{i,j=1}^n M_{ij}\bar\theta_i\theta_j + V(\bar\theta,\theta), \quad (7.31)$$

and the normalization $\mathcal{Z}$ is given by the integral

$$\mathcal{Z} = \int \prod_i d\bar\theta_i d\theta_i \; e^{E(\theta,\bar\theta)}, \quad (7.32)$$

one can expand in powers of the polynomial V and then calculate gaussian expectation values using Wick's theorem in the form (7.30).

Example. We consider the example

$$E(\theta,\bar\theta) = \sum_{i,j=1}^n M_{ij}\bar\theta_i\theta_j + \frac{1}{2}\sum_{i,j} V_{ij}\bar\theta_i\theta_i\bar\theta_j\theta_j,$$

where $V_{ij} = V_{ji}$. The first terms of the expansion in powers of V of $\ln \mathcal{Z}$, where $\mathcal{Z}$ is the normalization integral (7.32), has the form

$$\ln \mathcal{Z} - \ln \det \mathbf{M} = \tfrac{1}{2} \sum_{i,j} V_{ij} \left\langle \bar{\theta}_i \theta_i \bar{\theta}_j \theta_j \right\rangle_{0,c} + \tfrac{1}{8} \sum_{i,j,k,l} V_{ij} V_{kl} \left\langle \bar{\theta}_i \theta_i \bar{\theta}_j \theta_j \bar{\theta}_k \theta_k \bar{\theta}_l \theta_l \right\rangle_{0,c}$$
$$+ O(V^3),$$

where $\langle \bullet \rangle_{0,c}$ means connected gaussian expectation value. Using Wick's theorem, one obtains

$$\ln \mathcal{Z} - \ln \det \mathbf{M} = \tfrac{1}{2} \sum_{i,j} V_{ij} \left(\Delta_{ii} \Delta_{jj} - \Delta_{ji} \Delta_{ij} \right)$$
$$+ \tfrac{1}{4} \sum_{i,j,k,l} V_{ij} V_{kl} \left(-2 \Delta_{ki} \Delta_{ik} \Delta_{jj} \Delta_{ll} + 4 \Delta_{ki} \Delta_{il} \Delta_{lk} \Delta_{jj} - 2 \Delta_{ki} \Delta_{ij} \Delta_{jl} \Delta_{lk} \right.$$
$$\left. + \Delta_{ik} \Delta_{ki} \Delta_{lj} \Delta_{jl} - \Delta_{ki} \Delta_{il} \Delta_{lj} \Delta_{jk} \right) + O(V^3),$$

where the signs have an interpretation in terms of the parity of the number of fermion loops in Feynman diagrams (see Fig. 7.1).

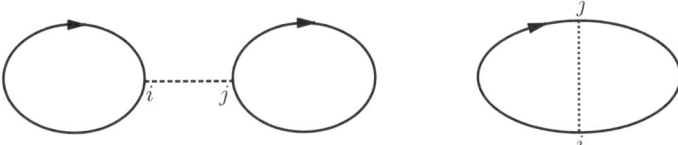

Fig. 7.1 Faithful Feynman diagrams: the contributions of order V, full lines corresponding to Δ (fermions), dotted lines to V.

Two-point expectation value. At order V^2,

$$\mathcal{Z} \left\langle \bar{\theta}_k \theta_\ell \right\rangle = \Delta_{\ell k} + \tfrac{1}{2} \sum_{i,j} V_{ij} \left\langle \bar{\theta}_k \theta_\ell \bar{\theta}_i \theta_i \bar{\theta}_j \theta_j \right\rangle_0$$
$$+ \tfrac{1}{8} \sum_{i,j,a,b} V_{ij} V_{ab} \left\langle \bar{\theta}_k \theta_\ell \bar{\theta}_i \theta_i \bar{\theta}_j \theta_j \bar{\theta}_a \theta_a \bar{\theta}_b \theta_b \right\rangle_0 + O(V^3),$$

where $\langle \bullet \rangle_0$ denotes gaussian expectation values.

Application of Wick's theorem, after cancellation of the normalization contributions, then leads to (the order V is displayed in Fig. 7.2)

$$\left\langle \bar{\theta}_k \theta_\ell \right\rangle = \Delta_{\ell k} + \sum_{i,j} V_{ij} \left(\Delta_{\ell j} \Delta_{ji} \Delta_{ik} - \Delta_{\ell i} \Delta_{ik} \Delta_{jj} \right)$$
$$+ \sum_{i,j,a,b} \left(\Delta_{\ell j} V_{jb} \Delta_{bb} \Delta_{ji} V_{ia} \Delta_{aa} \Delta_{ik} - \Delta_{\ell j} V_{jb} \Delta_{jb} \Delta_{ja} V_{ia} \Delta_{ai} \Delta_{ik} \right.$$
$$- \Delta_{\ell j} V_{jb} \Delta_{jb} \Delta_{bi} V_{ia} \Delta_{aa} \Delta_{ik} + \Delta_{\ell j} V_{jb} \Delta_{jb} \Delta_{ba} V_{ai} \Delta_{ai} \Delta_{ik}$$
$$+ \Delta_{\ell i} V_{ia} \Delta_{aj} V_{jb} \Delta_{bb} \Delta_{ja} \Delta_{ik} - \Delta_{\ell j} V_{ji} \Delta_{ja} V_{ab} \Delta_{bb} \Delta_{ai} \Delta_{ik}$$
$$- \Delta_{\ell i} V_{ia} \Delta_{aj} V_{jb} \Delta_{jb} \Delta_{ba} \Delta_{ik} + \Delta_{\ell j} V_{ji} \Delta_{ja} V_{ab} \Delta_{ab} \Delta_{bi} \Delta_{ik}$$
$$\left. - \Delta_{\ell j} V_{jb} \Delta_{ab} \Delta_{ba} V_{ai} \Delta_{ji} \Delta_{ik} + \Delta_{\ell j} V_{ja} \Delta_{jb} V_{bi} \Delta_{ba} \Delta_{ai} \Delta_{ik} \right) + O(V^3).$$

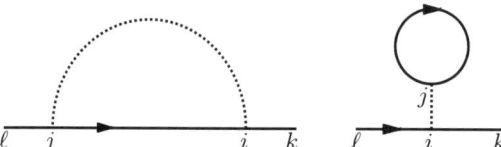

Fig. 7.2 Faithful Feynman diagrams: the contributions of order V, full lines corresponding to Δ (fermions), dotted lines to V.

Generating function. More general expectation values can also be inferred from the generating function

$$\mathcal{Z}(\eta,\bar{\eta}) = \int \left(\prod_i \mathrm{d}\bar{\theta}_i \mathrm{d}\theta_i\right) \mathrm{e}^{E(\theta,\bar{\theta},\bar{\eta},\eta)} \tag{7.33}$$

with

$$E(\theta,\bar{\theta},\bar{\eta},\eta) = \sum_{i,j=1}^{n} M_{ij}\bar{\theta}_i\theta_j + V(\bar{\theta},\theta) + \sum_{i=1}^{n} \left(\bar{\eta}_i\theta_i + \bar{\theta}_i\eta_i\right). \tag{7.34}$$

An expansion in a power series in V reduces its calculation to gaussian expectation values. Using the results (7.24, 7.25), one shows that the perturbative expansion then has the compact formal expression

$$\mathcal{Z}(\eta,\bar{\eta}) = \exp\left[V\left(-\frac{\partial}{\partial\eta},\frac{\partial}{\partial\bar{\eta}}\right)\right]\mathcal{Z}_{\mathrm{G}}(\eta,\bar{\eta}). \tag{7.35}$$

7.5 Fermion vector space and operators

The discussion of fermion systems by the Grassmann formalism that we have just described is quite parallel to the discussion of bosons in terms of the holomorphic representation. For example, we have shown in Section 6.1 how to define a scalar product between analytic functions (equation (6.11)). This scalar product then allows constructing a Hilbert space of analytic functions. From the quantum viewpoint, the coefficients of the Taylor series expansion correspond to the components of the wave function on states with a given number of particles (see Section 6.6). Quantum operators are then represented by multiplications and differentiations acting on these analytic functions.

Following an analogous scheme, we first define Grassmann 'analytic' functions. These functions form a vector space in which we define a scalar product. Quantum operators are then represented by elements of the algebra of the operators (7.11) acting on these functions.

7.5.1 Grassmann analytic functions and scalar product: one state

We first consider a Grassmann algebra $\mathfrak{A}$ with two generators $\theta, \bar\theta$. We define a Grassmann analytic function as an element of the algebra that depends only on the variable θ:
$$\frac{\partial f}{\partial \bar\theta} = 0.$$

Grassmann analytic functions form a subalgebra $\mathfrak{A}_{\text{an.}}$ of the algebra $\mathfrak{A}$. Analytic functions $f(\theta)$ are automatically affine functions:
$$f(\theta) = f_0 + f_1 \theta, \quad f_0, f_1 \in \mathbb{C}^2.$$

These functions span a complex two-dimensional vector space isomorphic to the space of the components (f_0, f_1). The components correspond directly to the two possible values 0 and 1 allowed by Fermi–Dirac statistics for the occupation number of one quantum state.

The function complex conjugate to the function f is (see equation (7.7))
$$\bar f(\bar\theta) \equiv \bar f_0 + \bar f_1 \bar\theta.$$

As for analytic functions of usual complex variables, one then defines a scalar product between functions of θ:
$$(f, g) = \int d\theta d\bar\theta \, e^{\bar\theta\theta} \, \overline{f(\theta)} \, g(\theta). \tag{7.36}$$

Parametrizing the function g as
$$g(\theta) = g_0 + g_1 \theta,$$

one verifies that the integral leads to the usual scalar product of the two corresponding complex vectors (f_0, f_1) and (g_0, g_1):
$$(f, g) = \bar f_0 g_0 + \bar f_1 g_1.$$

Note, for later purpose, that the same scalar product has an equivalent form:
$$(f, g) = \int d\bar\theta d\theta \, e^{\theta\bar\theta} \, g(\theta) \, \overline{f(\theta)}, \tag{7.37}$$

as one verifies by explicit calculation.

Operators and hermitian conjugation. We consider the scalar product
$$(f, \partial g/\partial \theta) = \int d\theta d\bar\theta \, e^{\bar\theta\theta} \, \overline{f(\theta)} \, \frac{\partial g(\theta)}{\partial \theta}. \tag{7.38}$$

Using the identity
$$\frac{\partial}{\partial \theta} e^{\bar\theta\theta} g(\theta) = e^{\bar\theta\theta} \left(\frac{\partial g(\theta)}{\partial \theta} - \bar\theta g \right),$$

and integrating by parts, one infers (the same method has been used in Section 6.2.2)

$$(f, \partial g/\partial \theta) = (\theta f, g),$$

which shows that the operators θ and $\partial/\partial \theta$ are hermitian conjugate.

Dirac's δ function. In Grassmann algebras, the role of Dirac's δ function is played by the function

$$\delta(\theta) \equiv \theta. \tag{7.39}$$

Indeed,

$$\int d\theta\, \theta f(\theta) = f(0),$$

where $f(0)$ means the constant part of the affine function $f(\theta)$. This δ function has a very useful integral representation:

$$\delta(\theta) = \int d\bar{\theta}\, e^{\bar{\theta}\theta}, \tag{7.40}$$

where $\bar{\theta}$ is an additional variable and, thus, an additional generator of the Grassmann algebra. This representation is analogous to the Fourier representation of the usual δ function. One verifies directly

$$f(0) = \int d\theta d\bar{\theta}\, e^{\bar{\theta}\theta}\, f(\theta).$$

7.5.2 General Grassmann algebra

The generalization to an arbitrary number n of generators θ_i is straightforward. We now consider a Grassmann algebra $\mathfrak{A}$ with generators $\{\theta_i, \bar{\theta}_i\}$. We define Grassmann analytic functions as elements of the algebra that depend only on the variables θ_i:

$$\frac{\partial f}{\partial \bar{\theta}_i} = 0, \ \forall i.$$

They form a subalgebra $\mathfrak{A}_{\text{an.}}$ of the algebra $\mathfrak{A}$.

The complex conjugation is defined as in (7.7) (i.e. as the hermitian conjugation of operators). Then, the scalar product of two functions f and g is defined by

$$(f, g) = \int \left(\prod_i d\theta_i d\bar{\theta}_i \right) \exp\left(\sum_i \bar{\theta}_i \theta_i \right) f^\dagger(\theta)\, g(\theta). \tag{7.41}$$

Since

$$\left(f^\dagger(\theta)\, g(\theta)\right)^\dagger = g^\dagger(\theta)\, f(\theta),$$

it follows from the remarks of Section 7.3.1, that

$$\overline{(f, g)} = (g, f).$$

To prove that the scalar product defines a positive norm $\|f\|$, one can expand all elements of the algebra $\mathfrak{A}_{\mathrm{an.}}$, considered as a complex vector space, on the basis of the 2^n distinct monomials $\psi_\nu(\boldsymbol{\theta})$, $\nu = 1, \ldots, 2^n$:

$$\{\psi_\nu(\boldsymbol{\theta})\} = \{1, \theta_{i_1}\theta_{i_2}\ldots\theta_{i_p}, \ \forall\ 1 \le p \le n \text{ and } \forall\ i_1 < i_2 < \cdots < i_p\}. \tag{7.42}$$

One verifies immediately that $(\psi_\mu, \psi_\nu) \propto \delta_{\mu\nu}$. Moreover,

$$(\theta_1\theta_2\ldots\theta_p, \theta_1\theta_2\ldots\theta_p) = \int \left(\prod_{i=1}^p \mathrm{d}\theta_i \mathrm{d}\bar\theta_i\right) \bar\theta_p\ldots\bar\theta_2\bar\theta_1\theta_1\theta_2\ldots\theta_p = 1.$$

The monomials $\psi_\nu(\boldsymbol{\theta})$ thus form an orthonormal basis. As a consequence, if one expands

$$f = \sum_\nu f_\nu \psi_\nu, \quad g = \sum_\nu g_\nu \psi_\nu,$$

where f_ν, g_ν are two complex vectors, one finds

$$(f, g) = \sum_\nu \bar f_\nu g_\nu,$$

that is the usual scalar product of the two vectors.

Again, the same results can be derived from the other form of the scalar product, which generalizes the definition (7.37):

$$(f, g) = \int \left(\prod_i \mathrm{d}\bar\theta_i \mathrm{d}\theta_i\right) \exp\left(\sum_i \theta_i \bar\theta_i\right) g(\theta) f^\dagger(\bar\theta). \tag{7.43}$$

7.5.3 Operators and kernels

Kernel of the identity. The scalar product (7.41) allows defining an orthonormal basis and, thus, a representation of the identity operator in the form of a kernel:

$$\mathcal{I}(\theta, \bar\theta) = \prod_i (1 + \theta_i \bar\theta_i) = \exp\left(-\sum_i \bar\theta_i \theta_i\right). \tag{7.44}$$

A direct verification is based on the representation (7.40) of the δ function. Indeed,

$$\int \prod_i \mathrm{d}\theta'_i \mathrm{d}\bar\theta'_i \, \mathcal{I}(\theta, \bar\theta') \exp\left(\sum_i \bar\theta'_i \theta'_i\right) f(\theta')$$

$$= \int \prod_i \mathrm{d}\theta'_i \mathrm{d}\bar\theta'_i \exp\left(\sum_i \bar\theta'_i(\theta'_i - \theta_i)\right) f(\theta') = f(\theta). \tag{7.45}$$

Operator algebra. We have already introduced (see (7.11)) the algebra of left-multiplication and differentiation operators acting on a Grassmann algebra $\mathfrak{A}_{\mathrm{an.}}$.

We have shown in Section 7.5.1 that the operators θ_i and $\partial/\partial\theta_i$ are hermitian conjugate.

Using the anti-commutation relations, one can write all elements of this operator algebra as a linear combination of monomials in which all differential operators are on the right: the *normal order*. One can then use equation (7.45) to represent all elements by kernels

$$\theta_{i_1}\theta_{i_2}\ldots\theta_{i_p}\frac{\partial}{\partial\theta_{j_1}}\frac{\partial}{\partial\theta_{j_2}}\ldots\frac{\partial}{\partial\theta_{j_q}}\mathcal{I}(\theta,\bar\theta) = \theta_{i_1}\theta_{i_2}\ldots\theta_{i_p}\bar\theta_{j_1}\bar\theta_{j_1}\ldots\bar\theta_{j_q}\mathcal{I}(\theta,\bar\theta), \quad (7.46)$$

which are general elements of the Grassmann algebra $\mathfrak{A}$.

In this representation, the action of an operator $O(\theta,\partial/\partial\theta)$ with kernel

$$\mathcal{O}(\theta,\bar\theta) = O(\theta,\bar\theta)\mathcal{I}(\theta,\bar\theta),$$

is given by

$$[\mathcal{O}f](\theta) = \int \prod_i d\theta'_i d\bar\theta'_i\, \mathcal{O}(\theta,\bar\theta')\exp\left(\sum_i \bar\theta'_i\theta'_i\right) f(\theta'). \quad (7.47)$$

As in the holomorphic formalism, we introduce here also the rather suggestive matrix element notation

$$\langle\theta|\,\mathcal{O}\,|\bar\theta\rangle \equiv \mathcal{O}(\theta,\bar\theta), \quad (7.48)$$

without defining precisely the corresponding bra and ket vectors.

Finally, the kernel corresponding to the product $\mathcal{O}_2\mathcal{O}_1$ is given by

$$\langle\theta|\,\mathcal{O}_2\mathcal{O}_1\,|\bar\theta\rangle = \int \prod_i d\theta'_i d\bar\theta'_i\, \langle\theta|\,\mathcal{O}_2\,|\bar\theta'\rangle \exp\left(\sum_i \bar\theta'_i\theta'_i\right) \langle\theta'|\,\mathcal{O}_1\,|\bar\theta\rangle. \quad (7.49)$$

All operators can be expressed in terms of the elements of the basis (7.42):

$$\langle\theta|\,\mathcal{O}\,|\bar\theta\rangle = \sum_{\mu,\nu}\mathcal{O}_{\mu\nu}\psi_\mu(\theta)\psi_\nu^\dagger(\theta), \quad (7.50)$$

where the coefficients $\mathcal{O}_{\mu\nu}$ are the matrix elements of $\mathcal{O}$ in this basis.

Trace. In terms of its kernel, the trace of an operator reads

$$\mathrm{tr}\,\mathcal{O} = \int \prod_i d\bar\theta_i d\theta_i \exp\left(-\sum_i \bar\theta_i\theta_i\right) \langle\theta|\,\mathcal{O}\,|\bar\theta\rangle. \quad (7.51)$$

Comparing expressions (7.49) and (7.51), one might be surprised by the interchange between the $\bar\theta_i$ and θ_i. The reason for this change can be directly related to the second form (7.43) of the scalar product. Moreover, the consistency of the definition with the cyclic property of the trace, which is not obvious, can be directly verified by replacing in equation (7.49) $\mathcal{O}_1$ and $\mathcal{O}_2$ by two arbitrary monomials and taking the trace of the equation.

Finally, using the expression (7.50), one finds

$$\operatorname{tr} \mathcal{O} = \sum_{\nu} \mathcal{O}_{\nu\nu}.$$

Hermitian conjugation and kernels. Comparing explicitly the scalar products $(f, \mathcal{O}g)$ and $(\mathcal{O}^\dagger f, g)$, where the action of a operator on a function is given by expression (7.47), one verifies directly that the kernel corresponding to the hermitian conjugate of an operator is its complex conjugate in the algebra, as defined in (7.7):

$$\mathcal{O} \mapsto \mathcal{O}(\theta, \bar{\theta}) \;\Rightarrow\; \mathcal{O}^\dagger \mapsto \bar{\mathcal{O}}^\dagger(\theta, \bar{\theta}).$$

As expected, the identity kernel $\mathcal{I}(\theta, \bar{\theta})$ corresponds to a hermitian operator.

7.6 One-state hamiltonian

We now represent fermion hamiltonians and fermion statistical operators following the strategy introduce in Section 6.6 for bosons. A noticeable difference is that the Hilbert space for an arbitrary number of particles obeying the Fermi–Dirac statistics (i.e. fermions) and that can occupy only one state of energy ω is really only a complex two-dimensional vector space, as a direct consequence of the Pauli principle: a state can only be empty (state with zero particle or *vacuum*) or occupied once. Therefore, the sophisticated Grassmann formalism is not really necessary to deal with this problem. However, the situation changes drastically when fermions can occupy a large number of states, and this justifies the discussion that follows.

7.6.1 Quantum states and operators

As discussed in Section 7.5.1, a vector corresponding to a linear combination of an empty state and a one-particle state can be represented by an affine function:

$$\psi(\theta) = \psi_0 + \psi_1 \theta, \quad (\psi_0, \psi_1) \in \mathbb{C}^2,$$

where, by choosing θ to be the generator of a Grassmann algebra $\mathfrak{A}$, one ensures that a state can be occupied only once since $\theta^2 = 0$. We then normalize the functions ψ by the scalar product (7.36).

The eigenvalues of a hamiltonian H_0 that conserves the number of particles can only be 0 for an empty state and another value, which we denote by ω, for the one-particle state. It can thus be represented by (in this first part we set $\hbar = 1$)

$$H_0 = \omega \theta \frac{\partial}{\partial \theta}, \quad \omega > 0, \tag{7.52}$$

and this is the most general hamiltonian with only one fermion state. The condition $\omega > 0$ ensures that the ground state is an empty state. Note that H_0 is hermitian with respect to the scalar product (7.36).

One verifies that
$$H_0^2 = \omega H_0. \tag{7.53}$$

The statistical operator $U_0(t) = \mathrm{e}^{-H_0 t}$ can then be inferred from the relation (7.53), for example, by using Lagrange formula:
$$U_0(t) = \mathrm{e}^{-H_0 t} = \left[\omega \mathbf{1} + (\mathrm{e}^{-\omega t} - 1) H_0\right]/\omega. \tag{7.54}$$

Matrix elements. The matrix elements of the hamiltonian H_0 are obtained by acting with the differential operator (7.52) on the identity:
$$\langle \theta | H_0 | \bar\theta \rangle = \omega \theta \frac{\partial}{\partial \theta} \mathrm{e}^{-\bar\theta \theta} = \omega \theta \bar\theta \, \mathrm{e}^{-\bar\theta \theta} = -\omega \bar\theta \theta.$$

Note that the preservation of the fermion character implies that in the reflection (7.3)
$$P(\langle \theta | H_0 | \bar\theta \rangle) = \langle \theta | H_0 | \bar\theta \rangle ,$$
and, thus, the matrix elements of H_0, as well as those of $U_0(t)$, belong to the commutative subalgebra of $\mathfrak{A}$.

The matrix elements of the operator $U_0(t)$ satisfy the equation
$$\frac{\partial}{\partial t} \langle \theta | U_0(t) | \bar\theta \rangle = -\omega \theta \frac{\partial}{\partial \theta} \langle \theta | U_0(t) | \bar\theta \rangle \tag{7.55}$$

with
$$\langle \theta | U_0(0) | \bar\theta \rangle = \mathrm{e}^{-\bar\theta \theta}.$$

One verifies that the solution is
$$\langle \theta | U_0(t) | \bar\theta \rangle = \mathrm{e}^{-\bar\theta \theta \, \mathrm{e}^{-\omega t}} = 1 + \theta \bar\theta \, \mathrm{e}^{-\omega t}, \tag{7.56}$$

in agreement with expression (7.54).

Replacing in (7.49) $\mathcal{O}_1$ and $\mathcal{O}_2$ by $U(t_1)$ and $U(t_2)$, respectively, one can also verify directly the semi-group property.

Moreover, notice that the kernels corresponding to the hermitian operators H_0 and $U_0(t)$ are indeed invariant under complex conjugation (7.7).

Using the explicit expression (7.56) and the definition (7.51) of the trace, one can calculate the partition function. One finds the expected result
$$\mathcal{Z}_0(\beta) = \mathrm{tr}\, U_0(\beta) = \int \mathrm{d}\bar\theta \mathrm{d}\theta \, \mathrm{e}^{-\bar\theta \theta} \, \mathrm{e}^{-\bar\theta \theta \, \mathrm{e}^{-\omega}} = 1 + \mathrm{e}^{-\omega\beta}. \tag{7.57}$$

Remark. The action of the operator U_0 on a function $f(\theta)$ is given by
$$[U_0(t) f](\theta) = \int \mathrm{d}\theta' \mathrm{d}\bar\theta' \, \mathrm{e}^{-\bar\theta' \theta \, \mathrm{e}^{-\omega t}} \, \mathrm{e}^{\bar\theta' \theta'} f(\theta')$$
$$= f(\mathrm{e}^{-\omega t} \theta), \tag{7.58}$$

where the equation (7.40) has been used. Note the analogy with equation (6.30).

7.6.2 Annihilation and creation operators

As in the case of bosons, the operators θ and $\partial/\partial\theta$ form a representation of the algebra of the so-called fermion annihilation and creation operators a and $a^\dagger$, with the correspondence
$$a^\dagger \mapsto \theta \,, \ a \mapsto \frac{\partial}{\partial \theta}\,,$$
Indeed, they are hermitian conjugate and satisfy the commutation relations .
$$a^2 = a^{\dagger 2} = 0\,, \ aa^\dagger + a^\dagger a = 1\,, \tag{7.59}$$
which, clearly, encode the Pauli principle.

In terms of annihilation and creation operators, the hamiltonian (7.52) reads
$$H_0 = \omega a^\dagger a\,. \tag{7.60}$$
The correspondence between the normalized eigenvectors $|0\rangle$, the ground state, and the occupied state $|1\rangle = a^\dagger |0\rangle$ is then
$$|0\rangle \mapsto 1\,, \ |1\rangle \mapsto \theta\,.$$

Creation–annihilation symmetry. The relations (7.11) exhibit a complete symmetry between the hermitian conjugate operators θ_i and $\partial/\partial\theta_i$. Therefore, θ could also be associated with fermion annihilation and $\partial/\partial\theta$ with fermion creation. Then, θ would become the empty state and 1 the occupied state. This remark becomes specially relevant when the parameter ω in the hamiltonian (7.52) is negative. Then, the ground state is the vector θ. One generally prefers identifying the ground state with the vacuum, and assign positive energies to particle excitations.

In the more general framework of Section 7.7.1, the vacuum would then correspond to the product of all generators.

7.7 Many-particle states. Partition function

We now generalize the preceding construction to the situation in which one-particle states belong to a finite-dimensional vector space. As we have already pointed out, due to the Pauli principle, this assumption is much more drastic for fermions than for bosons. We first define fermion state vectors and the action of hamiltonians on these states. We then define a generating function of state vectors and determine the corresponding representation of hamiltonians.

7.7.1 Fermion states. Hamiltonians

One-particle states. A fermion state is defined by a vector, which we denote by ψ_i, which belongs to a complex vector space $\mathfrak{H}_1$ of finite dimension N.

Many-particle states. An n-particle state is then described by a vector $\psi_{i_1 i_2 \ldots i_n}$, where the indices i_k take N values. The Pauli principle for fermions implies that the vector $\psi_{i_1 i_2 \ldots i_n}$ is antisymmetric in all permutations of the indices
$$\psi_{i_1 i_2 \ldots i_k i_{k+1} \ldots i_n} = -\psi_{i_1 i_2 \ldots i_{k+1} i_k \ldots i_n} \quad \forall k\,.$$

The vectors $\psi_{i_1 i_2 \ldots i_n}$ are, thus, antisymmetric tensors with n indices, and belong to a complex vector space $\mathfrak{H}_n$ of dimension $\binom{N-n+1}{n}$.

Independent particle hamiltonian. A one-particle (or one-body) hamiltonian $\mathbf{H}^{(1)}$ is defined by its action on a one-particle state: it is then represented by a hermitian $N \times N$ matrix $H^{(1)}_{ij}$, which can be diagonalized. We denote by ω_i its eigenvalues ($\omega_i > 0$). Then,

$$[\mathbf{H}^{(1)} \psi]_i = \omega_i \psi_i .$$

Its action on n-particle states is additive:

$$[\mathbf{H}^{(1)} \psi]_{i_1 i_2 \ldots i_n} = \sum_\ell \omega_{i_\ell} \psi_{i_1 i_2 \ldots i_n} .$$

When the complete hamiltonian reduces to a one-particle hamiltonian, the fermions do not interact: one then speaks of independent particles.

Pair interaction. A pair or two-body interaction $\mathbf{H}^{(2)}$ is defined by its action on a two-particle state:

$$[\mathbf{H}^{(2)} \psi]_{i_1 i_2} = \sum_{j_1, j_2} H^{(2)}_{i_1 i_2, j_1 j_2} \psi_{j_1 j_2} ,$$

where $H^{(2)}_{i_1 i_2, j_1 j_2}$ is a hermitian matrix that satisfies

$$H^{(2)}_{i_1 i_2, j_1 j_2} = H^{(2)}_{i_2 i_1, j_2 j_1} = \bar{H}^{(2)}_{j_j j_2, i_1 i_2} ,$$

and, thus, is an internal mapping in the vector space $\mathfrak{H}_2$ of antisymmetric tensors. It is, of course, possible following the same strategy to define many-particle interactions but we restrict the discussion here to the two-body interaction, for simplicity.

The action of $\mathbf{H}^{(2)}$ on an n-particle state then is given by

$$[\mathbf{H}^{(2)} \psi]_{i_1 i_2 \ldots i_n} = \frac{1}{2} \sum_{\ell \neq m} \sum_{j,k} H^{(2)}_{i_\ell i_m, jk} \psi_{i_1 i_2 \ldots i_{\ell-1} j i_{\ell+1} \ldots i_{m-1} k i_{m+1} \ldots i_n} .$$

7.7.2 Generating function of state vectors

We now consider the set of state vectors corresponding to an arbitrary number of particles that belong to the space $\oplus_n \mathfrak{H}_n$, $n = 0, 1, \ldots$, and associate to them a generating function. The tensors $\psi_{i_1 i_2 \ldots i_n}$ being antisymmetric, we must introduce a Grassmann algebra with N generators θ_i. A generating function of state vectors then has the form

$$\Psi(\theta) = \sum_{n=0}^{N} \frac{1}{n!} \sum_{i_1, i_2, \ldots, i_n} \theta_{i_1} \theta_{i_2} \ldots \theta_{i_n} \psi_{i_1 i_2 \ldots i_n} .$$

With our assumptions, $\Psi(\theta)$ is a polynomial of degree N. Notice, conversely, that the function $\Psi(\theta)$ can only generate antisymmetric tensors.

The function $\Psi(\theta)$ can be considered as a Grassmann analytic function. Such functions form a vector space that can be endowed with the scalar product (7.41).

As in the Bose case, we note that

$$\sum_j \theta_j \omega_j \frac{\partial \Psi(\theta)}{\partial \theta_j} = \sum_n \frac{1}{(n-1)!} \sum_j \theta_j \omega_j \sum_{i_2,i_3,\ldots,i_n} \theta_{i_2} \theta_{i_3} \ldots \theta_{i_n} \psi_{j i_2 \ldots i_n}$$

$$= \sum_{n=1}^{\infty} \frac{1}{n!} \sum_{i_1,i_2,\ldots,i_n} \theta_{i_1} \theta_{i_2} \ldots \theta_{i_n} \sum_\ell \omega_\ell \psi_{i_1 i_2 \ldots i_n} .$$

The representation of the one-particle hamiltonian $\mathbf{H}^{(1)}$ acting on the generating functions $\Psi(\theta)$ thus is

$$\mathbf{H}^{(1)} \equiv \sum_i \theta_i \omega_i \frac{\partial}{\partial \theta_i} .$$

An analogous calculation shows that the two-body is represented by

$$\mathbf{H}^{(2)} = \frac{1}{2} \sum_{i_1,i_2,j_1,j_2} \theta_{i_1} \theta_{i_2} H^{(2)}_{i_1 i_2, j_1 j_2} \frac{\partial^2}{\partial \theta_{j_1} \partial \theta_{j_2}} .$$

The total hamiltonian

$$\mathbf{H} = \mathbf{H}^{(1)} + \mathbf{H}^{(2)}$$

is hermitian with respect to the scalar product (7.41). It has a representation analogous to the hamiltonians of the holomorphic representation and the general strategy described in Sections 6.6 and 6.7 can again be followed here.

Particle number operator. From the viewpoint of statistical physics, the preceding formalism leads to a grand canonical formulation, that is a formulation in which the number of particles is fixed only on average. The hamiltonian $\mathbf{H}$ conserves the number of particles. One verifies immediately that the representation of the particle number operator $\mathbf{N}$, when acting on $\Psi(\theta)$, is

$$\mathbf{N} = \sum_i \theta_i \frac{\partial}{\partial \theta_i} .$$

It commutes with $\mathbf{H}$:

$$[\mathbf{N}, \mathbf{H}] = 0 .$$

To vary the average number of particles, one then introduces a *chemical potential* μ coupled to $\mathbf{N}$. This leads to the substitution

$$\mathbf{H} \mapsto \mathbf{H} - \mu \mathbf{N} .$$

Here, this simply amounts to modifying $\mathbf{H}^{(1)}$.

Remark. As in the example with only one generator, the operators θ_i and $\partial/\partial\theta_i$, because they are hermitian conjugates and satisfy the commutation relations (7.11), form a representation of fermion annihilation and creation operators $a_i, a_i^\dagger$ with the correspondence

$$a_i^\dagger \mapsto \theta_i , \quad a_i \mapsto \partial/\partial\theta_i .$$

Indeed,

$$a_i^\dagger a_j^\dagger + a_j^\dagger a_i^\dagger = a_i a_j + a_j a_i = 0 \quad \text{and} \quad a_i^\dagger a_j + a_j a_i^\dagger = \delta_{ij} . \tag{7.61}$$

7.8 Path integral: one-state problem

We first construct a path integral representation of the statistical operator in the case of the one-state problem. Since the statistical operator then reduces to a 2×2 matrix, its calculation from a path integral may appear as an unnecessary complication. However, the path integral representation is useful because it can easily be generalized to an arbitrary number of available states. This justifies its introduction even to solve this elementary problem.

Thus, we now construct a path integral representation for the matrix elements of the quantum statistical operator $U_0(t) = e^{-tH_0}$. The method, in the case of a fermion hamiltonian, follows, rather closely, the method of Section 6.4, the main difference being that complex variables are replaced by Grassmann variables.

7.8.1 Gaussian path integrals

To construct a path integral, one needs the expansion of the statistical operator (7.56) to first order for $t \to 0$:

$$\langle \theta | U_0(t) | \bar{\theta} \rangle = \exp\left[-\bar{\theta}\theta(1 - \omega t) + O\left(t^2\right) \right]. \tag{7.62}$$

Using the semi-group property $U_0(t) = [U_0(t/n)]^n$ expressed in the form (7.49), one can then write the statistical operator at finite euclidean time as

$$\langle \theta'' | U_0(t'', t') | \bar{\theta}' \rangle = \lim_{n \to \infty} \int \prod_{k=1}^{n-1} d\theta_k d\bar{\theta}_k \, \exp\left[-\bar{\theta}_0 \theta_0 - \mathcal{S}_\varepsilon(\theta, \bar{\theta}) \right] \tag{7.63}$$

with

$$\mathcal{S}_\varepsilon(\theta, \bar{\theta}) = \sum_{k=1}^{n} \left[\bar{\theta}_{k-1}(\theta_k - \theta_{k-1}) - \omega\varepsilon\bar{\theta}_{k-1}\theta_k \right], \tag{7.64}$$

$\varepsilon = (t'' - t')/n$, and the definitions

$$\bar{\theta}_0 = \bar{\theta}', \quad \theta_n = \theta''. \tag{7.65}$$

The formal limit $n \to \infty$ yields a generalized path integral, with Grassmann paths. The matrix elements of $U_0(t)$ are given by

$$\langle \theta'' | U_0(t'', t') | \bar{\theta}' \rangle = \int_{\bar{\theta}(t') = \bar{\theta}'}^{\theta(t'') = \theta''} [d\theta(t) d\bar{\theta}(t)] \exp\left[-\bar{\theta}(t')\theta(t') - \mathcal{S}_0(\theta, \bar{\theta}) \right] \tag{7.66}$$

with

$$\mathcal{S}_0(\theta, \bar{\theta}) = \int_{t'}^{t''} dt \, \bar{\theta}(t) \left(\dot{\theta}(t) - \omega\theta(t) \right). \tag{7.67}$$

Generating function of fermion correlation functions. To calculate a generating function of fermion correlation functions with the weight $e^{-\mathcal{S}_0}/\mathcal{Z}_0$, it is convenient to immediately consider the more general integral obtained by adding to the action $\mathcal{S}_0$ linear terms corresponding to a coupling to external Grassmann sources $\bar{\eta}(t)$ and $\eta(t)$:

$$\langle \theta'' | U_0(t'',t';\eta,\bar{\eta}) | \bar{\theta}' \rangle = \int_{\bar{\theta}(t')=\bar{\theta}'}^{\theta(t'')=\theta''} \left[\mathrm{d}\theta(t)\mathrm{d}\bar{\theta}(t) \right] \mathrm{e}^{-\bar{\theta}(t')\theta(t')} \mathrm{e}^{-\mathcal{S}_\mathrm{G}(\theta,\bar{\theta};\eta,\bar{\eta})} \quad (7.68)$$

with

$$\mathcal{S}_\mathrm{G}(\theta,\bar{\theta};\eta,\bar{\eta}) = \mathcal{S}_0(\theta,\bar{\theta}) - \int_{t'}^{t''} \mathrm{d}t \left[\bar{\eta}(t)\theta(t) + \bar{\theta}(t)\eta(t) \right], \quad (7.69)$$

where now the independent (infinite) sets $\{\theta(t)\}, \{\bar{\theta}(t)\}, \{\eta(t)\}$ and $\{\bar{\eta}(t)\}$ form a set of generators of the Grassmann algebra.

Calculation of the gaussian integral. The integral (7.68) is gaussian and can be calculated exactly. The 'saddle point' equation obtained by varying $\theta(t)$ yields

$$\dot{\bar{\theta}}(t) + \omega\bar{\theta}(t) + \bar{\eta}(t) = 0$$

and, thus, taking into account boundary conditions,

$$\bar{\theta}(t) = \bar{\theta}_\mathrm{s}(t) \equiv \mathrm{e}^{-\omega(t-t')} \bar{\theta}' - \int_{t'}^{t} \mathrm{e}^{-\omega(t-u)} \bar{\eta}(u)\mathrm{d}u. \quad (7.70)$$

In the same way, the variation of $\bar{\theta}(t)$ yields

$$\dot{\theta}(t) - \omega\theta(t) - \eta(t) = 0$$

and, thus,

$$\theta(t) = \theta_\mathrm{s}(t) \equiv \mathrm{e}^{-\omega(t''-t)} \theta'' - \int_{t}^{t''} \mathrm{e}^{-\omega(u-t)} \eta(u)\mathrm{d}u. \quad (7.71)$$

Translating $\theta(t)$ and $\bar{\theta}(t)$ by the solutions of the 'classical' equations: $\theta \mapsto \theta_\mathrm{s} + \theta$, $\bar{\theta} \mapsto \bar{\theta}_\mathrm{s} + \bar{\theta}$, one then obtains

$$\langle \theta'' | U(t'',t';\eta,\bar{\eta}) | \bar{\theta}' \rangle = \mathcal{N}(t',t'') \exp \left[-\bar{\theta}(t')\theta(t') - \mathcal{S}_\mathrm{c}(\theta'',\bar{\theta}';\bar{\eta},\eta) \right] \quad (7.72)$$

with

$$\bar{\theta}(t')\theta(t') = \bar{\theta}'\theta'' \mathrm{e}^{-\omega(t''-t')} - \bar{\theta}' \int_{t'}^{t''} \mathrm{d}t \, \mathrm{e}^{-\omega(t-t')} \eta(t)$$

and

$$\mathcal{S}_\mathrm{c} = -\int_{t'}^{t''} \bar{\eta}(t)\theta_\mathrm{s}(t)\mathrm{d}t$$

$$= -\int_{t'}^{t''} \mathrm{d}t \, \bar{\eta}(t) \mathrm{e}^{-\omega(t''-t)} \theta'' + \int_{t'}^{t''} \mathrm{d}t \int_{t}^{t''} \mathrm{d}u \, \bar{\eta}(t) \mathrm{e}^{-\omega(u-t)} \eta(u).$$

To calculate the normalization

$$\mathcal{N}(t', t'') = \int_{\bar{\theta}(t')=0}^{\theta(t'')=0} [d\theta(t)d\bar{\theta}(t)] \exp\left[-\int_{t'}^{t''} dt\, \bar{\theta}(t)\left(\dot{\theta}(t) - \omega\theta(t)\right)\right],$$

one again changes variables, setting

$$\bar{\theta}(t) = e^{-\omega t}\bar{\zeta}(t), \qquad \theta(t) = e^{\omega t}\zeta(t).$$

The jacobian equals 1. After this change, the dependence on ω has disappeared. One finds $\mathcal{N} = 1$ in the mixed representation evaluated for $\theta = \bar{\theta} = 0$. Thus $\mathcal{N} = 1$.

As in the boson case, correlation functions are obtained by differentiating expression (7.72) with respect to η and $\bar{\eta}$.

7.8.2 Partition function

The generalized partition function $\operatorname{tr} U(\beta/2, -\beta/2; \eta, \bar{\eta})$ can be derived from expression (7.72), the trace being defined by equation (7.51):

$$\operatorname{tr} U(\beta/2, -\beta/2; \eta, \bar{\eta}) = \int d\bar{\theta}\, d\theta\, e^{-\bar{\theta}\theta} \langle \theta | U(\beta/2, -\beta/2; \eta, \bar{\eta}) | \bar{\theta} \rangle. \tag{7.73}$$

A simple evaluation of the Grassmann integrals yields

$$\operatorname{tr} U(\beta/2, -\beta/2; \eta, \bar{\eta}) = \mathcal{Z}_0(\beta) \exp\left[-\int_{-\beta/2}^{\beta/2} du\, dt\, \bar{\eta}(u)\Delta(t-u)\eta(t)\right], \tag{7.74}$$

where $\mathcal{Z}_0(\beta)$ is the partition function:

$$\mathcal{Z}_0(\beta) = 1 + e^{-\omega\beta},$$

and

$$\Delta(t) = \tfrac{1}{2} e^{-\omega t} \left[\operatorname{sgn}(t) + \tanh(\omega\beta/2)\right], \tag{7.75}$$

where $\operatorname{sgn}(t)$ is the sign function, $\operatorname{sgn}(t) = 1$ for $t > 0$, $\operatorname{sgn}(t) = -1$ for $t < 0$.

The function $\Delta(t)$ is the solution of the differential equation

$$\dot{\Delta} + \omega\Delta = \delta(t),$$

with, in contrast with the boson case (equations (2.52, 6.44)), *anti-periodic* boundary conditions:

$$\Delta(\beta/2) = -\Delta(-\beta/2).$$

In the limit $\beta \to \infty$, it reduces to

$$\Delta(t) = \tfrac{1}{2} e^{-\omega t} \left(\operatorname{sgn}(t) + 1\right).$$

This expression is identical to the one obtained in the case of holomorphic integrals (see equation (6.45)).

Let us point out that, despite the non-trivial way the trace is defined, the result obtained here is identical to the one given by a path integral with *anti-periodic boundary conditions*:

$$\operatorname{tr} U_G(\beta/2, -\beta/2; \eta, \bar\eta) = \int [\mathrm{d}\theta(t)\mathrm{d}\bar\theta(t)] \exp\left[-\mathcal{S}_G(\theta, \bar\theta; \eta, \bar\eta)\right] \tag{7.76}$$

with $\theta(-\beta/2) = -\theta(\beta/2)$, $\bar\theta(-\beta/2) = -\bar\theta(\beta/2)$ and

$$\mathcal{S}_G(\theta, \bar\theta; \eta, \bar\eta) = \int_{-\beta/2}^{\beta/2} \mathrm{d}t\, \left\{\bar\theta(t)\left[\dot\theta(t) - \omega\theta(t)\right] - \bar\eta(t)\theta(t) - \bar\theta(t)\eta(t)\right\}. \tag{7.77}$$

Instead, the integral with periodic boundary conditions is related to the calculation of $\operatorname{tr}(-1)^F \mathrm{e}^{-\beta H}$, where F is the fermion number, and is obtained by integrating expression (7.72) with $\mathrm{e}^{\theta\bar\theta}$.

The two-point correlation function calculated with the gaussian measure $\mathrm{e}^{-\mathcal{S}_0}/\mathcal{Z}_0$ is obtained by differentiating twice $\operatorname{tr} U(\beta/2, -\beta/2; \eta, \bar\eta)$. One finds

$$\langle \bar\theta(t)\theta(u) \rangle = \Delta(t-u). \tag{7.78}$$

Remark. The derivative of $\ln \mathcal{Z}_0$ with respect to ω is related to the two-point function (7.75):

$$\frac{\mathrm{d}\ln \mathcal{Z}_0}{\mathrm{d}\omega} = \int_{-\beta/2}^{\beta/2} \mathrm{d}t\, \langle \bar\theta(t)\theta(t)\rangle = \beta\Delta(0) = \tfrac{1}{2}\beta[\operatorname{sgn}(0) + \tanh(\omega\beta/2)],$$

and, thus,

$$\mathcal{Z}_0(\beta) = \mathrm{e}^{\omega\beta(\operatorname{sgn}(0)+1)/2} + \mathrm{e}^{\omega\beta(\operatorname{sgn}(0)-1)/2}. \tag{7.79}$$

One faces the problem already encountered in the holomorphic case. The choice $\operatorname{sgn}(0) = -1$, in contrast, corresponds to the normal order (7.52), but leads to the same kind of difficulties as in the commutative case. The choice $\operatorname{sgn}(0) = 0$ leads to the energy eigenvalues $\pm\omega/2$ and corresponds to the hamiltonian $\omega[a^\dagger, a]/2$, which is the 'symmetrized' version of the operator (7.52). A way to ensure that the choice $\operatorname{sgn}(0) = 0$ yields the exact result, is to insert in the path integral the average of the normal and anti-normal ordered forms (see the discussion of Section 6.5.3).

7.9 Path integrals: Generalization

The preceding formalism can be generalized to a Grassmann algebra with an arbitrary number of generators θ_i, as the calculation of the partition function in the second quantization formalism of Section 7.7 requires.

7.9.1 General hamiltonian

A general hamiltonian is represented by a differential operator $H(\boldsymbol{\theta}, \partial/\partial\boldsymbol{\theta}; t)$ acting on functions of θ_i's. To a hamiltonian written in normal form with all differentiations on the right, one can also associate a kernel (Section 7.5.3)

$$\langle \boldsymbol{\theta} | H | \bar{\boldsymbol{\theta}} \rangle = H(\boldsymbol{\theta}, \bar{\boldsymbol{\theta}}; t) \mathcal{I}(\boldsymbol{\theta}, \bar{\boldsymbol{\theta}}).$$

An important restriction is that the matrix elements of the hamiltonian must belong to the commutative subalgebra,

$$P[H] = H,$$

that is, do not mix fermions and bosons. On the other hand, the hamiltonian does not necessarily conserve the number of fermions. Fermion number conservation implies that in each monomial contributing to the kernel, the numbers of θ and $\bar{\theta}$ factors are equal.

The generalized form of equation (7.62) is

$$\langle \boldsymbol{\theta} | U(t+\varepsilon, t) | \bar{\boldsymbol{\theta}} \rangle - \exp\left[\sum_i \theta_i \bar{\theta}_i - \varepsilon\, H(\boldsymbol{\theta}, \bar{\boldsymbol{\theta}}; t) + O\left(\varepsilon^2\right) \right]. \qquad (7.80)$$

Following the method of Section 7.8.1, one then derives a path integral representation for $U(t'', t')$ at finite euclidean time difference. One obtains

$$\langle \boldsymbol{\theta}'' | U(t'', t') | \bar{\boldsymbol{\theta}}' \rangle = \int_{\bar{\boldsymbol{\theta}}(t')=\bar{\boldsymbol{\theta}}'}^{\boldsymbol{\theta}(t'')=\boldsymbol{\theta}''} [\mathrm{d}\boldsymbol{\theta}(t)\mathrm{d}\bar{\boldsymbol{\theta}}(t)]\, \mathrm{e}^{-\bar{\boldsymbol{\theta}}(t')\cdot\boldsymbol{\theta}(t')} \exp\left[-\mathcal{S}(\boldsymbol{\theta}, \bar{\boldsymbol{\theta}})\right] \qquad (7.81)$$

with

$$\mathcal{S}(\boldsymbol{\theta}, \bar{\boldsymbol{\theta}}) = \int_{t'}^{t''} \mathrm{d}t\, \left\{ \bar{\boldsymbol{\theta}}(t)\cdot\dot{\boldsymbol{\theta}}(t) + H\left[\boldsymbol{\theta}(t), \bar{\boldsymbol{\theta}}(t); t\right] \right\}. \qquad (7.82)$$

We have shown in Section 7.4 how to calculate gaussian integrals and expectation values of polynomials. The same method can be used here to calculate the path integral (7.84) perturbatively. Writing a hamiltonian as the sum of a quadratic term and an interaction:

$$H(\boldsymbol{\theta}, \bar{\boldsymbol{\theta}}) = -\sum_i \omega_i \bar{\theta}_i \theta_i + H_{\mathrm{I}}(\boldsymbol{\theta}, \bar{\boldsymbol{\theta}}),$$

one expands the integral in powers of H_{I} and calculates the successive terms, for example, using Wick's theorem (7.30). One verifies that the perturbative expansion of general matrix elements has the formal compact representation

$$\langle \boldsymbol{\theta}'' | U(t'', t') | \bar{\boldsymbol{\theta}}' \rangle$$
$$= \exp\left[-\int \mathrm{d}t\, H_{\mathrm{I}}(\partial/\partial\bar{\eta}, -\partial/\partial\eta)\right] \langle \boldsymbol{\theta}'' | U_{\mathrm{G}}(\eta; t'', t') | \bar{\boldsymbol{\theta}}' \rangle \big|_{\eta=\bar{\eta}=0}, \qquad (7.83)$$

where U_G is the product of integrals (7.68) corresponding to the different pairs of generators $\theta_i, \bar\theta_i$.

In the case of a time-independent hamiltonian, the corresponding partition then reads

$$\mathcal{Z}(\beta) = \operatorname{tr} U(\beta/2, -\beta/2) = \int [\mathrm{d}\boldsymbol{\theta}(t)\mathrm{d}\bar{\boldsymbol{\theta}}(t)] \exp\left[-\mathcal{S}(\boldsymbol{\theta}, \bar{\boldsymbol{\theta}})\right]. \tag{7.84}$$

Moreover, the path integral has to be calculated with *anti-periodic boundary conditions*:

$$\boldsymbol{\theta}(-\beta/2) = -\boldsymbol{\theta}(\beta/2), \quad \bar{\boldsymbol{\theta}}(-\beta/2) = -\bar{\boldsymbol{\theta}}(\beta/2).$$

This follows, for example, from the representation (7.76) and perturbation theory. The trace of the r.h.s. of equation (7.83) involves only $\operatorname{tr} U_G$, which can be calculated from the path integral with anti-periodic boundary conditions (equation (7.76)).

Remark. In perturbative calculations, problems due to operator ordering appear here also, as in the case of the holomorphic path integral. Indeed, perturbative calculations involve $\operatorname{sgn}(0)$. The ansatz consistent with the normal-order construction, is again to set $\operatorname{sgn}(0_-) = -1$, but generates some difficulties. It is more convenient to substitute to $H(\boldsymbol{\theta}, \bar{\boldsymbol{\theta}})$ the equivalent expression consistent with $\operatorname{sgn}(0) = 0$.

7.9.2 Fermion systems with pair interactions

We now specialize the expressions of Section 7.9 to the many-fermion systems described in Section 7.7. The partition function $\mathcal{Z}(\beta, \mu)$ is then given by a Grassmann path integral of the form (7.84) where the action is

$$\mathcal{S}(\boldsymbol{\theta}, \bar{\boldsymbol{\theta}}) = \int_{-\beta/2}^{\beta/2} \mathrm{d}t \left[\bar{\boldsymbol{\theta}}(t) \cdot (\dot{\boldsymbol{\theta}}(t) - \mu\boldsymbol{\theta}(t)) + H(\boldsymbol{\theta}(t), \bar{\boldsymbol{\theta}}(t))\right], \tag{7.85}$$

$$H(\boldsymbol{\theta}, \bar{\boldsymbol{\theta}}) = \sum_i \theta_i \omega_i \bar\theta_i + \frac{1}{2} \sum_{i_1, i_2, j_1, j_2} \theta_{i_1}\theta_{i_2} H^{(2)}_{i_1 i_2, j_1 j_2} \bar\theta_{j_1} \bar\theta_{j_2}, \tag{7.86}$$

a form analogous to the boson expression (6.77).

Equation of state. Differentiating the path integral (7.84), one verifies that the equation state can be written as

$$\langle \mathbf{N} \rangle = \frac{1}{\beta} \frac{\partial \ln \mathcal{Z}}{\partial \mu} = \frac{1}{\beta} \sum_i \int_{-\beta/2}^{\beta/2} \mathrm{d}t \, \langle \theta_i(t)\bar\theta_i(t)\rangle = \sum_i \langle \theta_i(0)\bar\theta_i(0)\rangle, \tag{7.87}$$

where in the second expression the expectation value is calculated with the weight $\mathrm{e}^{-\mathcal{S}}/\mathcal{Z}$.

In the case of independent particles, expectation values involve only the gaussian two-point function (7.75). One recovers the standard expression (with the choice $\operatorname{sgn}(0) = -1$)

$$\langle \mathbf{N} \rangle = \sum_i \frac{1}{\mathrm{e}^{\beta(\omega_i - \mu)} + 1} = \operatorname{tr} \frac{1}{\mathrm{e}^{\beta(H^{(1)} - \mu)} + 1}, \tag{7.88}$$

where $H^{(1)}$ is the one-particle hamiltonian.

At low temperature, that is for $\beta \to \infty$,
$$\langle \mathbf{N} \rangle \sim \sum_i \theta(\mu - \omega_i),$$

where $\theta(s)$ is the step function. The chemical potential can be identified with the Fermi energy: at zero temperature, all states below the Fermi energy are occupied; all states above the Fermi energy are empty. At low temperature, when interactions between fermions are added, only states with energies close to the Fermi energy are relevant.

7.10 Quantum Fermi gas

We first generalize the preceding results to independent fermions when the available quantum states belong to a Hilbert space, instead of a finite-dimensional complex vector space. We then extend the analysis to interacting fermions. We show in this section how a rather straightforward generalization of the path integral of Section 7.9 allows expressing the partition function of non-relativistic fermion systems as a field or functional integral (one then integrates over Grassmann fields).

7.10.1 Independent fermions: Hilbert space

As we have pointed out, the assumption of a finite-dimensional vector space of one-particle states is quite restrictive for fermions since the total number of fermions is then bounded. More interesting applications require replacing, for one-particle states, finite dimensional vector spaces by Hilbert spaces. Then, the equation of state can still be written in the form (7.88),

$$\langle \mathbf{N} \rangle = \text{tr}\, \frac{1}{e^{\beta(H^{(1)} - \mu)} + 1},$$

but now $H^{(1)}$ is a general one-particle quantum hamiltonian.

As an illustration, we consider a gas of free fermions in a box of equal size L in all dimensions and, thus, of volume L^d in dimension d. The one-particle quantum hamiltonian is simply
$$H^{(1)} = \hat{\mathbf{p}}^2 / 2m.$$

In a box, momenta are quantized, the precise form depending on the boundary conditions. Assuming periodic boundary conditions for convenience, but this plays no role in the discussion, one finds
$$\mathbf{p} = 2\pi \hbar \mathbf{n}/L, \quad \mathbf{n} \in \mathbb{Z}^d,$$

the corresponding energy being $E = \mathbf{p}^2 / 2m$. The derivation of the equation of state in d space dimensions then follows from the arguments presented in Section 6.8.2. In the infinite volume limit $L \to \infty$, one finds for the density

$$\rho(\beta, \mu) \underset{L \to \infty}{=} \frac{\langle \mathbf{N} \rangle}{L^d} = \frac{1}{(2\pi\hbar)^d} \int \frac{d^d p}{e^{\beta(\mathbf{p}^2/2m - \mu)} + 1}. \tag{7.89}$$

In particular, in isotropic space the Fermi surface thus is the sphere $\mathbf{p}^2/2m = \mu$.

7.10.2 Interacting fermions: field integral

We now consider the interacting hamiltonian **H** of Section 6.9 with the form (6.85) in the n-particle space and we follow the same strategy. The only difference is that we deal here with fermion systems and, thus, the wave functions ψ_n are antisymmetric. Therefore, the arguments of a generating functional must be functions $\varphi(x)$ that are generators of an infinite dimensional Grassmann algebra and satisfy

$$\varphi(x)\varphi(x') + \varphi(x')\varphi(x) = 0.$$

We define the functional

$$\Psi(\varphi) = \sum_{n=0}^{\infty} \frac{1}{n!} \left(\int \prod_i \mathrm{d}^d x_i\, \varphi(x_i) \right) \psi_n(x_1, \ldots, x_n). \tag{7.90}$$

The derivation of a field integral representation then follows closely the arguments already presented in the boson case in Section 6.9, except that it is necessary to carefully keep track of the ordering of factors in field products and of the signs.

The scalar product of two generating functionals is defined in terms of a *Grassmann field* integral, which generalizes expression (7.41):

$$(\Psi_1, \Psi_2) = \int [\mathrm{d}\varphi(x)\mathrm{d}\bar\varphi(x)]\Psi_1^\dagger(\varphi)\Psi_2(\varphi) \exp\left[\int \mathrm{d}^d x\, \bar\varphi(x)\varphi(x)\right].$$

The corresponding identity kernel, which generalizes expression (7.44), is

$$\mathcal{I}(\varphi, \bar\varphi) = \exp\left[-\int \mathrm{d}^d x\, \bar\varphi(x)\varphi(x)\right]. \tag{7.91}$$

The formal expression of the kinetic term is the same as in the boson case (see Section 6.9). The potential term remains also the same, but with a specific order of the fields in products. With the conventions of Section 7.8, the hamiltonian has the kernel representation

$$\langle \varphi | \mathbf{H} | \bar\varphi \rangle = H(\varphi, \bar\varphi)\mathcal{I}(\varphi, \bar\varphi),$$

with

$$H(\varphi, \bar\varphi) = -\frac{\hbar^2}{2m} \int \mathrm{d}^d x\, \varphi(x)\nabla_x^2 \bar\varphi(x) + \frac{1}{2} \int \mathrm{d}^d x\, \mathrm{d}^d y\, \varphi(x)\varphi(y)V(x-y)\bar\varphi(x)\bar\varphi(y). \tag{7.92}$$

A field integral representation of the partition function of the Fermi gas follows:

$$\mathcal{Z}(\tau/\hbar) = \mathrm{tr}\, \mathbf{U}(\tau/2, -\tau/2) = \int [\mathrm{d}\varphi(t,x)\mathrm{d}\bar\varphi(t,x)] \exp[-\mathcal{S}(\bar\varphi, \varphi)/\hbar] \tag{7.93}$$

with anti-periodic boundary conditions:

$$\varphi(\tau/2, x) = -\varphi(-\tau/2, x), \quad \bar\varphi(\tau/2, x) = -\bar\varphi(-\tau/2, x),$$

and the euclidean action

$$\mathcal{S}(\bar\varphi, \varphi) = \int \mathrm{d}t\, \mathrm{d}^d x\, \bar\varphi(t,x) \left(\hbar \frac{\partial}{\partial t} + \frac{\hbar^2}{2m} \nabla_x^2 + \mu \right) \varphi(t,x)$$
$$+ \frac{1}{2} \int \mathrm{d}t\, \mathrm{d}^d x\, \mathrm{d}^d y\, \bar\varphi(t,x)\varphi(t,x)V(x-y)\bar\varphi(t,y)\varphi(t,y). \tag{7.94}$$

7.10.3 The equation of state

We first verify that, in free field theory, the equation of state reduces to the equation (7.89) of free fermions, and then comment briefly about the effect of interactions.

The free theory. The action of the free theory reduces to

$$S(\bar\varphi, \varphi) = \int dt\, d^d x\, \bar\varphi(t,x) \left(\hbar \frac{\partial}{\partial t} + \frac{\hbar^2}{2m}\nabla_x^2 + \mu \right) \varphi(t,x).$$

A convenient representation is obtained by introducing the Fourier representation of the field:

$$\varphi(t,x) = \int d^d p\, e^{ipx/\hbar}\, \tilde\varphi(t,p), \quad \bar\varphi(t,x) = \int d^d p\, e^{-ipx/\hbar}\, \tilde\varphi^\dagger(t,p).$$

The quadratic form in the action is then diagonalized and the action becomes

$$S(\bar\varphi, \varphi) = (2\pi\hbar)^d \int dt\, d^d p\, \tilde\varphi^\dagger(t,p) \left(\hbar \frac{\partial}{\partial t} - \frac{p^2}{2m} + \mu \right) \tilde\varphi(t,p).$$

In a free (gaussian) theory, all quantities can be expressed in terms of the two-point function. Using the result (7.75), one obtains

$$\langle \tilde\varphi^\dagger(t,p) \tilde\varphi(t',p') \rangle = \frac{1}{(2\pi\hbar)^d} \Delta(t-t',p) \delta^d(p-p')$$

with

$$\Delta(t,p) = \frac{1}{2} e^{-\omega(p)t/\hbar} \left[\operatorname{sgn}(t) + \frac{\sinh[\omega(p)\tau/2\hbar]}{\cosh[\omega(p)\tau/2\hbar]} \right],$$

where $\omega(p) = p^2/2m - \mu$.

The equation of state is obtained by differentiating the partition function (6.94). Assuming a periodic box of linear size L, one finds the density

$$\rho(\beta,\mu) = \frac{1}{\beta L^d} \frac{\partial \ln \mathcal{Z}}{\partial \mu} = \frac{1}{\beta L^d \hbar} \int dt\, d^d x\, \langle \varphi(t,x)\bar\varphi(t,x) \rangle = \langle \varphi(0,0)\bar\varphi(0,0) \rangle,$$

where translation invariance in space and time has been used. Introducing the field Fourier representation, one obtains

$$\rho(\beta,\mu) = -\int d^d p\, d^d p'\, \langle \tilde\varphi^\dagger(0,p)\tilde\varphi(0,p') \rangle = -\frac{1}{(2\pi\hbar)^d} \int d^d p\, \Delta(0,p)$$

$$= -\frac{1}{(2\pi\hbar)^d} \int d^d p\, \frac{1}{2} \left[\operatorname{sgn}(0) + \frac{\sinh[\beta\omega(p)/2]}{\cosh[\beta\omega(p)/2]} \right].$$

This expression coincides with the result (7.89) obtained directly when $\operatorname{sgn}(0)$ is chosen to be -1, and otherwise differs by an infinite constant that has to be removed by adding a constant linear in μ to the action.

Interactions: The δ function potential. An interesting interacting example is provided by the two-body pseudo-potential

$$V(x) = g\delta(x).$$

The action then becomes local, in the sense that it becomes the integral of a lagrangian density depending only on the field and its derivatives. In the case of fermions without internal degrees of freedom, the two-body interaction then vanishes since it involves the squares of Grassmann variables, and the fermions are free.

A more interesting example is provided by systems where fermions have an internal degree of freedom with two possible values (like the spin for an electron). The action then depends on two pairs of fields $\varphi_\alpha(t,x)$, $\alpha = 1, 2$ and the interaction no longer vanishes

$$\mathcal{S}(\bar{\varphi}, \varphi) = \int dt\, d^d x \left[\sum_\alpha \bar{\varphi}_\alpha(t,x) \left(\hbar \frac{\partial}{\partial t} + \frac{\hbar^2}{2m} \nabla_x^2 + \mu \right) \varphi_\alpha(t,x) \right.$$
$$\left. + g \bar{\varphi}_1(t,x) \varphi_1(t,x) \bar{\varphi}_2(t,x) \varphi_2(t,x) \right]. \tag{7.95}$$

The action and the corresponding field integral are then invariant under unitary transformations

$$\varphi_\alpha \mapsto \sum_\beta U_{\alpha\beta} \varphi_\beta, \quad \bar{\varphi}_\alpha \mapsto \sum_\beta U^\dagger_{\alpha\beta} \bar{\varphi}_\beta$$

with

$$UU^\dagger = \mathbf{1}.$$

Indeed, the kinetic term is a complex scalar product, and for the interaction term one obtains $|\det U|^2 = 1$. This symmetry is a combination of the $U(1)$ particle number conservation and the $SU(2)$ spin group symmetry.

The one-dimensional quantum system is completely integrable, in the sense that all eigenstates of the hamiltonian are linear combinations of a finite number of plane waves (Bethe ansatz).

Finally, note that this system has a relativistic generalization, the Thirring model, which is also integrable in one space dimension.

Mean-field approximation. Interesting physics is associated to an attractive interaction, that is $g < 0$. However, unlike for bosons, the steepest descent method yields no direct insight into this problem. A possible strategy relies on introducing an auxiliary boson field $\chi(t,x)$ and rewriting the quartic fermion interaction as an integral over χ with an action quadratic in the fermions. The fermion integral becomes gaussian and can be performed. The remaining χ integral can be evaluated by the steepest descent method.

7.11 Real gaussian integrals. Wick's theorem

The gaussian integrals that we have calculated in Section 7.4 have properties analogous to the complex integrals of the holomorphic formalism of Section 6.1. When the fermion number is not conserved, more general integrals become relevant, in particular, gaussian integrals of the more general form

$$\mathcal{Z}(\mathbf{A}) = \int \mathrm{d}\theta_{2n} \ldots \mathrm{d}\theta_2 \mathrm{d}\theta_1 \, \exp\left(\frac{1}{2} \sum_{i,j=1}^{2n} \theta_i A_{ij} \theta_j \right). \tag{7.96}$$

Since the product $\theta_i \theta_j$ is antisymmetric in (ij), the matrix A_{ij} can be chosen antisymmetric:

$$A_{ij} + A_{ji} = 0. \tag{7.97}$$

In contrast with the integrals considered so far, these integrals have properties somewhat analogous to the gaussian integrals (1.4). In particular, generically they lead to real results only if the Grassmann algebra is defined on real numbers.

Expanding the exponential in a power series, one notices that only the term of order n, which contains all products of degree $2n$ in θ, gives a non-vanishing contribution:

$$\mathcal{Z}(\mathbf{A}) = \frac{1}{2^n n!} \int \mathrm{d}\theta_{2n} \ldots \mathrm{d}\theta_1 \, (\theta_i A_{ij} \theta_j)^n. \tag{7.98}$$

In the expansion of the product, only the terms containing a permutation of $\theta_1 \ldots \theta_{2n}$ do not vanish. Commuting Grassmann generators to cast all products into the standard form $\theta_1 \theta_2 \ldots \theta_{2n}$, one finds

$$\mathcal{Z}(\mathbf{A}) = \frac{1}{2^n n!} \sum_{\substack{\text{permutations } P \\ \text{of } \{i_1 \ldots i_{2n}\}}} \epsilon(P) A_{i_1 i_2} A_{i_3 i_4} \ldots A_{i_{2n-1} i_{2n}}, \tag{7.99}$$

where $\epsilon(P) = \pm 1$ is the signature of the permutation. This expression can be further reduced by noting that the only distinct terms correspond to all pairings of the indices $1, 2, \ldots, 2n$. Thus,

$$\mathcal{Z}(\mathbf{A}) = \sum_{\substack{\text{pairings } P \\ \text{of } \{i_1 \ldots i_{2n}\}}} \epsilon(P) A_{i_1 i_2} A_{i_3 i_4} \ldots A_{i_{2n-1} i_{2n}}, \tag{7.100}$$

where $\epsilon(P)$ is the signature of the permutation of the indices.

The quantity on the r.h.s. is called the *pfaffian* of the antisymmetric matrix $\mathbf{A}$, and we use the notation

$$\mathcal{Z}(\mathbf{A}) = \mathrm{Pf}\,(\mathbf{A}). \tag{7.101}$$

Pfaffian and determinant. The techniques of Grassmann integration allow deriving the classical algebraic identity

$$\mathrm{Pf}^2(\mathbf{A}) = \det \mathbf{A}, \tag{7.102}$$

which implies

$$\int d\theta_{2n}\ldots d\theta_2 d\theta_1 \exp\left(\frac{1}{2}\sum_{i,j=1}^{2n} \theta_i A_{ij}\theta_j\right) = \pm\sqrt{\det \mathbf{A}},$$

a result reminiscent of the analogous result for real gaussian integrals.

To prove the identity, we consider

$$\mathcal{Z}^2(\mathbf{A}) = \int d\theta_{2n}\ldots d\theta_1\, d\theta'_{2n}\ldots d\theta'_1 \exp\left[\frac{1}{2}\sum_{i,j}\left(\theta_i A_{ij}\theta_j + \theta'_i A_{ij}\theta'_j\right)\right]. \quad (7.103)$$

We change variables, setting

$$\eta_k = \frac{1}{\sqrt{2}}(\theta_k + i\theta'_k), \qquad \bar{\eta}_k = \frac{1}{\sqrt{2}}(\theta_k - i\theta'_k). \quad (7.104)$$

The jacobian is $(-1)^n$. Moreover,

$$\theta_i\theta_j + \theta'_i\theta'_j = \bar{\eta}_i\eta_j - \bar{\eta}_j\eta_i, \quad (7.105)$$

$$d\eta_{2n}\ldots d\eta_1 d\bar{\eta}_{2n}\ldots d\bar{\eta}_1 = (-1)^{n^2}\prod_i d\eta_i d\bar{\eta}_i. \quad (7.106)$$

Using the antisymmetry of the matrix $\mathbf{A}$, one finds

$$\mathrm{Pf}^2(\mathbf{A}) = \int d\eta_1 d\bar{\eta}_1\ldots d\eta_{2n} d\bar{\eta}_{2n} \exp\left(\sum_{i,j}\bar{\eta}_i A_{ij}\eta_j\right) = \det \mathbf{A}. \quad (7.107)$$

Wick's theorem. One can also derive another form of Wick's theorem for gaussian expectation values with the measure $\exp[\sum_{i,j}\theta_i A_{ij}\theta_j/2]/\mathcal{Z}$. One finds

$$\langle \theta_{i_1}\theta_{i_2}\ldots\theta_{i_{2p}}\rangle = \sum_{\substack{\text{all pairings}\\ \text{of } (1,2,\ldots,2p)}} \epsilon(P)\langle \theta_{i_{P_1}}\theta_{i_{P_2}}\rangle\cdots\langle \theta_{i_{P_{2p-1}}}\theta_{i_{P_{2p}}}\rangle, \quad (7.108)$$

where $\epsilon(P)$ is the signature of the permutation P. Note that this expression differs from Wick's theorem (1.17) only by signs.

7.12 Mixed change of variables: Berezinian and supertrace

One often meets integrals with both commuting and anti-commuting integration variables (bosons and fermions). Occasionally, it may be useful to perform mixed changes of variables. For completeness, we thus describe some properties of such mixed integrals.

We denote by θ, θ' and x, x' the anti-commuting and commuting variables, respectively, and set (respecting reflection symmetry)

$$x_a = x_a(x', \theta') \in \mathfrak{A}_+(\theta'), \qquad \theta_i = \theta_i(x', \theta') \in \mathfrak{A}_-(\theta'). \qquad (7.109)$$

We introduce the matrix of partial derivatives

$$\mathbf{M} = \begin{pmatrix} \mathbf{A} & \mathbf{B} \\ \mathbf{C} & \mathbf{D} \end{pmatrix}$$

with

$$\mathbf{A}_{ab} = \frac{\partial x_a}{\partial x'_b}, \quad \mathbf{B}_{ai} = \frac{\partial x_a}{\partial \theta'_i}, \quad \mathbf{C}_{ia} = \frac{\partial \theta_i}{\partial x'_a}, \quad \mathbf{D}_{ij} = \frac{\partial \theta_i}{\partial \theta'_j},$$

and, thus,

$$\mathbf{A} \text{ and } \mathbf{D} \in \mathfrak{A}_+, \ \mathbf{B} \text{ and } \mathbf{C} \in \mathfrak{A}_-.$$

It is then convenient to change variables in two steps:

(i) One first changes from (θ, x) to (θ, x'). This step generates the jacobian

$$J_1 = \det \left. \frac{\partial x_a}{\partial x'_b} \right|_\theta = \det \left(\mathbf{A} - \mathbf{B} \mathbf{D}^{-1} \mathbf{C} \right). \qquad (7.110)$$

(ii) One then changes from (θ, x') to (θ', x'). This second step generates, as shown before, the jacobian

$$J_2 = (\det \mathbf{D})^{-1}. \qquad (7.111)$$

The complete jacobian J, also called the *berezinian* of the matrix of derivatives, is thus

$$J \equiv \frac{D(x, \theta)}{D(x', \theta')} = J_1 J_2 = \operatorname{Ber} \mathbf{M} \equiv \det \left(\mathbf{A} - \mathbf{B} \mathbf{D}^{-1} \mathbf{C} \right) (\det \mathbf{D})^{-1}. \qquad (7.112)$$

For the jacobian to be non-singular, the matrices $\mathbf{A}$ and $\mathbf{D}$ must be invertible (and thus the matrix of the terms of degree zero in θ' must be invertible). It follows from the construction that

$$\operatorname{Ber} \mathbf{M}_1 \operatorname{Ber} \mathbf{M}_2 = \operatorname{Ber}[\mathbf{M}_1 \mathbf{M}_2].$$

Trace of mixed matrices. In the case of an integration over usual commuting variables, a change of variables infinitesimally close to the identity,

$$x_a = x'_a + \varepsilon f_a(x'),$$

as a consequence of the relation $\ln \det = \operatorname{tr} \ln$, leads to the jacobian

$$J = \det \frac{\partial x_a}{\partial x'_b} = 1 + \varepsilon \operatorname{tr} \frac{\partial f_a}{\partial x'_b} + O\left(\varepsilon^2\right) = 1 + \varepsilon \sum_a \frac{\partial f_a}{\partial x'_a} + O\left(\varepsilon^2\right).$$

We now consider the mixed case

$$x_a = x'_a + \varepsilon f_a(x', \theta'), \quad \theta_i = \theta'_i + \varepsilon \varphi_i(x', \theta'). \tag{7.113}$$

Then, setting

$$\mathbf{M} = 1 + \varepsilon \mathbf{M}_1 + O\left(\varepsilon^2\right), \quad \mathbf{M}_1 = \begin{pmatrix} \mathbf{A}_1 & \mathbf{B}_1 \\ \mathbf{C}_1 & \mathbf{D}_1 \end{pmatrix},$$

we find, as a consequence of identity (7.112),

$$J = 1 + \varepsilon \left(\operatorname{tr} \mathbf{A}_1 - \operatorname{tr} \mathbf{D}_1\right) + O\left(\varepsilon^2\right), \quad \operatorname{tr} \mathbf{A}_1 - \operatorname{tr} \mathbf{D}_1 = \sum_a \frac{\partial f_a}{\partial x_a} - \sum_i \frac{\partial \varphi_i}{\partial \theta_i}. \tag{7.114}$$

To generalize the relation between a jacobian for an infinitesimal change of variables and the trace, one is thus led to define the *supertrace* of a mixed matrix, denoted by Str, as the difference between two traces:

$$\operatorname{Str} \mathbf{M}_1 = \operatorname{tr} \mathbf{A}_1 - \operatorname{tr} \mathbf{D}_1. \tag{7.115}$$

One verifies directly that the supertrace satisfies the cyclic property

$$\operatorname{Str}[\mathbf{M}_1 \mathbf{M}_2] = \operatorname{Str}[\mathbf{M}_2 \mathbf{M}_1].$$

Indeed, it follows from the definition that

$$\operatorname{Str}[\mathbf{M}_1 \mathbf{M}_2] = \operatorname{tr}(\mathbf{A}_1 \mathbf{A}_2 + \mathbf{B}_1 \mathbf{C}_2) - \operatorname{tr}(\mathbf{C}_1 \mathbf{B}_2 + \mathbf{D}_1 \mathbf{D}_2).$$

Then,

$$\operatorname{tr}(\mathbf{A}_1 \mathbf{A}_2) = \operatorname{tr}(\mathbf{A}_2 \mathbf{A}_1), \quad \operatorname{tr}(\mathbf{D}_1 \mathbf{D}_2) = \operatorname{tr}(\mathbf{D}_2 \mathbf{D}_1),$$

but

$$\operatorname{tr}(\mathbf{B}_1 \mathbf{C}_2) = -\operatorname{tr}(\mathbf{C}_2 \mathbf{B}_1), \quad \operatorname{tr}(\mathbf{C}_1 \mathbf{B}_2) = -\operatorname{tr}(\mathbf{B}_2 \mathbf{C}_1),$$

where the signs result from the cyclic property of the usual trace and from the anti-commutation between $\mathbf{B}$ and $\mathbf{C}$.

This allows proving a generalization of the usual relation $\operatorname{tr} \ln = \ln \det$, which now becomes

$$\operatorname{Str} \ln = \ln \operatorname{Ber}.$$

Exercises

Exercise 7.1

Calculate the expectation value of

$$\langle \bar{\theta}_1 \theta_1 \bar{\theta}_2 \theta_2 \rangle ,$$

where $\bar{\theta}_1, \theta_1, \bar{\theta}_2, \theta_2$ are Grassmann variables, with the measure

$$d\theta_1 d\bar{\theta}_1 d\theta_2 d\bar{\theta}_2 \exp\left[\bar{\theta}_1 \theta_2 + \bar{\theta}_2 \theta_1\right].$$

Solution. Writing the argument of the exponential as $\sum_{i,j} \bar{\theta}_i M_{ij} \theta_j$, one finds

$$\det M = -1 \quad M^{-1} = \begin{pmatrix} 0 & 1 \\ 1 & 0 \end{pmatrix}, \quad \langle \bar{\theta}_1 \theta_1 \bar{\theta}_2 \theta_2 \rangle = (-1) \times 1 = -1.$$

Exercise 7.2

Calculate the scalar product (f, f) where f is the function of the Grassmann variables θ_+, θ_-:

$$f(\theta_+, \theta_-) = 2\theta_+ + 3\theta_+ \theta_- ,$$

the scalar product (f, g) of two functions f and g being defined by

$$(f, g) = \int d\theta_+ d\bar{\theta}_+ d\theta_- d\bar{\theta}_- \ e^{\bar{\theta}_+ \theta_+ + \bar{\theta}_- \theta_-} \ \overline{f(\theta)} g(\theta). \tag{7.116}$$

(One is reminded that Grassmann conjugation has the form of a hermitian conjugation.)

Solution.

$$(f, f) = 13.$$

Exercise 7.3

With the same scalar product (7.116), calculate the scalar product (f, g) of the two functions

$$f(\theta_+, \theta_-) = 5\theta_+ + 3\theta_- + 2\theta_+ \theta_- ,$$
$$g(\theta_+, \theta_-) = \theta_+ - 3\theta_- + 2\theta_+ \theta_- .$$

Solution.

$$(f, g) = 0.$$

Exercise 7.4

One considers the normalized expectation values ($\langle 1 \rangle = 1$)

$$\mathcal{N}(\mathbf{M}) \langle \bar\theta_{i_1}\theta_{j_1}\bar\theta_{i_2}\theta_{j_2}\ldots\bar\theta_{i_p}\theta_{j_p}\rangle$$
$$= \int \left(\prod_i \mathrm{d}\theta_i \mathrm{d}\bar\theta_i\right) \bar\theta_{i_1}\theta_{j_1}\ldots\bar\theta_{i_p}\theta_{j_p} \exp\left(\sum_{i,j=1}^n M_{ij}\bar\theta_i\theta_j\right). \quad (7.117)$$

Determine the normalization constant $\mathcal{N}$ using only Wick's theorem (7.30).

The same question in the case of the measure $\exp[\tfrac{1}{2}\sum_{i,j}\theta_i A_{ij}\theta_j]$.

Solution. The solutions in both cases are similar. For example, in the first case, one calculates the expectation value of the product

$$\prod_{i=1}^n \bar\theta_i \theta_i .$$

In this case, the factor $\exp(\sum_{i,j=1}^n M_{ij}\bar\theta_i\theta_j)$ in the integrand can be replaced by 1, the integral gives 1, and Wick's theorem reconstructs the determinant of the matrix $(\mathbf{M})^{-1}$ in the l.h.s..

Exercise 7.5

Use the formalism based on Grassmann integrals to obtain an expansion in powers of λ, up to order λ^3 included, of the determinant $\det(M+\lambda)$ where the matrix M is invertible. It will be convenient to set $\Delta = M^{-1}$. It is suggested to use Wick's theorem.

Solution. Using the notation $t_n = \operatorname{tr}\Delta^n$, one finds

$$\det(M+\lambda) = \det M \left[1 + t_1\lambda + \tfrac{1}{2}(t_1^2 - t_2)\lambda^2 + \tfrac{1}{6}\left(t_1^3 - 3t_1 t_2 + 2t_3\right)\lambda^3\right] + O(\lambda^4).$$

Exercise 7.6

Generalize the method to expand $\det(M+\Lambda)$ where Λ is a diagonal matrix $\Lambda_{ij} = \lambda_i \delta_{ij}$, in powers of λ_i.

Solution.

$$\det[M+\Lambda] = \det M \left[\sum_{n=0} \frac{1}{n!} \sum_{i_1,i_2,\ldots,i_n} \lambda_{i_1}\lambda_{i_2}\ldots\lambda_{i_n} \det \Delta_{i_l i_k}\right].$$

Exercise 7.7

One site electron problem. Electrons living on one site are still characterized by the two possible values of their spin. A site thus can be empty, occupied by one electron with spin up or down, or two electrons with spin up and down.

One denotes below by θ_+, θ_- the Grassmann variables associated with spin up and down electrons, respectively. Hermiticity is then defined with respect to the scalar product (7.116).

Write, in normal-ordered form, the most general hermitian hamiltonian invariant under spin reversal and conserving total electron number. One then considers the hamiltonian

$$H = \omega\left(\theta_+ \frac{\partial}{\partial \theta_+} + \theta_- \frac{\partial}{\partial \theta_-}\right) + v\theta_-\theta_+ \frac{\partial}{\partial \theta_+}\frac{\partial}{\partial \theta_-}, \quad \omega > 0, \quad v \in \mathbb{R}. \quad (7.118)$$

What are its symmetries? Determine its spectrum. Infer the partition function at temperature $1/\beta$. Show that the situation $v < -2\omega$ has a natural interpretation if the role of creation and annihilation operators is interchanged.

Solution. In addition to the terms appearing in (7.118), one possible additional term is

$$\theta_+ \frac{\partial}{\partial \theta_-} + \theta_- \frac{\partial}{\partial \theta_+}.$$

It is eliminated if one demands separate conservation of up and down electron numbers.

The four eigenvectors are $1, \theta_+, \theta_-, \theta_-\theta_+$ and the corresponding eigenvalues are $E_0 = 0$, $E_1 = E_2 = \omega$, $E_3 = 2\omega + v$, respectively. The partition function thus is

$$\mathcal{Z}(\beta) = 1 + 2\,\mathrm{e}^{-\omega\beta} + \mathrm{e}^{-(2\omega+v)\beta}.$$

When $2\omega + v$ is negative, the ground state is no longer the empty state but, instead, the doubly-occupied state. The description in terms of electron excitations is no longer convenient. The excitations now correspond to remove electrons. A formal way to implement this idea is to introduce the operators

$$\zeta_- = \frac{\partial}{\partial \theta_+}, \quad \zeta_+ = \frac{\partial}{\partial \theta_-}, \quad \theta_+ = \frac{\partial}{\partial \zeta_-}, \quad \theta_- = \frac{\partial}{\partial \zeta_+}.$$

This transformation is consistent with the commutation relations and hermitian conjugation. The hamiltonian can then be rewritten as

$$H = E_0 - (\omega + v)\left(\zeta_+ \frac{\partial}{\partial \zeta_+} + \zeta_- \frac{\partial}{\partial \zeta_-}\right) + v\zeta_-\zeta_+ \frac{\partial}{\partial \zeta_+}\frac{\partial}{\partial \zeta_-}, \quad E_0 = 2\omega + v,$$

where $2\omega + v < 0$ implies $-\omega - v > \omega > 0$.

Exercise 7.8

Write the path integral representation of the corresponding partition function. Expand and calculate it up to second order in v (the gaussian two-point function is given by equation (7.75)).

Solution.
$$\mathcal{Z}(\beta) = \int [d\theta(t)d\bar\theta(t)] \exp\left[-\mathcal{S}(\theta,\bar\theta)\right],$$

where the action takes the form (7.84):
$$\mathcal{S}(\theta,\bar\theta) = \int_{-\beta/2}^{\beta/2} dt \left[\sum_\pm \bar\theta_\pm(t)\dot\theta_\pm(t) + H(\theta(t),\bar\theta(t))\right],$$

$$H(\theta,\bar\theta) = \omega \sum_\pm \theta_\pm \bar\theta_\pm + v\,\theta_-\theta_+\bar\theta_+\bar\theta_-\,,$$

and the boundary conditions are anti-periodic:
$$\theta_\pm(-\beta/2) = -\theta_\pm(\beta/2),\ \bar\theta_\pm(-\beta/2) = -\bar\theta_\pm(\beta/2).$$

The perturbative expansion takes the form

$$\mathcal{Z}(\beta)/\mathcal{Z}_0(\beta) = 1 - v \int_{-\beta/2}^{\beta/2} dt \left\langle \theta_-(t)\theta_+(t)\bar\theta_+(t)\bar\theta_-(t) \right\rangle$$

$$+ \tfrac{1}{2}v^2 \int_{-\beta/2}^{\beta/2} dt\,du\, \left\langle \theta_-(t)\theta_+(t)\bar\theta_+(t)\bar\theta_-(t)\theta_-(u)\theta_+(u)\bar\theta_+(u)\bar\theta_-(u) \right\rangle + O(v^3),$$

$$\mathcal{Z}(\beta)/\mathcal{Z}_0(\beta)$$

$$= 1 - v \int_{-\beta/2}^{\beta/2} dt \left\langle \theta_-(t)\theta_+(t)\bar\theta_+(t)\bar\theta_-(t) \right\rangle$$

$$+ \tfrac{1}{2}v^2 \int_{-\beta/2}^{\beta/2} dt\,du\, \left\langle \theta_-(t)\theta_+(t)\bar\theta_+(t)\bar\theta_-(t)\theta_-(u)\theta_+(u)\bar\theta_+(u)\bar\theta_-(u) \right\rangle + O(v^3),$$

where $\mathcal{Z}_0$ is the square of the partition function (7.79). One finds

$$\mathcal{Z}(\beta)/\mathcal{Z}_0(\beta) = 1 - v\beta\Delta^2(0) + \tfrac{1}{2}v^2\beta \int_{-\beta/2}^{\beta/2} dt\,\left(\Delta^2(0) - \Delta(t)\Delta(-t)\right)^2 + O(v^3)$$

$$= 1 - \tfrac{1}{4}v\beta\left(\mathrm{sgn}(0) + \tanh(\omega\beta/2)\right)^2$$
$$+ \tfrac{1}{32}v^2\beta^2\left(1 + 2\,\mathrm{sgn}(0)\tanh(\omega\beta/2) + \mathrm{sgn}^2(0)\right)^2 + O(v^3).$$

Setting $\epsilon = \mathrm{sgn}(0)$, one concludes
$$E_0 = -(1+\epsilon)\omega + \tfrac{1}{4}(1+\epsilon)^2 v\,,$$
$$E_1 = E_2 = -\epsilon\omega - \tfrac{1}{4}(1-\epsilon^2)v\,,$$
$$E_3 = (1-\epsilon)\omega + \tfrac{1}{4}(1-\epsilon)^2 v\,.$$

For $\mathrm{sgn}(0) = -1$, one recovers the spectrum of the initial hamiltonian. To obtain the same result with the convention $\mathrm{sgn}(0) = 0$, one must substitute $\omega \mapsto \omega' = \omega + v/2$ and shift the empty state energy by $E_0 = -\omega' + v/4$.

Exercise 7.9

One now adds to the hamiltonian the term

$$\gamma\left(\theta_-\theta_+ + \frac{\partial}{\partial\theta_+}\frac{\partial}{\partial\theta_-}\right) \tag{7.119}$$

with γ real. This models the interaction with a medium that can absorb and emit electron pairs with equal probability. Determine the eigenvectors and spectrum of the total hamiltonian.

Solution. The new interaction mixes only the spinless states 1 and $\theta_-\theta_+$. The energies of the other states are unchanged. In the $1, \theta_-\theta_+$ subspace the hamiltonian reads

$$\begin{pmatrix} 0 & \gamma \\ \gamma & 2\omega + v \end{pmatrix}$$

with eigenvalues $\omega + v/2 \pm \sqrt{(\omega + v/2)^2 + \gamma^2}$. The ground state energy thus is $\omega + v/2 - \sqrt{(\omega + v/2)^2 + \gamma^2}$. Shifting the spectrum, in such a way that the ground state has zero energy, one obtains

$$0, \; -v/2 + \sqrt{(\omega + v/2)^2 + \gamma^2}, \; 2\sqrt{(\omega + v/2)^2 + \gamma^2}.$$

Exercise 7.10

In what follows one sets $v = 0$ and $\rho = \sqrt{\omega^2 + \gamma^2}$, $\omega = \rho\cos\varphi$, $\gamma = \rho\sin\varphi$. Show that the spectrum then has an interpretation in terms of two independent quasi-particles by setting

$$\theta_+ = a\eta_+ + b\frac{\partial}{\partial\eta_-}, \quad \theta_- = d\eta_- + c\frac{\partial}{\partial\eta_+}, \quad a, b, c, d \in \mathbb{R},$$

where η_+, η_- are two generators of a Grassmann algebra, and the corresponding operators $\eta_\pm$ and $\partial/\partial\eta_\pm$ are hermitian conjugate in the same way as $\theta_\pm$ and $\partial/\partial\theta_\pm$.

First, express $\partial/\partial\theta_\pm$ in terms of the η type operators. Then, show that the consistency of these transformations with the commutation relations (7.11) imply three conditions on the coefficients a, b, c, d that express that a, b, c, d are the four entries of an orthogonal matrix. Finally determine the coefficients to reduce the hamiltonian to the form

$$H = E_0 + \Omega\left(\eta_+\frac{\partial}{\partial\eta_+} + \eta_-\frac{\partial}{\partial\eta_-}\right), \quad \Omega > 0.$$

Solution. The form of the hamiltonian is obtained for the choice

$$\begin{pmatrix} a & b \\ c & d \end{pmatrix} = \begin{pmatrix} \cos(\varphi/2) & \sin(\varphi/2) \\ -\sin(\varphi/2) & \cos(\varphi/2) \end{pmatrix},$$

and $E_0 = \omega - \rho$, $\Omega = \rho$.

Exercise 7.11

Write the corresponding path integral representation of the partition function and perform the change of variables

$$\theta_+ = a\eta_+ + b\bar\eta_-, \quad \theta_- = d\eta_- + c\bar\eta_+,$$
$$\bar\theta_+ = a\bar\eta_+ + b\eta_-, \quad \bar\theta_- = d\bar\eta_- + c\eta_+,$$

using the values found in the preceding exercise. Show that the resulting path integral is consistent with the spectrum.

Solution. After the change of variables, one finds a path integral corresponding to the hamiltonian

$$H = \Omega\left(\eta_+ \frac{\partial}{\partial \eta_+} + \eta_- \frac{\partial}{\partial \eta_-}\right),$$

up to an additive constant.

Exercise 7.12

Spin group: fermion representation. One now considers the three operators

$$\tau_1 = \frac{1}{2}\left(\theta_+ \frac{\partial}{\partial \theta_-} + \theta_- \frac{\partial}{\partial \theta_+}\right), \quad \tau_2 = \frac{i}{2}\left(\theta_- \frac{\partial}{\partial \theta_+} - \theta_+ \frac{\partial}{\partial \theta_-}\right),$$
$$\tau_3 = \frac{1}{2}\left(\theta_+ \frac{\partial}{\partial \theta_+} - \theta_- \frac{\partial}{\partial \theta_-}\right).$$

In what follows, one sets $\hbar = 1$.

Verify that the operators are hermitian. Calculate the products $\tau_i \tau_j$ and $\tau^2 = \sum_i \tau_i^2$. Infer the commutators $[\tau_i, \tau_j]$. Find the eigenvectors and eigenvalues of τ_3 and τ^2. Calculate the commutators of τ_i with the hamiltonian (7.118) and the interaction (7.119). Comment.

Solution. The operators are generators of the Lie algebra of the spin group $SU(2)$:

$$\tau_i \tau_j = \tfrac{1}{3}\tau^2 \delta_{ij} + \tfrac{1}{2}i\epsilon_{ijk}\tau_k,$$
$$\tau^2 = \frac{3}{4}\left(\theta_+ \frac{\partial}{\partial \theta_+} + \theta_- \frac{\partial}{\partial \theta_-} - 2\theta_+ \frac{\partial}{\partial \theta_+}\theta_- \frac{\partial}{\partial \theta_-}\right).$$

The eigenvectors of τ_3 are $1, \theta_+, \theta_-, \theta_+\theta_-$ with eigenvalues $0, 1/2, -1/2, 0$, respectively, that is the spin components of the corresponding states. The corresponding eigenvalues of τ^2 are $0, 3/4, 3/4, 0$, as expected for spin $1/2$ particles. Finally, $\theta_+ \partial/\partial \theta_+$ and $\theta_- \partial/\partial \theta_-$ commute, and introducing

$$h = \theta_+ \frac{\partial}{\partial \theta_+} + \theta_- \frac{\partial}{\partial \theta_-},$$

one verifies

$$\tau_i h = h\tau_i = \tau_i.$$

The commutation of the hamiltonian (7.118) with the generators of the $SU(2)$ group follows from these relations and the commutations of the τ_i's with τ^2. One verifies that the interaction (7.119) is also $SU(2)$ invariant, which explains the form of the spectrum.

Exercise 7.13

Random hermitian matrices, in the limit of large size, can model some properties of complex quantum hamiltonians (in particular, those corresponding to a chaotic classical motion). A simple problem is then the determination of the spectrum of such random matrices in the limit of infinite size. As a related problem, we suggest here to calculate the expectation value of the characteristic polynomial of random hermitian matrices with a gaussian probability distribution, invariant under unitary transformations.

One thus considers the set of hermitian random $N \times N$ matrices with the probability distribution

$$\mathrm{d}\rho(M) = \mathcal{N}^{-1} \mathrm{d}^{N^2} M \, \mathrm{e}^{-N \operatorname{tr} M^2/2}, \quad \mathrm{d}^{N^2} M \equiv \prod_i \mathrm{d} M_{ii} \prod_{i<j} \mathrm{d} \operatorname{Re} M_{ij} \, \mathrm{d} \operatorname{Im} M_{ij}. \tag{7.120}$$

The normalization constant $\mathcal{N}$ is determined by the condition that the expectation value of 1 is 1 and, thus,

$$\mathcal{N} = \int \mathrm{d}^{N^2} M \, \mathrm{e}^{-N \operatorname{tr} M^2/2}.$$

The measure (7.120) is invariant under unitary transformations,

$$M \mapsto U^\dagger M U, \quad U U^\dagger = 1$$

and, thus, gives equal probability to all matrices having the same spectrum. This set of random matrices is called the gaussian unitary ensemble (GUE). One denotes the expectation value of a function $F(M)$ with the measure (7.120) by

$$\langle F(M) \rangle = \mathcal{N}^{-1} \int \mathrm{d}^{N^2} M \, \mathrm{e}^{-N \operatorname{tr} M^2/2} F(M).$$

(i) Justify that if X is an arbitrary $N \times N$ matrix

$$\left\langle \mathrm{e}^{\operatorname{tr} XM} \right\rangle = \mathrm{e}^{\operatorname{tr} X^2/2N}.$$

(ii) Represent the characteristic polynomial $\det(M - z)$ of a matrix M as a Grassmann integral.

(iii) Calculate then the expectation value $\mathcal{H}_N(z) = \langle \det(M - z) \rangle$ for the GUE.

(iv) The integrand takes the form of the exponential of a quartic polynomial in the Grassmann variables. Verify that the integrand can be written as a gaussian integral (over one real variable) of a gaussian function of the Grassmann variables.

(v) Evaluate then the gaussian integral over the Grassmann variables. The result is a integral over one real variable.

(vi) Generalize the method to calculate the expectation value of $[\det(M - z)]^n$ (the solution is not required for what follows).

(vii) One proves, quite generally, that the expectation value of $\det(M-z)$ with the measure $e^{-\operatorname{tr} V(M)}$, yields the orthogonal polynomials with respect to the measure $e^{-V(z)}$. Verify this property explicitly here by showing that the functions

$$\psi_N(z) = e^{-Nz^2/4}\mathcal{H}_N(z),$$

satisfy a Schrödinger equation. (the solution is not required for what follows.)

(viii) Evaluate $\mathcal{H}_N$ for $N \to \infty$ by applying the steepest descent method to the integral obtained in (v). What can be said about the support of the zeros of $\mathcal{H}_N(z)$? Determine the zeros in the large N limit.

(ix) How should the calculation be modified for *real, symmetric* matrices with gaussian $O(N)$ invariant measure (the GOE of the literature)?

Solution.

$$\det(M-z) = \int \prod_i d\bar{\theta}_i d\theta_i \, e^{\operatorname{tr} XM - z\operatorname{tr} X}$$

with

$$X_{ij} = \bar{\theta}_i \theta_j.$$

Then,

$$\operatorname{tr} X^2 = \sum_{i,j} \bar{\theta}_i \theta_j \bar{\theta}_j \theta_i = -\left(\sum_i \bar{\theta}_i \theta_i\right)^2 = -(\operatorname{tr} X)^2.$$

One notices

$$e^{\operatorname{tr} X^2/2N} = \frac{1}{\sqrt{2\pi}} \int ds \, e^{-s^2/2 + is \sum_i \bar{\theta}_i \theta_i/\sqrt{N}}.$$

It is convenient to change variables $s \mapsto s\sqrt{N}$. The integral over Grassmann variables is straightforward and yields the representation of Hermite polynomials:

$$\mathcal{H}_N(z) = \sqrt{N/2\pi} \int_{-\infty}^{+\infty} ds \, e^{-Ns^2/2}(is-z)^N. \tag{7.121}$$

A representation of $[\det(M-z)]^n$ is obtained by introducing an $n \times n$ hermitian matrix S and the expression above is replaced by

$$\langle [\det(M-z)]^n \rangle = (N/2\pi)^{n^2/2} \int dS \, e^{-N \operatorname{tr} S^2/2} \, [\det(iS-z)]^N.$$

The functions $\psi_N(z)$ satisfy the Schrödinger equation of the harmonic oscillator (with an N-dependent potential)

$$-\psi_N'' + \tfrac{1}{8}N^2 z^2 \psi_N = N(N+\tfrac{1}{2})\psi_N.$$

Alternative version: One introduces the polynomials

$$H_N(z) = \int_{-\infty}^{+\infty} ds \, e^{-s^2}(is-z)^N \propto \mathcal{H}_N(z\sqrt{2/N}).$$

The eigenfunctions of the harmonic oscillator then are
$$\psi_N(z) = e^{-z^2/2} H_N(z).$$

After the change of variables $s + iz = s'$, one finds
$$\psi_N(z) = e^{z^2/2} \int_{-\infty}^{+\infty} ds\, e^{-s^2 + 2izs} (is)^N.$$

One then verifies immediately
$$(x - d/dx)\psi_N(z) = 2\psi_{N+1}(z),$$

and, thus,
$$\psi_N(z) = \sqrt{\pi} 2^{-N} (x - d/dx)^N e^{-z^2/2}.$$

One recognizes creation operators acting on the ground state of the harmonic oscillator.

For $N \to \infty$, the integral (7.121) can be calculated by the steepest descent method (for details see exercise 1.6). The saddle points are
$$s = \frac{1}{s + iz} \Rightarrow s_\pm = -\tfrac{1}{2}iz \pm \sqrt{1 - z^2/4}.$$

One finds two saddle points. The zeros are given by the condition that the contributions of the two saddle points cancel. This implies first that the integrands have the same value and, thus, that z is real (this result is expected from general arguments that imply that the zeros of eigenfunctions of a hermitian hamiltonian of the form $p^2 + V(q)$ are real). Moreover, $\sqrt{1 - z^2/4}$ must be real, and this implies that the support of the zeros is the interval $-2 \leq z \leq 2$. Then, at leading order,
$$e^{-s_+^2/2}(is_+ - z) = e^{2i\pi k/N} e^{-s_-^2/2}(is_- - z),$$

where the r.h.s. comes from the Nth root and $1 \leq k \leq N$. It is convenient to set $z = 2\cos\varphi$, $0 \leq \varphi \leq \pi$. Then,
$$s_\pm = -i\, e^{\pm i\varphi}$$

and, thus,
$$2\varphi - \sin(2\varphi) = 2\pi k/N \quad (\text{mod } 2\pi).$$

One infers that, for $N \to \infty$, the density $w(\varphi)$ of the values φ is given by the derivative of the l.h.s.:
$$w(\varphi) = \frac{N}{2\pi}(1 - \cos(2\varphi)).$$

The density of eigenvalues of the matrix then is
$$w(z) = \left|\frac{d\varphi}{dz}\right| w(\varphi) = \frac{N}{\pi}\sqrt{1 - z^2/4}.$$

This classical result bears the name of Wigner's semi-circle law.

For real symmetric matrices, the only change comes from from the matrix X that must be symmetric and, thus,

$$X_{ij} = \tfrac{1}{2}\left(\bar\theta_i\theta_j + \bar\theta_j\theta_i\right),$$

$$\operatorname{tr} X^2 = -\tfrac{1}{2}\left(\sum_i \bar\theta_i\theta_i\right)^2.$$

The calculation then follows the same lines.

8 BARRIER PENETRATION: SEMI-CLASSICAL APPROXIMATION

A classical particle is always reflected by a potential barrier if its energy is lower than the potential. In contrast, a quantum particle has a non-vanishing probability to tunnel through a barrier, a property also called barrier penetration.

This chapter is devoted to a study of various physical manifestations of barrier penetration, in the semi-classical approximation, using the path integral formalism. Two specific problems are discussed: we evaluate, in the semi-classical limit $\hbar \to 0$, the splitting between two classically degenerate energy levels corresponding to two symmetric minima of a potential; we calculate in the same limit the decay rate, and thus the lifetime, of metastable states.

Since no classical trajectory can be associated to barrier penetration, one may wonder how it is possible to evaluate such effects, even in the semi-classical limit. Actually, it has been noticed that, formally, barrier penetration has a semi-classical interpretation in terms of classical particles moving in imaginary time (see also Section 9.5.3). The euclidean formalism used so far, based on calculating $e^{-\beta H}$, describes formally an evolution in imaginary time. We show, in this chapter, that it allows evaluating physical quantities related to barrier penetration.

Although the methods can be generalized, we mainly discuss properties of the ground state or close excited energy levels and, thus, for example, the partition function for $\beta \to \infty$. Our tool is the steepest descent method applied to the path integral, but in this problem the saddle points correspond to solutions of the equations of the classical motion that are no longer constants. These solutions satisfy one condition: the difference between their action and the action of the minimal constant solution remains finite when $\beta \to \infty$. One associates to such solutions the name *instanton*.

To calculate instanton contributions, one must master two problems that are increasingly difficult: find the saddle points by solving classical equations, expand the integrand around the saddle point and evaluate the path integral at leading order by integrating over gaussian fluctuations.

Finally, note that calculations based on the steepest descent method lead to semi-classical evaluations that can also be obtained by solving the Schrödinger equation in the WKB approximation, but the steepest descent method can be generalized much more easily to barrier penetration problems in quantum field theory.

8.1 Quartic double-well potential and instantons

We study a first family of quantum systems where tunneling plays a role: the hamiltonian has a discrete space symmetry, but the potential has minima at points that are not group invariant. The positions of the degenerate minima are then related by symmetry group transformations.

Classically, the minimal energy solutions correspond to particles at rest in any one of the minima of the potential. The position of the particle breaks (spontaneously) the symmetry of the system. In contrast, for a quantum system the ground state cannot be degenerate, as we have shown in Section 3.6. We thus expect the ground state to correspond to a symmetric wave function, its modulus being maximal near each of the minima of the potential. This phenomenon is a consequence of barrier penetration, a particle initially in one of the minima having a non-vanishing probability to tunnel into the others. We intend, here, to evaluate the energy splitting between the classically degenerate energy levels, in the semi-classical limit.

8.1.1 The quartic double-well potential

A simple example of a such a situation is provided by a reflection-symmetric hamiltonian with a quartic potential of the form of a symmetric double-well:

$$H = \tfrac{1}{2}\hat{p}^2 + \tfrac{1}{2}\left(\hat{x}^2 - \tfrac{1}{4}\right)^2.$$

This hamiltonian is clearly invariant under the reflection $x \mapsto -x$ and, thus, $p \mapsto -p$. To this reflection is associated an operator P that acts on wave functions as

$$[P\psi](x) = \psi(-x). \tag{8.1}$$

reflection symmetry, then, is expressed by the commutation of the quantum hamiltonian H with the reflection operator P:

$$[P, H] = 0.$$

The two operators P and H can thus be diagonalized simultaneously: the eigenfunctions ψ_n of H are even or odd functions:

$$[P\psi_{n,\pm}](x) = \psi_{n,\pm}(-x) = \pm \psi_{n,\pm}(x).$$

We have shown quite generally that the ground state wave function can be chosen positive (Section 3.6). The ground state wave function $\psi_{0,+}$, with energy $E_{0,+}$, is thus necessarily even:

$$\psi_{0,+}(x) = \psi_{0,+}(-x).$$

General arguments also imply that the wave function $\psi_{0,-}(x)$ of the first excited state, with energy $E_{0,-}$, vanishes once and is thus odd:

$$\psi_{0,-}(x) = -\psi_{0,-}(-x).$$

In the limit $\hbar \to 0$, one expects that if $\psi_{0,+}$ and $\psi_{0,-}$ have the same normalization, the wave functions $\psi_{0,+} \pm \psi_{0,-}$ correspond to particles almost entirely localized in each of the wells.

Perturbation theory. The properties of the ground state, in the semi-classical limit, can be inferred from the partition function $\mathcal{Z}(\tau/\hbar)$ in the limit $\hbar \to 0$ and then $\tau \to \infty$ (see the discussion of Sections 2.9 and 3.1). The partition function is given by the path integral

$$\mathcal{Z}(\tau/\hbar) = \int_{x(-\tau/2)=x(\tau/2)} [\mathrm{d}x(t)] \exp\left[-\mathcal{S}(x)/\hbar\right] \tag{8.2}$$

with

$$\mathcal{S}(x) = \int_{-\tau/2}^{\tau/2} \left[\tfrac{1}{2}\dot{x}^2(t) + \tfrac{1}{2}\left(x^2(t) - \tfrac{1}{4}\right)^2\right] \mathrm{d}t. \tag{8.3}$$

The potential has two degenerate minima $x = \pm 1/2$. Thus, for $\hbar \to 0$, the integrand has two saddle points that correspond to the two constant functions $x(t) = \pm 1/2$ and that, for symmetry reasons, yield identical contributions. To calculate the contribution of one saddle point, for example, $x(t) = -1/2$, one can set

$$x(t) = -\tfrac{1}{2} + q(t)\sqrt{\hbar}.$$

The action becomes

$$\mathcal{S}(q)/\hbar = \int_{-\tau/2}^{\tau/2} \left[\tfrac{1}{2}\dot{q}^2(t) + \tfrac{1}{2}q^2(t)\left(1 - q(t)\sqrt{\hbar}\right)^2\right] \mathrm{d}t. \tag{8.4}$$

One then expands in powers of $\hbar$. The existence of the two symmetric saddle points yields a factor 2, which simply indicates the presence of two states of energy E_0 degenerate to all orders in $\hbar$, corresponding to two wave functions located in each of the two wells of the potential.

Level splitting. The expected energy difference $E_{0,-} - E_{0,+}$ thus vanishes faster than any power of $\hbar$, and cannot easily be inferred from a calculation of $\mathrm{tr}\, \mathrm{e}^{-\tau H/\hbar}$. Indeed, in the double limit $\hbar \to 0$ then $\tau \to \infty$, one finds

$$\mathrm{tr}\, \mathrm{e}^{-\tau H/\hbar} \sim \mathrm{e}^{-\tau E_{0,+}/\hbar} + \mathrm{e}^{-\tau E_{0,-}/\hbar}$$

$$\sim 2\, \mathrm{e}^{-\tau(E_{0,+}+E_{0,-})/2\hbar} \cosh[\tau(E_{0,+} - E_{0,-})/2\hbar]. \tag{8.5}$$

The partition function, thus, is dominated by the perturbative expansion of the half-sum $E_0 = \tfrac{1}{2}(E_{0,+} + E_{0,-})$, and depends on the non-perturbative difference between $E_{0,+}$ and $E_{0,-}$ only at order $(E_{0,+} - E_{0,-})^2$.

To calculate the difference $E_{0,+} - E_{0,-}$, it is much more convenient to evaluate the quantity $\mathrm{tr}\, P\mathrm{e}^{-\tau H/\hbar}$. The eigenvalue of $\mathrm{tr}\, P\mathrm{e}^{-\tau H/\hbar}$ corresponding to the eigenvector $\psi_{n,\pm}(x)$ is then $\pm\mathrm{e}^{-\tau E_{n,\pm}/\hbar}$. In the same limits $\hbar \to 0$ then $\tau \to \infty$, one finds

$$\mathrm{tr}\, P\mathrm{e}^{-\tau H/\hbar} = \sum_{\pm,n} \pm \mathrm{e}^{-\tau E_{n,\pm}/\hbar}$$

$$\sim \mathrm{e}^{-\tau E_{0,+}/\hbar} - \mathrm{e}^{-\tau E_{0,-}/\hbar}$$

$$\sim -2\sinh[\tau(E_{0,+} - E_{0,-})/2\hbar]\, \mathrm{e}^{-\tau(E_{0,+}+E_{0,-})/2\hbar}. \tag{8.6}$$

Since $E_{0,+} - E_{0,-}$ vanishes to all orders in $\hbar$, at leading order (since $E_\pm \sim \tfrac{1}{2}\hbar$)

$$\operatorname{tr} P e^{-\tau H/\hbar} \sim -\tau e^{-\tau/2}\,\frac{E_{0,+} - E_{0,-}}{\hbar}. \tag{8.7}$$

Actually, it is convenient to consider the ratio between the quantities (8.5) and (8.6):

$$\operatorname{tr} P e^{-\tau H/\hbar} \Big/ \operatorname{tr} e^{-\tau H/\hbar} \sim -\frac{\tau}{2\hbar}(E_{0,+} - E_{0,-}). \tag{8.8}$$

The path integral representation of $\operatorname{tr} P e^{-\tau H/\hbar}$ differs from the representation of the partition function only in the boundary conditions. Indeed, in terms of matrix elements

$$\langle x|P|x'\rangle = \delta(x + x').$$

Therefore, for all operators U,

$$\operatorname{tr} PU = \int dy\, dx\, \delta(y+x)\,\langle x|U|y\rangle = \int dx\,\langle x|\,U\,|-x\rangle.$$

Applying this remark to the path integral, one infers

$$\operatorname{tr} P e^{-\tau H/\hbar} = \int_{x(-\tau/2)=-x(\tau/2)} [dx(t)]\exp\left[-\mathcal{S}(x)/\hbar\right] \tag{8.9}$$

with the same action (8.3).

8.1.2 Instantons

Notation. In what follows, we restrict the discussion to the two lowest energy eigenvalues and, thus, omit the subscript 0 on E.

While the path integral representing $\operatorname{tr} e^{-\tau H/\hbar}$ is dominated by the constant saddle points $x(t) = \pm 1/2$, these paths do not contribute to the integral (8.9) because they do not satisfy the corresponding boundary conditions. This is not too surprising since we have shown that the difference $E_+ - E_-$ vanishes faster than any power of $\hbar$. One must thus look for non-constant solutions of the equation of the euclidean classical motion. Moreover, the action of these solutions must have a finite limit in the relevant limit $\tau \to \infty$, otherwise they do not contribute. One associates to such solutions the name *instanton*, as if they would correspond to particles.

Since both the kinetic term and the potential are positive, this condition implies that both vanish for $|t| \to \infty$. In particular,

$$x(-\infty) = \pm\tfrac{1}{2} \text{ and } x(+\infty) = \mp\tfrac{1}{2}.$$

The splitting between the two energy levels thus depends on the existence of instanton solutions joining the two symmetric minima of the potential (Fig. 8.1).

The saddle point equation, which is the equation of the classical motion in euclidean or imaginary time, is

$$-\ddot{x}(t) + 2x(t)\left(x^2(t) - \tfrac{1}{4}\right) = 0. \tag{8.10}$$

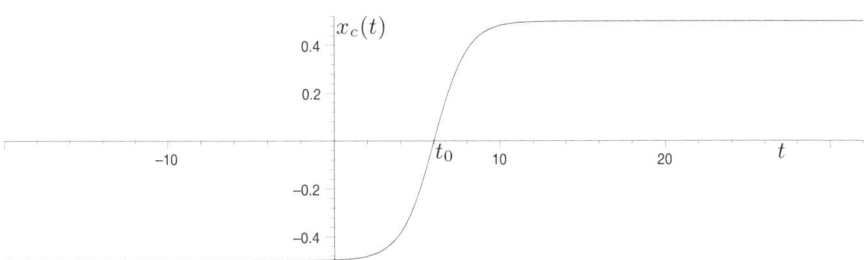

Fig. 8.1 Instanton-type solution.

In the limit $\tau \to \infty$, the equation has two families of solutions with finite action:
$$x_c^{\pm}(t) = \pm \tfrac{1}{2} \tanh\bigl((t - t_0)/2\bigr), \tag{8.11}$$

where t_0 is an integration constant, reflection of time-translation invariance for τ infinite.

The corresponding value of the action is
$$\mathcal{S}(x_c) = \tfrac{1}{6}. \tag{8.12}$$

Once the saddle point is identified, the corresponding contribution to the path integral is, in general, given at leading order by a gaussian integration. Here, the integration involves a rather subtle problem that we discuss later. However, note that we have found two families of degenerate saddle points, which depend on the parameter t_0. Since for τ large but finite, t_0 varies in an interval of size τ, the sum over all saddle points generates a factor τ, consistent with expression (8.8).

A complete calculation then yields
$$\operatorname{tr} P \,\mathrm{e}^{-\tau H/\hbar} \sim 2\sqrt{\frac{\hbar}{\pi}} \tau \,\mathrm{e}^{-\tau/2}\, \mathrm{e}^{-1/6\hbar}. \tag{8.13}$$

Comparing with expression (8.8), one obtains the asymptotic behaviour of $E_+ - E_-$ for $\hbar \to 0$:
$$E_- - E_+ \underset{\hbar \to 0}{=} 2\sqrt{\frac{\hbar}{\pi}} \,\mathrm{e}^{-1/6\hbar}\bigl(1 + O(\hbar)\bigr). \tag{8.14}$$

The difference decreases exponentially for $\hbar \to 0$ and, thus, faster than any power of $\hbar$, a result consistent with the perturbative discussion of Section 8.1.1.

8.2 Degenerate minima: semi-classical approximation

To evaluate the contribution of instantons, at leading order, it is convenient to consider a general potential V with similar properties: the potential is a regular, even function; it is minimum at two symmetric points $\pm x_0 \neq 0$ where it vanishes:
$$V(x) = V(-x),\ V(\pm x_0) = 0,\ V \geq 0.$$

8.2.1 Instantons

Following the analysis of Section 8.1.2, we calculate the twisted partition function

$$\operatorname{tr} P \mathrm{e}^{-\tau H/\hbar} = \int_{x(\tau/2) = -x(-\tau/2)} [\mathrm{d}x(t)] \exp\left[-\mathcal{S}(x)/\hbar\right] \tag{8.15}$$

with

$$\mathcal{S}(x) = \int_{-\tau/2}^{\tau/2} \left[\tfrac{1}{2}\dot{x}^2(t) + V(x(t))\right] \mathrm{d}t, \tag{8.16}$$

for $\hbar \to 0$ and $\tau \to \infty$.

The saddle point equation, obtained by varying the euclidean action, is identical to the equation of the usual classical motion (i.e. in real time) in the potential $-V(x)$:

$$-\ddot{x} + V'(x) = 0 \tag{8.17}$$

with the boundary conditions $x(-\infty) = -x(\infty)$.

Again, since both the kinetic term and the potential are positive, a solution, whose action remains finite when $\tau \to \infty$, must interpolate between the minima of the potential. In the limit $\tau \to \infty$, for finite action solutions, the integration of the equation yields

$$\tfrac{1}{2}\dot{x}_c^2(t) - V(x_c(t)) = 0. \tag{8.18}$$

Moreover, if $x_c(t)$ is a solution, $x_c(t - t_0)$ is a solution. For τ large but finite, the parameter t_0 varies in an interval of size τ.

Integrating equation (8.18), one can write the solutions as

$$t - t_0 = \pm \int_0^{x_c} \frac{\mathrm{d}y}{\sqrt{2V(y)}}.$$

Moreover, one infers from equation (8.18) that the corresponding action can be written as

$$A = \int_{-\infty}^{+\infty} \mathrm{d}t\, \dot{x}_c^2(t). \tag{8.19}$$

8.2.2 Gaussian integration and zero mode

We now expand the action $\mathcal{S}(x)$ around the saddle point, setting

$$x(t) = x_c(t) + r(t), \quad r(\tau/2) = -r(-\tau/2).$$

To second order in r, the expansion takes the form

$$\mathcal{S}(x_c + r) = A + \int_{-\tau/2}^{\tau/2} \mathrm{d}t\, \left[\tfrac{1}{2}\dot{r}^2(t) + V''(x_c(t))r^2(t)\right] + O(r^3).$$

8.2 Barrier penetration: semi-classical approximation

The quadratic form in r can be written as

$$\Sigma(r) = \int_{-\tau/2}^{\tau/2} dt \left[\tfrac{1}{2}\dot{r}^2(t) + \tfrac{1}{2}V''(x_c(t))r^2(t)\right]$$
$$= \frac{1}{2}\int dt_1 dt_2\, r(t_1)M(t_1,t_2)r(t_2),$$

where

$$M(t_1,t_2) = \frac{\delta^2 \mathcal{S}}{\delta x_c(t_1)\delta x_c(t_2)} = \left[-d_{t_1}^2 + V''(x_c(t_1))\right]\delta(t_1-t_2). \tag{8.20}$$

The differential operator M acts on a function $r(t)$ as

$$\int dt'\, M(t,t')r(t') = \frac{\delta}{\delta r(t)}\Sigma(r) = -\ddot{r}(t) + V''(x_c(t))r(t). \tag{8.21}$$

It has the form of a hermitian quantum hamiltonian, t playing the role of a position variable and $V''(x_c(t))$ being the potential. All its eigenvalues are real as well as its determinant.

Note that in the limit $\tau \to \infty$, only the trajectories that satisfy $r(\pm\infty) = 0$ contribute to the path integral in such a way that the boundary conditions are automatically satisfied.

Naively, the gaussian integral over $r(t)$ then leads to

$$\operatorname{tr} P\mathrm{e}^{-\tau/\hbar} \propto \mathrm{e}^{-A/\hbar}\int [dr(t)]\exp(-\Sigma(r)) \propto \frac{\mathrm{e}^{-A/\hbar}}{\sqrt{\det(M/\hbar)}},$$

an expression that must be evaluated in the limit $\tau \to \infty$.

The zero mode. Differentiating the equation of motion (8.17) with respect to t, one finds

$$\left[-d_t^2 + V''(x_c(t))\right]\dot{x}_c(t) = 0. \tag{8.22}$$

Comparing with equation (8.21), one recognizes the action of M on $\dot{x}_c$. Since the function $\dot{x}_c(t)$ is square integrable (equation (8.19)), the equation implies that $\dot{x}_c(t)$ is an eigenvector of M with vanishing eigenvalue:

$$M\dot{x}_c = 0. \tag{8.23}$$

The gaussian integration yields a result proportional to $(\det M)^{-1/2}$, which is thus infinite!

The problem should have been anticipated: as we have already pointed out, due to time-translation invariance, one finds a one-parameter family of degenerate saddle points related by continuous time-translations. The action is thus invariant under an infinitesimal variation of $x_c(t)$, which corresponds to a variation of the parameter t_0 and, thus, is proportional to $\dot{x}_c$. The problem that we face here is by no means specific to path integrals, as the example of an ordinary integral will show. Its solution requires the introduction of collective coordinates associated to the continuous symmetries of the integrand.

Another remark is important here. One infers from the general theory of orthogonal functions that the number of zeros of eigenfunctions of the hamiltonian M is directly related to the hierarchy of eigenvalues: the ground state of M has no zero, the first excited state has one zero... In the present example, the eigenfunction $\dot{x}_c(t)$ does not vanish (see Fig. 8.1): thus, it corresponds to the ground state, and all other eigenvalues of M are positive.

8.3 Collective coordinates and gaussian integration

To investigate the problem of the zero mode, we first consider an ordinary integral in which the integrand is invariant under some continuous group of transformations, here rotations in the plane. We show how the problem can be solved by introducing collective coordinates, using the so-called Faddeev–Popov method, a method that we then generalize to path integrals.

8.3.1 Zero modes in simple integrals

We consider a double integral of the general form (3.33):

$$I(g) = \int d^2\mathbf{x}\, e^{-S(\mathbf{x})/g}, \quad S(\mathbf{x}) = -\mathbf{x}^2/2 + (\mathbf{x}^2)^2/4, \qquad (8.24)$$

where $\mathbf{x}$ is the two-component vector (x_1, x_2), and the integrand is a function only of $\mathbf{x}^2$.

For $g \to 0_+$, this integral can be calculated by the steepest descent method. A naive approach is the following: the saddle points are solutions of the equation

$$\frac{\partial S}{\partial x_\mu} = -x_\mu(1 - \mathbf{x}^2) = 0. \qquad (8.25)$$

The origin $\mathbf{x} = \mathbf{0}$, which corresponds to a relative maximum, is not a relevant saddle point. The minima correspond to

$$|\mathbf{x}| = 1. \qquad (8.26)$$

Due to the rotation invariance of the integrand, one finds here also a one-parameter family of degenerate saddle points belonging to a circle, since only the size of the vector $\mathbf{x}$ is determined by the saddle point equation. If one chooses one particular saddle point and evaluates its contribution in the gaussian approximation, one finds a result that involves the determinant of the matrix

$$M_{\mu\nu} = \left.\frac{\partial^2 S}{\partial x_\mu \partial x_\nu}\right|_{|\mathbf{x}|=1} = 2x_\mu x_\nu. \qquad (8.27)$$

The matrix is a projector on $\mathbf{x}$. The vector orthogonal to $\mathbf{x}$ corresponds to a flat direction for the integrand and, thus, is an eigenvector with a vanishing eigenvalue.

Here, the problem has a straightforward solution: the integral over the angular variable that parametrizes the set of all saddle points, also called the *collective coordinate*, must be calculated exactly; only the integral over the length of the vector can be evaluated by the steepest descent method.

8.3.2 Faddeev–Popov's method

Since it is not always so easy to factorize the integration measure in terms of variables that parametrize the saddle points, the collective coordinates, and variables that can be handled by the steepest descent method, the so-called Faddeev–Popov's method (which plays an essential role in the quantization of non-abelian gauge theories), can often be used. One first introduces a function of the integration variables that is not invariant, that is, here, not invariant under rotations, and one averages it over the symmetry group. In example (8.24), we can choose the function $\delta(x_2)$ where δ is Dirac's function. Then,

$$\int_0^{2\pi} d\theta\, \delta(x_1 \sin\theta + x_2 \cos\theta) = \frac{2}{|\mathbf{x}|} \Leftrightarrow \frac{|\mathbf{x}|}{2} \int_0^{2\pi} d\theta\, \delta(x_1 \sin\theta + x_2 \cos\theta) = 1,$$

The result of the integration is an invariant function; in the example it depends only on the length of the vector $\mathbf{x}$. One introduces the latter identity into the initial integral in the form

$$I(g) = \frac{1}{2} \int d^2\mathbf{x} \int_0^{2\pi} d\theta\, |\mathbf{x}|\, \delta(x_1 \sin\theta + x_2 \cos\theta)\, e^{-S(\mathbf{x})/g}.$$

One then interchanges the order between integrations and one changes variables, $\mathbf{x} \mapsto \mathbf{y}$, setting

$$y_1 = x_1 \cos\theta - x_2 \sin\theta, \quad y_2 = x_1 \sin\theta + x_2 \cos\theta,$$

a change of variables that is also a rotation. The function e^{-S}/g and the integration measure are not affected. Thus,

$$I(g) = \frac{1}{2} \int_0^{2\pi} d\theta \int d^2\mathbf{y}\, |\mathbf{y}|\, \delta(y_2)\, e^{-S(\mathbf{y})/g}$$

$$= \frac{1}{2} \int_0^{2\pi} d\theta \int_{-\infty}^{+\infty} dy_1\, |y_1|\, e^{-S(y_1)/g}$$

$$= 2\pi \int_0^{+\infty} y_1 dy_1\, e^{-S(y_1)/g}.$$

One recognizes the radial integral that can then be evaluated by the steepest descent method.

8.3.3 Collective coordinates in path integrals

In the case of a path integral also, it is necessary to integrate exactly over the variables that parametrize the saddle points, the so-called *collective coordinates*. In the example of the instanton solutions of equation (8.18), the time-translation parameter is the collective coordinate. Again, one must explicitly factorize the integration over the collective time parameter in the integration measure. This is the idea of the method of collective coordinates. The problem is slightly more subtle than in the example (8.24) because the number of integration variables is infinite.

Collective coordinates and Faddeev–Popov's method. To factorize the integration over to the collective time parameter (the collective coordinate), we adapt Faddeev–Popov's method, explained in Section 8.3.2, to this new situation.

We denote now by $x_c(t)$ a particular solution of the saddle point equation (8.18) corresponding to $t_0 = 0$ and the general solution then is $x_c(t - t_0)$.

We start from the identity

$$1 = \frac{1}{\sqrt{2\pi\xi}} \int_{-\infty}^{+\infty} d\lambda\, e^{-\lambda^2/2\xi},$$

where ξ is an arbitrary constant. We then change variables, $\lambda \mapsto t_0$, with

$$\lambda = \int dt\, \dot{x}_c(t)\bigl(x(t+t_0) - x_c(t)\bigr).$$

We obtain a new identity:

$$\frac{1}{\sqrt{2\pi\xi}} \int dt_0 \left[\int dt\, \dot{x}_c(t)\dot{x}(t+t_0)\right] \exp\left\{-\frac{1}{2\xi}\left[\int dt\, \dot{x}_c(t)\bigl(x(t+t_0) - x_c(t)\bigr)\right]^2\right\}$$
$$= 1. \tag{8.28}$$

The constant ξ has been introduced partially for cosmetic reasons, but is considered to be of order $\hbar$.

We insert identity (8.28) into the path integral (8.15):

$$\operatorname{tr} P\,e^{-\tau H/\hbar} = \frac{1}{\sqrt{2\pi\xi}} \int dt_0 \int [dx(t)] \left[\int dt\, \dot{x}_c(t)\dot{x}(t+t_0)\right] \exp\left[-\mathcal{S}_\xi(x)/\hbar\right],$$

where the total action

$$\mathcal{S}_\xi(x) = S(x) + \frac{\hbar}{2\xi}\left[\int dt\, \dot{x}_c(t)\bigl(x(t+t_0) - x_c(t)\bigr)\right]^2$$

is no longer invariant under time-translations because time appears explicitly through the function $x_c(t)$.

The function $x(t+t_0)$ can now be renamed $x(t)$. This affects $S(x)$, but we change variables, $t - t_0 \mapsto t$, in the action. Then, for $\tau = \infty$, one recovers the initial action because the integration domain is not modified.

The integrand then no longer depends on the variable t_0 and the integration over t_0 is immediate. For $\tau \to \infty$,

$$\operatorname{tr} P\,e^{-\tau H/\hbar} \sim \frac{\tau}{\sqrt{2\pi\xi}} \int [dx(t)] \left[\int dt\, \dot{x}_c(t)\dot{x}(t)\right] \exp\left[-\mathcal{S}_\xi(x)/\hbar\right]$$

with

$$\mathcal{S}_\xi(x) = S(x) + \frac{\hbar}{2\xi}\left[\int dt\, \dot{x}_c(t)\bigl(x(t) - x_c(t)\bigr)\right]^2.$$

8.3.4 Gaussian integration

The saddle point equation becomes

$$\frac{\delta S}{\delta x(t)} + \frac{\hbar}{\xi}\dot{x}_c(t)\int dt'\, \dot{x}_c(t')\bigl(x(t') - x_c(t')\bigr) = 0. \tag{8.29}$$

Clearly, the solution of this equation is $x(t) = x_c(t)$. The second functional derivative of the action at the saddle point is then modified by an additional contribution:

$$\frac{\delta^2 S}{\delta x_c(t_1)\delta x_c(t_2)} \mapsto M_\xi(t_1,t_2) \equiv \frac{\delta^2 S}{\delta x_c(t_1)\delta x_c(t_2)} + \frac{\hbar}{\xi}\dot{x}_c(t_1)\dot{x}_c(t_2).$$

The additional operator is a projector on to the eigenvector of $\delta^2 S/\delta x_c\delta x_c$ corresponding to the vanishing eigenvalue. The modified operator, thus, has the same eigenvectors and the same eigenvalues as the initial operator $\delta^2 S/\delta x_c\delta x_c$, with one exception: the eigenvalue corresponding to the eigenvector $\dot{x}_c$ is now

$$\mu = \hbar\|\dot{x}_c\|^2/\xi \tag{8.30}$$

instead of 0. Therefore, the determinant of the operator M_ξ no longer vanishes and the problem of the zero mode is solved.

The normalization of the path integral can be inferred by comparing it to the partition function $\mathcal{Z}_0(\tau/\hbar)$ of the harmonic oscillator:

$$\mathcal{Z}_0(\tau/\hbar) = \int_{x(-\tau/2)=x(\tau/2)} [\mathrm{d}x(t)]\exp\left\{-\frac{1}{2\hbar}\int_{-\tau/2}^{\tau/2} dt\, \left[\dot{x}^2(t) + x^2(t)\right]\right\}, \tag{8.31}$$

which for $\tau \to \infty$ reduces to $\mathrm{e}^{-\tau/2}$. In this limit, the gaussian integral can be expressed in terms of the operator

$$M_0(t_1,t_2) = \left[-(\mathrm{d}_{t_1})^2 + 1\right]\delta(t_1 - t_2). \tag{8.32}$$

As we show later, what can be evaluated easily is the determinant of the product of operators $(M + \varepsilon)(M_0 + \varepsilon)^{-1}$, where ε is an arbitrary constant. For $\varepsilon \to 0$, this expression vanishes linearly in ε and we thus set

$$\lim_{\varepsilon \to 0}\frac{1}{\varepsilon}\det(M+\varepsilon)(M_0+\varepsilon)^{-1} \equiv \det' M M_0^{-1}. \tag{8.33}$$

On the other hand, what is needed here is (the factors $\hbar$ cancel in the ratio of gaussian integrals)

$$\det M_\xi = \det(M + \mu\,|0\rangle\langle 0|)M_0^{-1}$$
$$= \lim_{\varepsilon \to 0}\det(M+\varepsilon+\mu\,|0\rangle\langle 0|)(M_0+\varepsilon)^{-1},$$

where $|0\rangle$ is the vector proportional to $\dot{x}_c$ with unit norm and $\mu = \|\dot{x}_c\|^2 \hbar/\xi$. Then, after some simple algebra,

$$\det\left(M + \varepsilon + \mu |0\rangle \langle 0|\right)(M_0 + \varepsilon)^{-1} = \det\left(M + \varepsilon\right)(M_0 + \varepsilon)^{-1}$$
$$\times \det\left[1 + \mu |0\rangle \langle 0| (M + \varepsilon)^{-1}\right]$$
$$= (1 + \mu/\varepsilon) \det\left(M + \varepsilon\right)(M_0 + \varepsilon)^{-1}.$$

In the limit $\varepsilon \to 0$, one thus finds

$$\det{}' M M_0^{-1} \|\dot{x}_c\|^2 \hbar/\xi.$$

This result leads to the factor

$$\frac{1}{\sqrt{2\pi\xi}} \tau \|\dot{x}_c\|^2.$$

Collecting all factors, one concludes that the gaussian integration over the configurations in the neighbourhood of the saddle point yields a factor

$$\frac{\tau}{\sqrt{2\pi\hbar}} \|\dot{x}_c\| (\det{}' M M_0^{-1})^{-1/2} e^{-\tau/2}.$$

As expected, the dependence on ξ has cancelled.

Taking into account the two families of saddle points and the ratio 2 between $\mathcal{Z}_0(\tau/\hbar)$ and $\operatorname{tr} e^{-\tau H/\hbar}$ for $\tau \to \infty$, one obtains

$$\operatorname{tr} P e^{-\tau H/\hbar} / \operatorname{tr} e^{-\tau H/\hbar} \sim \frac{\tau}{\sqrt{2\pi\hbar}} \|\dot{x}_c\| \left[\det{}' M (\det M_0)^{-1}\right]^{-1/2} e^{-A/\hbar} \qquad (8.34)$$

and, thus, using the result (8.8), the splitting of levels

$$E_- - E_+ \sim 2\sqrt{\frac{\hbar}{2\pi}} \|\dot{x}_c\| \left[\det{}' M (\det M_0)^{-1}\right]^{-1/2} e^{-A/\hbar}$$

8.3.5 Application: the double-well potential

A solution of the classical equation of motion is

$$x_c(t) = \tfrac{1}{2} \tanh(t/2).$$

Moreover, (equation (8.19))

$$\|x_c\| = \sqrt{A} = \frac{1}{\sqrt{6}}.$$

Finally,

$$M = -\mathrm{d}_t^2 + 1 - \frac{3}{2\cosh^2(t/2)}.$$

8.3   *Barrier penetration: semi-classical approximation*   237

The operator M has the form of a hamiltonian of Bargmann–Wigner's type: the corresponding Schrödinger equation can be solved explicitly and quantum scattering is reflectionless (the S-matrix is given in Exercise 9.2; its poles yield the spectrum of the hamiltonian). The determinant can also be calculated explicitly. Quite generally, for

$$M = -d_t^2 + 1 - \frac{\lambda(\lambda+1)\omega^2}{\cosh^2(\omega t)},$$

one finds

$$\det(M+\varepsilon)(M_0+\varepsilon)^{-1} = \frac{\Gamma(1+z)\Gamma(z)}{\Gamma(1+\lambda+z)\Gamma(z-\lambda)} \quad (8.35)$$

with

$$z = \sqrt{1+\varepsilon}/\omega.$$

Here $\omega = 1/2$, $\lambda = 2$ and thus $z - 2 \sim \varepsilon$. Then,

$$\det(M+\varepsilon)(M_0+\varepsilon) = \frac{(z-2)(z-1)}{(z+1)(z+2)} \underset{\varepsilon\to 0}{\sim} \frac{\varepsilon}{12}.$$

One infers

$$E_- - E_+ \sim 2\sqrt{\frac{\hbar}{\pi}} e^{-1/6\hbar}.$$

Remarks.

(i) We have calculated the instanton contribution only in the $\tau = \infty$ limit, in which the action has boundary conditions invariant under time-translations. The calculation for τ large but finite, involves a few additional subtleties.

(ii) It is possible to study the semi-classical effects to all orders in an expansion in powers of $e^{-1/6\hbar}$. This has led to a conjecture, later proved, which generalizes the usual Bohr–Sommerfeld's formula to the situation of potentials with degenerate minima. The energy eigenvalues E of the hamiltonian are solutions of a secular equation that can be written, in the case of the quartic double-well potential, as

$$\Gamma^2\left(\tfrac{1}{2} - B(E,\hbar)\right) \left(-\frac{2}{\hbar}\right)^{2B(E,\hbar)} e^{-A(E,\hbar)} + 2\pi = 0 \quad (8.36)$$

with

$$B(E,\hbar) = -B(E,-\hbar) = \frac{E}{\hbar} + \sum_{k=1}^{\infty} \hbar^k b_{k+1}(E/\hbar), \quad (8.37)$$

$$A(E,\hbar) = -A(E,-\hbar) = \frac{1}{3\hbar} + \sum_{k=1}^{\infty} \hbar^k a_{k+1}(E/\hbar). \quad (8.38)$$

The coefficients $a_k(s)$ and $b_k(s)$ are even or odd polynomials in s according to the degree k.

The perturbative expansion for $\hbar \to$ applies to energy eigenvalues $E = O(\hbar)$, while the semi-classical expansion assumes $E = O(1)$. This amounts to summing the terms of largest degree in E to all orders in $\hbar$.

8.4 Instantons and metastable states

We now study another situation in which quantum tunneling plays a role: the decay of metastable states. We assume a quantum particle initially located in the well of a potential that corresponds to a local but not absolute minimum. An example of such a potential is exhibited in Fig. 8.2, where the origin does not correspond to an absolute minimum of the potential. As a consequence of quantum tunneling, a quantum particle has a finite probability per unit time to leave the well and this is the probability we now want to determine in the limit $\hbar \to 0$.

As a restriction, we discuss only initial states localized deep in the well, that is close to the pseudo-ground state in the well (the equivalent of a classical particle almost at rest). We will show that, as for the perturbative calculation, one can derive the decay rate from the partition function $\mathcal{Z}(\tau/\hbar) = \operatorname{tr} e^{-\tau H/\hbar}$ for $\tau \to \infty$ (see Sections 2.9 and 3.1).

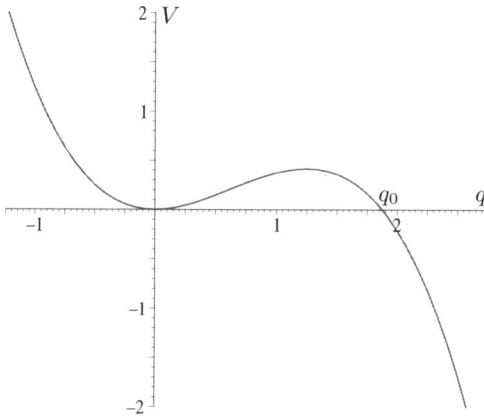

Fig. 8.2 Potential well leading to metastability.

Quantum metastability. In the case of a potential of the type exhibited in Fig. 8.2, the origin is not the absolute minimum of the potential. A state corresponding to a wave function $\psi(t)$, localized at initial time $t = 0$ (t is here the *real physical time* of the Schrödinger equation) in the well of the potential around $q = 0$, decays through barrier penetration. In order to understand how to calculate the decay rate, we vary continuously a parameter in the potential in order to pass continuously from a situation where the origin is an absolute minimum to a situation where it becomes only a relative minimum. In the stable situation, the solution of the time-dependent Schrödinger equation, corresponding to the ground state energy E_0, behaves as

$$\psi_0(t) \sim e^{-iE_0 t/\hbar}.$$

After analytic continuation, E_0 becomes complex and, thus, $\psi_0(t)$ decreases exponentially with time:

$$|\psi_0(t)| \underset{t \to +\infty}{\sim} e^{-|\operatorname{Im} E_0| t/\hbar}.$$

8.4 Barrier penetration: semi-classical approximation

The parameter $|\hbar/\operatorname{Im} E_0|$ is the lifetime of a now metastable state with wave function $\psi(t)$. Let us point out that the decay of a state receives contributions from the continuation of all excited states. However, one expects, for intuitive reasons, that when the real part of the energy increases the corresponding contribution decreases faster with time, a property that can, indeed, be verified in examples. Thus, for large times, only the component corresponding to the pseudo-ground state survives. We show now how to calculate $\operatorname{Im} E_0$ for $\hbar \to 0$.

8.4.1 A simple integral

We first study a simple integral that shares some of the properties of the path integral:
$$I(\lambda, \varepsilon) = \int_{-\infty}^{\infty} \mathrm{d}x \, \mathrm{e}^{-S(x)},$$
$$S(x) = \tfrac{1}{2}x^2 + \tfrac{1}{3}\lambda x^3 + \tfrac{1}{4}\varepsilon x^4.$$

Initially, the two parameters ε and λ are positive. For $\varepsilon > \lambda^2/4$, the function $S(x)$ is minimum at $x = 0$. The function $I(\lambda, \varepsilon)$ can be expanded in powers of λ:
$$I(\lambda, \varepsilon) = \sum_{k=0}^{\infty} \frac{(-\lambda)^k}{3^k k!} \int_{-\infty}^{\infty} \mathrm{d}x \, x^{3k} \, \mathrm{e}^{-x^2/2 - \varepsilon x^4/4},$$

and the expansion is convergent.

Each term of the expansion has a finite limit when $\varepsilon \to 0$ and, formally,
$$I(\lambda, 0) = \sum_{k=0}^{\infty} \frac{(-\lambda)^k}{3^k k!} \int_{-\infty}^{+\infty} \mathrm{d}x \, x^{3k} \, \mathrm{e}^{-x^2/2}$$
$$= \sum_{k=0}^{\infty} \frac{2^{3k-1/2} \Gamma(3k+1/2)}{3^{2k}(2k)!} \lambda^{2k}. \tag{8.39}$$

However, the limiting function $I(\lambda, 0)$ is given by an integral that no longer converges, and the interpretation of the expansion is not immediate. Moreover, the limiting series (8.39) diverges for all values of λ, and no longer defines an analytic function.

Analytic continuation. If instead of taking the limit $\varepsilon \to 0$ with ε real positive, one performs an analytic continuation to complex values
$$\pi/3 < \operatorname{Arg} \varepsilon < \pi/2 \text{ or } -\pi/2 < \operatorname{Arg} \varepsilon < -\pi/3,$$
then the initial integration contour can be deformed into the contours
$$C_{\pm} : \operatorname{Im} x = 0 \text{ for } \operatorname{Re} x > 0, \ \operatorname{Arg} x = \pi \pm 5\pi/24 \text{ for } \operatorname{Re} x < 0,$$

respectively, and the final integrals always converge but define two different limiting functions
$$I_{\pm}(\lambda) = \oint_{C_{\pm}} \mathrm{e}^{-x^2/2 - \lambda x^3/3} \, \mathrm{d}x,$$

depending on the choice of contour. The two functions are complex conjugate:

$$I_-(\lambda) = \bar{I}_+(\lambda).$$

When $\lambda \to 0$, these integrals can be evaluated by the steepest descent method. The saddle points are given by

$$S'(x) = x + \lambda x^2 = 0 \Rightarrow x = 0 \text{ or } x = -1/\lambda.$$

The corresponding values of $S(x)$ are

$$S(0) = 0, \ S(-1/\lambda) = 1/6\lambda^2.$$

For $\lambda \to 0$, the leading saddle point is always $x = 0$ and the expansion (8.39) in powers of λ is the expansion generated by the steepest descent method. To all orders in λ, the two functions have identical real expansions. The second saddle point is sub-leading and gives negligible contributions.

The divergent series (8.39) is an asymptotic expansion of both functions $I_\pm$. An asymptotic series does not define, in general, a unique analytic function but may, nevertheless, provide a very good approximation when the expansion parameter is small enough and the summation is limited to a finite number of terms (see also the discussion of Section 1.5).

The difference between the two integrals is purely imaginary. It is given by the contour $C_+ - C_-$, which can be deformed into a contour that avoids the origin. The resulting contour integral is dominated for $\lambda \to 0$ by the second saddle point. The imaginary parts are invisible in the expansion (8.39) because they decrease faster than any power. A calculation of the contribution of the saddle point then yields, at leading order,

$$I_+(\lambda) - I_-(\lambda) = 2i \operatorname{Im} I_+(\lambda) \sim i\sqrt{2\pi}\, e^{-1/6\lambda^2}.$$

8.4.2 Path integral and steepest descent method: instantons

We now apply an analogous strategy to the path integral. We begin with a situation, where in the hamiltonian

$$H = \frac{1}{2m}p^2 + V(q),$$

the potential has an absolute minimum at the origin where

$$V(q) = \tfrac{1}{2}m\omega^2 q^2 + O(q^3).$$

By an analytic continuation in a parameter of V, we pass to a situation where the minimum of the potential at $q = 0$ is only relative and, thus, there exist values of q for which $V < 0$. We do not consider below the special situation where the minimum of the potential is degenerate because some aspects of this problem have already been discussed in Sections 8.1-8.3.

8.4 Barrier penetration: semi-classical approximation

Here, we calculate the imaginary part of $\mathcal{Z}(\tau/\hbar) = \operatorname{tr} e^{-\tau H/\hbar}$ for $\tau \to \infty$. The result is expected to have to form

$$\operatorname{Im} \mathcal{Z}(\tau/\hbar) \sim \operatorname{Im} e^{-\tau E_0/\hbar} \sim -\frac{\tau}{\hbar} \operatorname{Im} E_0 \, e^{-\tau \operatorname{Re} E_0/\hbar}.$$

At leading order in $\hbar$, $\operatorname{Re} E_0$ can be replaced by the value its assumes in the harmonic approximation and, thus,

$$\operatorname{Im} \mathcal{Z}(\tau/\hbar) \sim -\frac{\omega \tau}{\hbar} e^{-\omega \tau/2} \operatorname{Im} E_0. \tag{8.40}$$

Instantons. We look for non-trivial saddle points of the path integral. The saddle point equation, obtained by varying the euclidean action, is

$$-m\ddot{q} + V'(q) = 0 \tag{8.41}$$

with $q(-\tau/2) = q(\tau/2)$.

The functions

$$q(t) = q_{\text{ext.}} = \text{const.}, \tag{8.42}$$

where $q_{\text{ext.}}$ corresponds to an extremum of the potential, are clearly solutions. We do not take into account the saddle points with $V < 0$ for the same reasons as in the case of the simple integral: the analytic continuation leads to integration domains that avoid such saddle points. On the other hand, the contributions of saddle points corresponding to extrema where $V > 0$ are of order $e^{-\tau V_{\text{ext.}}/\hbar}$ and, thus, negligible for $\tau \to \infty$ and $\hbar \ll 1$ since we consider only energy eigenvalues of order $\hbar$.

Therefore, we look for solutions that have an action that has a finite limit when $\tau \to +\infty$, that is, *instanton-type* solutions.

The solutions of equation (8.41) with periodic boundary conditions correspond to periodic motions in *real time* in the potential $-V(q)$. It is clear that trajectories can be found that oscillate around the minima of $-V$. A first integration of the equation of motion (8.41) yields

$$\tfrac{1}{2}m\dot{q}^2 - V(q) = \epsilon$$

with $\epsilon < 0$. Denoting by $q_- < q_+$ the two points where the velocity $\dot{q}$ vanishes, we obtain for the period of such a solution

$$\tau = 2\sqrt{m} \int_{q_-}^{q_+} \frac{dq}{\sqrt{2V(q) + 2\epsilon}}.$$

The period τ diverges only for constants ϵ such that $V(q) + \epsilon$ has a double zero at q_- or q_+, and this implies that $V'(q)$ vanishes. Moreover, the action remains finite in this limit only if $V(q(t))$ and $\dot{q}$ vanish for $|t| \to \infty$. These conditions are compatible only if ϵ and thus, for example, q_- vanish. The corresponding classical trajectory comes increasingly closer to the origin. Then, the limit $q_0 > 0$ of q_+ is the point on the trajectory where the velocity vanishes. In the limit $\tau \to \infty$, the classical solution is given by (t_0 is an integration constant)

$$t - t_0 = \sqrt{m} \int_{q_0}^{q} \frac{dq'}{\sqrt{2V(q')}} \quad \text{for } t < t_0,$$

$$t - t_0 = \sqrt{m} \int_{q}^{q_0} \frac{dq'}{\sqrt{2V(q')}} \quad \text{for } t > t_0.$$

The instanton action. A remark in the spirit of the virial theorem, interesting because it can be generalized to more complicated examples, can be used here. If $q_c(t)$ is a finite action solution on the interval $t \in (-\infty, +\infty)$, then $q_c(\lambda t)$ has also a finite action. After the change of variables $\lambda t = t'$, the action corresponding to $q_c(\lambda t)$ becomes

$$\mathcal{S}(\lambda) = \tfrac{1}{2} m \lambda \int dt\, \dot q_c^2(t) + \frac{1}{\lambda} \int dt\, V(q_c(t)).$$

Since $\mathcal{S}(q)$ is stationary for $q(t) = q_c(t)$, the derivative $d\mathcal{S}/d\lambda$ must vanish at $\lambda = 1$. One infers

$$\tfrac{1}{2} m \int dt\, \dot q_c^2(t) = \int dt\, V(q_c(t))$$

and, thus, the corresponding classical action

$$\mathcal{S}(q_c) \equiv A = m \int_{-\infty}^{+\infty} dt\, \dot q_c^2(t) = 2 \int_0^{q_0} \sqrt{2V(q)}\, dq \qquad (8.43)$$

is positive. The instanton thus gives a contribution of the order of $e^{-A/\hbar}$, which decreases exponentially for $\hbar/A \to 0$.

Remarks.

(i) One may wonder whether it makes sense to take into account such small contributions, since E_0 is first dominated by an expansion to all orders in $\hbar$. Actually, if one starts from a stable situation and proceeds by analytic continuation, one can obtain two complex conjugate results. Each result is indeed dominated by the same trivial saddle point $q(t) \equiv 0$, from which originates the perturbation series whose terms are all real. In contrast, if one calculates the difference between the two continuations, the contributions of the leading saddle point cancel and the difference is dominated by the instanton. As a consistency check, one must thus verify that the instanton contribution is purely imaginary.

(ii) Since the euclidean action is invariant under time-translations, the classical solution depends on an arbitrary parameter t_0, which for finite τ, varies in the interval $[-\tau/2, \tau/2]$. As in the example of Section 8.2, one finds a one-parameter family of degenerate saddle points. In the calculation of the contribution of a saddle point the dependence on t_0 disappears, and thus all saddle points give the same contribution.

(iii) One could have also considered trajectories that oscillate n times around the maximum of the potential in a time interval τ. It is easy to verify that the corresponding action in the limit $\tau \to \infty$ becomes

$$S(q_c) = nA, \qquad (8.44)$$

and yields a contribution of order $\mathrm{e}^{-nA/\hbar}$. For $\hbar \to 0$, the $n = 1$ contribution thus dominates the imaginary part of the path integral.

Leading order contribution: gaussian approximation. The arguments of Section 8.2 apply also here. The naive steepest descent method with gaussian integration involves the determinant of the operator

$$M(t_1, t_2) = \frac{\delta^2 S}{\delta q_c(t_1)\delta q_c(t_2)} = \left[-m\mathrm{d}_{t_1}^2 + V''\!\left(q_c(t_1)\right)\right]\delta(t_1 - t_2). \qquad (8.45)$$

A differentiation with respect to time of the equation of motion (8.41) yields

$$\left[-m\mathrm{d}_t^2 + V''\!\left(q_c(t)\right)\right]\dot{q}_c(t) \equiv M\dot{q}_c = 0. \qquad (8.46)$$

Thus, $\dot{q}_c$ (which is square integrable, see equation (8.43)) is an eigenvector of the hermitian operator M and the corresponding eigenvalue vanishes.

However, let us point out one important difference between this problem and the problem of degenerate minima. As we have already mentioned, the general theory of orthogonal functions shows that the number of zeros of an eigenfunction of the hamiltonian M is directly related to the hierarchy of eigenvalues: the ground state of M has no zero, the first excited state has one zero... Thus, the eigenfunction $\dot{q}_c(t)$, which vanishes exactly once, for $t = t_0$, corresponds to the first excited state, and this implies that in this problem M has one negative eigenvalue. The product $\det' M$ of the non-vanishing eigenvalues of M is negative and $\sqrt{\det' M}$ imaginary, as expected.

8.5 Collective coordinates: alternative method

Again, due to the existence of the time zero mode, it is necessary to introduce a time collective coordinate, and one can use the gaussian approximation only for the modes orthogonal to $\dot{q}_c$. The method of Section 8.3.3 can be adapted to this new situation, but it is instructive to describe an alternative solution to the same problem.

We now denote by $q_c(t)$ the particular solution of the saddle point equation (8.41) corresponding to $t_0 = 0$ and, thus, the general solution is $q_c(t - t_0)$.

To introduce an integration variable associated with time-translations, we set

$$q(t) = q_c(t - t_0) + r(t - t_0)\sqrt{\hbar}, \qquad (8.47)$$

where t_0 is no longer a simple parameter, but forms, together with the path $r(t)$ a new set of integration variables. However, an infinitesimal variation of t_0 adds to

$q(t)$ a contribution proportional to $\dot q_c$. In order for the new set $\{t_0, r(t)\}$ to include only independent variables, $r(t)$ must satisfy one constraint. We choose

$$\int \dot q_c(t - t_0) r(t - t_0) \mathrm{d}t = 0. \tag{8.48}$$

We then expand $r(t)$ on the orthonormal basis formed by the eigenvectors $f_n(t)$ of the hermitian operator (8.45):

$$r(t) = \sum_{n=0}^{\infty} r_n f_n(t).$$

The set $\{f_n\}$ includes the normalized eigenvector $\dot q_c / \|\dot q_c\|$, which we can identify with f_0. The condition (8.48) then implies that the component r_0 of f_0 vanishes.

In terms of the components f_n, the argument of the exponential in the gaussian integral takes the simple form

$$\frac{1}{2} \int \mathrm{d}t_1 \mathrm{d}t_2 \, r(t_1) M(t_1, t_2) r(t_2) = \frac{1}{2} \sum_{n>0} m_n r_n^2, \tag{8.49}$$

where $\{m_n\}$ is the set of all non-vanishing eigenvalues of M.

It is useful at this point to remember that the functional measure $[\mathrm{d}q(t)]$ can also be defined in the continuum as the flat integration measure over the components c_m of the expansion of $q(t)$ on an orthonormal basis of square-integrable functions (see the discussion of Section 2.7):

$$q(t) = \sum_{m=0}^{\infty} c_m g_m(t), \qquad g_m(t) \in \mathcal{L}^2,$$

$$[\mathrm{d}q(t)] = \mathcal{N} \prod_{m=0}^{\infty} \mathrm{d}c_m.$$

The jacobian of the transformation that relates the set $\{c_m\}$ to the set $\{t_0, \{r_n\}\}$ is given, at leading order in $\hbar$, by (see Section 8.6)

$$J \sim \|\dot q_c\|/\sqrt{\hbar} = \frac{1}{\sqrt{\hbar}} \left[\int \dot q_c^2(t) \mathrm{d}t \right]^{1/2} = \sqrt{A/m\hbar}. \tag{8.50}$$

Since the integrand does not depend on t_0, the integration over the collective coordinate t_0 yields simply a factor τ. The integration over the variables r_n yields $(\det' M)^{-1/2}$, where $\det' M$ is the product of all non-vanishing eigenvalues of M, which is also the determinant of M when restricted to the subspace orthogonal to $\dot q_c$.

Normalization. To normalize the path integral, we compare it to its limit at $\hbar = 0$ (a harmonic oscillator), which in the limit $\tau \to \infty$ reduces to $e^{-\omega\tau/2}$. For $\hbar \to 0$, the operator M tends toward the operator

$$M_0(t_1, t_2) = \left[-m d_{t_1}^2 + m\omega^2\right] \delta(t_1 - t_2),$$
$$\int dt_2 \, M_0(t_1, t_2) f(t_2) = -m \ddot{f}(t_1) + m\omega^2 f(t_1). \tag{8.51}$$

In the comparison between the contribution of the instanton and the reference path integral corresponding to the harmonic oscillator, one must take into account that in the case of the instanton one gaussian mode has been excluded. The two path integrals differ by one gaussian integration. It is thus necessary to divide the instanton contribution by the factor

$$\int_{-\infty}^{+\infty} e^{-\lambda^2/2} \, d\lambda = (2\pi)^{1/2}.$$

Dividing by a factor $2i$, one then obtains the imaginary part of $\mathcal{Z}(\tau/\hbar)$ in the form (8.40).

Collecting all factors, one obtains

$$\mathrm{Im}\, \mathcal{Z}(\tau/\hbar) \sim \frac{1}{2i} \left[\det'(MM_0^{-1})\right]^{-1/2} \sqrt{\frac{A}{m\hbar}} \frac{\tau}{\sqrt{2\pi}} e^{-\omega\tau/2} e^{-A/\hbar},$$

and, finally,

$$\mathrm{Im}\, E_0 \sim \frac{1}{2i} \left[\det'(MM_0^{-1})\right]^{-1/2} \sqrt{\frac{A\hbar}{2\pi m}} e^{-A/\hbar}. \tag{8.52}$$

The result is finite and real since, as we have indicated, M has one negative eigenvalue. It is dimensionally correct since one eigenvalue of M has a dimension mass/time^2 and one eigenvalue of M has been suppressed.

8.6 The jacobian

We now calculate the jacobian of the change of variables from $q(t)$ to the set $\{t_0, r_n\}$. To avoid a proliferation of factors $\hbar$, we change $q(t) \mapsto q(t)\sqrt{\hbar}$ in such a way that in what follows $q_c(t)$ corresponds to $q_c(t)/\sqrt{\hbar}$, where q_c is a special solution of equation (8.41).

We expand $q(t)$ on a complete orthonormal basis (in the $\mathcal{L}^2$ sense) of real periodic functions with period τ:

$$q(t) = \sum_{m=0}^{\infty} c_m g_m(t), \tag{8.53}$$

$$\delta_{nm} = \int_0^{\tau} dt \, g_m(t) g_n(t). \tag{8.54}$$

In the explicit evaluation of Section 2.4 (see equation (2.69)), we have defined the functional measure as

$$[dq(t)] = \mathcal{N} \prod_{m=0}^{\infty} dc_m, \tag{8.55}$$

where $\mathcal{N}$ is the usual normalization constant. We now change variables:

$$q(t) = q_c(t - t_0) + \sum_{n=1}^{\infty} r_n f_n(t - t_0), \tag{8.56}$$

where the set $\{\dot{q}_c(t)/\|\dot{q}_c\|, f_n(t)\}$ forms an orthonormal basis and the new integration variables are t_0 and the set $\{r_n\}$.

The variables c_m can be expressed in terms of the new variables:

$$c_m = \int dt \, g_m(t) q_c(t - t_0) + \sum_{n=1}^{\infty} r_n \int dt \, g_m(t) f_n(t - t_0). \tag{8.57}$$

The jacobian of the transformation is the determinant of the matrix

$$\left[\frac{\partial c_m}{\partial t_0}, \frac{\partial c_m}{\partial r_n} \right] \tag{8.58}$$

with

$$\frac{\partial c_m}{\partial t_0} = -\int dt \, g_m(t) \dot{q}_c(t - t_0) - \sum_{n=1}^{\infty} r_n \int dt \, g_m(t) \dot{f}_n(t - t_0),$$

$$\frac{\partial c_m}{\partial r_n} = \int dt \, g_m(t) f_n(t - t_0). \tag{8.59}$$

In a leading order calculation, one can neglect the dependence of the jacobian in the $\{r_n\}$. Since the set $\{\dot{q}_c/\|\dot{q}_c\|, f_n\}$ forms an orthonormal basis, the matrix

$$\left[\int dt \, g_m(t) \frac{\dot{q}_c(t - t_0)}{|\dot{q}_c|}, \int dt \, g_m(t) f_n(t - t_0) \right] \tag{8.60}$$

is orthogonal and its determinant is 1. Thus, the jacobian of the transformation is simply

$$\|\dot{q}_c\| = \left\{ \int dt \, [\dot{q}_c(t)]^2 \right\}^{1/2}. \tag{8.61}$$

8.7 Instantons: the quartic anharmonic oscillator

We now apply the preceding results to a rather simple example: the quartic anharmonic potential in which the sign of the quartic term is changed from positive to negative values. The corresponding hamiltonian is

$$H = -\tfrac{1}{2}(\mathrm{d}/\mathrm{d}q)^2 + \tfrac{1}{2}q^2 + \tfrac{1}{4}gq^4. \tag{8.62}$$

In what follows we set $\hbar = 1$ because, as we verify below, the parameter g plays here the role of $\hbar$.

We infer the eigenvalues of H from a calculation of the partition function

$$\mathcal{Z}(\beta) = \operatorname{tr} \mathrm{e}^{-\beta H} = \int_{q(-\beta/2)=q(\beta/2)} [\mathrm{d}q(t)] \exp\left[-\mathcal{S}(q)\right], \tag{8.63}$$

where $\mathcal{S}(q)$ is the euclidean action,

$$\mathcal{S}(q) = \int_{-\beta/2}^{\beta/2} \left[\tfrac{1}{2}\dot{q}^2(t) + \tfrac{1}{2}q^2(t) + \tfrac{1}{4}gq^4(t)\right] \mathrm{d}t. \tag{8.64}$$

A generalization of the arguments applicable to integrals over a finite number of variables indicates that the path integral (8.63) defines a function of g that is analytic in the half-plane $\operatorname{Re}(g) > 0$. In this domain, the integral is dominated for $g \to 0$ by the saddle point $q(t) \equiv 0$. Thus, it can be calculated by expanding the integrand in powers of g and integrating the successive terms. This leads to a perturbative expansion of the partition function, from which one can derive an expansion of the ground state energy $E_0(g)$ in the limit $\beta \to \infty$.

Remark.
After the change

$$q(t) \mapsto q(t)g^{-1/2},$$

the parameter g factorizes in front of the action:

$$\mathcal{S}(q) = \frac{1}{g}\mathcal{S}(q\sqrt{g}). \tag{8.65}$$

The coupling constant g thus plays the role of $\hbar$ from the viewpoint of the perturbative expansion.

Negative coupling. For all $g < 0$, the hamiltonian is no longer bounded from below. Therefore, the energy eigenvalues, considered as analytic functions of g, have a singularity at $g = 0$ and the perturbation series is always divergent.

Again, to understand how to define and evaluate $E_0(g)$ for g negative, we first study a simple integral that illustrates some aspects of the problem.

8.7.1 The simple quartic integral

We consider the integral

$$I(g) = \frac{1}{\sqrt{2\pi}} \int_{-\infty}^{+\infty} e^{-(x^2/2 + gx^4/4)} \, dx, \tag{8.66}$$

which counts the number of Feynman diagrams contributing to the partition function (8.63). For g positive and small, the integral is dominated by the saddle point at the origin and thus

$$I(g) = 1 + O(g). \tag{8.67}$$

The function $I(g)$ is analytic in a cut plane. To continue the integral analytically to $g < 0$, it is necessary to rotate the integration contour C when one changes the phase of g, for example, like

$$C: \operatorname{Arg} x = -\tfrac{1}{4} \operatorname{Arg} g \pmod{\pi}.$$

Then, $\operatorname{Re}(gx^4)$ always remains positive. Therefore, one obtains two different, complex conjugate, expressions $I_\pm(g)$, according to the orientation of the rotation in the g plane:

$$\text{for } g = -|g| + i0: \quad I_+(g) = \frac{1}{\sqrt{2\pi}} \int_{C_+} e^{-(x^2/2 + gx^4/4)} \, dx$$

$$\text{with } C_+: \quad \operatorname{Arg} x = -\frac{\pi}{4} \pmod{\pi}, \tag{8.68}$$

$$\text{for } g = -|g| - i0: \quad I_-(g) = \frac{1}{\sqrt{2\pi}} \int_{C_-} e^{-(x^2/2 + gx^4/4)} \, dx$$

$$\text{with } C_-: \quad \operatorname{Arg} x = \frac{\pi}{4} \pmod{\pi}. \tag{8.69}$$

For $g \to 0_-$, the two integrals are still dominated by the saddle point at the origin since the contributions of the other saddle points,

$$x + gx^3 = 0 \Rightarrow x^2 = -1/g, \tag{8.70}$$

is of order

$$e^{-(x^2/2 + gx^4/4)} \sim e^{1/4g} \ll 1. \tag{8.71}$$

However, the discontinuity of $I(g)$ across the cut is given by the difference between the two integrals:

$$I_+(g) - I_-(g) = 2i \operatorname{Im} I(g) = \frac{1}{\sqrt{2\pi}} \int_{C_+ - C_-} e^{-(x^2/2 + gx^4/4)} \, dx. \tag{8.72}$$

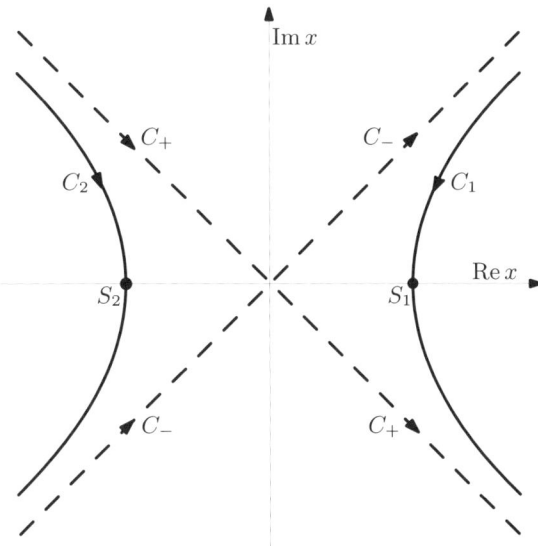

Fig. 8.3 The integration contours C_+, C_-, C_1 and C_2.

It corresponds to the contour $C_+ - C_-$, which, as Fig. 8.3 shows, can be deformed into the sum of contours C_1 and C_2 that avoid the leading saddle point but contain the non-trivial saddle points S_1 and S_2: $x = \pm 1/\sqrt{-g}$. This indicates that the contributions of the saddle point at the origin cancel, and that the integral is now dominated by the saddle points S_1 and S_2. Evaluating their contributions, one then finds

$$\operatorname{Im} I(g) \sim 2^{-1/2}\, \mathrm{e}^{1/4g}. \tag{8.73}$$

As a consequence, for g negative and small, while the real part of the integral is given by the perturbative expansion, the leading contributions to the imaginary part, which is exponentially small, come from the non-trivial saddle points.

We now generalize this strategy to the path integral (8.63).

8.7.2 Path integral

Inspired by the preceding example, we rotate the integration domain in functional $q(t)$ space while we change the phase of g to go from positive to negative values:

$$q(t) \mapsto q(t)\, \mathrm{e}^{-i\theta},$$

where θ is independent of time. Returning to the definition of the path integral as a limit of integrals with discrete times (see Chapter 2), one intuits that this is a meaningful procedure.

However, there is one major difference with the case of a simple integral: the domain must satisfy $\operatorname{Re}\left[\dot{q}^2(t)\right] > 0$, because, as we have emphasized in Section 2.2, the kinetic term $\int \dot{q}^2(t)\mathrm{d}t$ selects paths that are sufficiently regular and thus ensures the existence of a continuum limit of the discretized path integral.

For g negative, the two conditions

$$\operatorname{Re}\left[gq^4(t)\right] > 0, \ \operatorname{Re}\left[\dot{q}^2(t)\right] > 0, \tag{8.74}$$

are satisfied if one integrates over a domain satisfying

$$\operatorname{Arg} q(t) = -\theta \pmod{\pi}, \ \pi/8 < \theta < \pi/4 \text{ or } -\pi/4 < \theta < -\pi/8. \tag{8.75}$$

For $g \to 0$, the two path integrals corresponding to the two analytic continuations are here also dominated by the saddle point at the origin

$$q(t) = 0,$$

but in the difference between the two integrals, these contributions cancel.

The contribution of the other saddle points corresponding to the constant functions

$$q^2(t) = -1/g,$$

is of the order of $e^{\beta/4g}$ and is thus negligible for $\beta \to \infty$.

We then look for saddle points that are non-trivial solutions of the euclidean equation of motion for $g < 0$:

$$-\ddot{q}(t) + q(t) + gq^3(t) = 0 \tag{8.76}$$

with

$$q(-\beta/2) = q(\beta/2). \tag{8.77}$$

We are only interested in *instanton*-type solutions, whose action remains finite when $\beta \to +\infty$.

8.7.3 Instantons

The solutions of equation (8.76) with the periodic condition (8.77) have an interpretation as describing a classical periodic motion, in real time, in the potential

$$-V(q) = -\tfrac{1}{2}q^2 - \tfrac{1}{4}gq^4. \tag{8.78}$$

It is clear that the equation of motion has solutions that correspond to oscillations around the minima of $-V$, $q = \pm\sqrt{-1/g}$. Integrating once equation (8.76), one obtains

$$\tfrac{1}{2}\dot{q}^2 - \tfrac{1}{2}q^2 - \tfrac{1}{4}gq^4 = \epsilon$$

with $\epsilon < 0$. Calling q_- and q_+ the points with $q > 0$ where the velocity $\dot{q}$ vanishes, one finds for the period of such a solution

$$\beta = 2 \int_{q_-}^{q_+} \frac{dq}{\sqrt{q^2 + \tfrac{1}{2}gq^4 + 2\epsilon}}.$$

8.7 Barrier penetration: semi-classical approximation

β diverges only if the constant ϵ and thus q_- go to zero. The classical trajectory then passes increasingly closer to the origin. In the infinite β limit, the classical solutions become

$$q_c(t) = \pm \left(-\frac{2}{g}\right)^{1/2} \frac{1}{\cosh(t-t_0)}. \tag{8.79}$$

The corresponding value of the classical action is

$$\mathcal{S}(q_c) = -\frac{4}{3g} + O\left(e^{-\beta}/g\right). \tag{8.80}$$

Since the euclidean action is time-translation invariant, the classical solution depends on one arbitrary parameter t_0, which for β finite, varies in an interval of size β. We thus find two families of degenerate saddle points that depend on one parameter.

Leading order contribution. The operator second functional derivative of the action is given by

$$M(t_1, t_2) = \frac{\delta^2 \mathcal{S}}{\delta q_c(t_1) \delta q_c(t_2)} = \left[-\left(\frac{d}{dt_1}\right)^2 + 1 + 3g q_c^2(t_1)\right] \delta(t_1 - t_2). \tag{8.81}$$

One verifies that the function $\dot{q}_c(t)$ is square integrable and, therefore, M has a zero mode corresponding to the eigenvector $\dot{q}_c$.

Taking into account the two families of saddle points, the zero mode and collecting all factors, one obtains

$$\operatorname{Im} \operatorname{tr} e^{-\beta H} \sim \frac{2}{2i} \left[\det' M M_0^{-1}\right]^{-1/2} J \frac{\beta}{\sqrt{2\pi}} e^{-\beta/2} e^{4/3g}, \tag{8.82}$$

where J is the jacobian (8.50). Moreover, it is easy to calculate the eigenvalues of M analytically because M is a hamiltonian with a Bargmann–Wigner-type potential. The determinant can then be inferred from the general expression (8.35). One finally obtains

$$\operatorname{Im} E_0(g) = \frac{4}{\sqrt{2\pi}} \frac{e^{4/3g}}{\sqrt{-g}} \left[1 + O(g)\right], \quad g \to 0_-. \tag{8.83}$$

Exercises

Exercise 8.1

One considers the integral

$$Z(g) = \int d^2q \, \exp\left[-\frac{1}{2g} \mathbf{q}^2 \left(\mathbf{q}^2 - 1\right)^2\right],$$

where $\mathbf{q}$ is a two-component vector (q_1, q_2). Calculate the integral for $g \to 0_+$ by the steepest descent method.

Solution. One finds several saddle points corresponding to the same value: $\mathbf{q} = 0$ and the circle $|\mathbf{q}| = 1$. In contrast, the circle of saddle points $|\mathbf{q}| = 1/3$ corresponds to local minima and, thus, does not contribute. The saddle point $\mathbf{q} = 0$ can be dealt with by the standard steepest descent method and yields $2\pi g$. In the case of the circle $|\mathbf{q}| = 1$, one must introduce polar coordinates and one finds $\pi\sqrt{2\pi g}$, which is the leading contribution for $g \to 0$.

Exercise 8.2

One considers the hamiltonian ($[\hat{q}, \hat{p}] = i\hbar$)

$$H = \tfrac{1}{2}\hat{p}^2 + \tfrac{1}{2}(\hat{q}^4 - 1)^2.$$

Calculate the action $\mathcal{S}_c$ of the instanton associated, in the semi-classical limit, to the energy splitting between the ground state and the first excited state.

Solution.
$$\mathcal{S}_c = 8/5.$$

Exercise 8.3

Calculate the imaginary part of the energy of the pseudo-ground state of the hamiltonian

$$H = -\tfrac{1}{2}(\mathrm{d}/\mathrm{d}q)^2 + \tfrac{1}{2}q^2 + \tfrac{1}{2}gq^{2N}$$

for g negative. The result (8.35) will again be useful.

Solution. A few elements of the solution are
(i) The classical solution

$$|q_c|^{N-1}(t) = \frac{1}{\sqrt{-g}} \frac{1}{\cosh[(N-1)(t-t_0)]}.$$

(ii) The classical action

$$\mathcal{S}_c = A(N)/(-g)^{1/(N-1)},$$

$$A(N) = \frac{\sqrt{\pi}\,\Gamma(N/(N-1))}{2\Gamma((3N-1)/2(N-1))} = 4^{1/(N-1)} \frac{\Gamma^2(N/(N-1))}{\Gamma(2N/(N-1))}.$$

(iii) The operator second derivative of the action at the saddle point

$$M = -\mathrm{d}_t^2 + 1 - \frac{N(2N-1)}{\cosh^2[(N-1)t]}.$$

The determinants are then given by equation (8.35) with

$$\lambda = N/(N-1), \qquad z = \sqrt{1+\epsilon}/(N-1).$$

It follows
$$\det(M+\varepsilon)(M_0+\varepsilon)^{-1} \sim -2^{-(N+1)/(N-1)} A(N)\varepsilon.$$

(iv) The operator M has a vanishing eigenvalue associated with time-translations. The jacobian J generated by the introduction of time as a collective coordinate is given by
$$J^2 = \int dt\, \dot{q}^2(t) = A/(-g)^{1/(N-1)}.$$

Using equation (8.82) and collecting all factors, one obtains the imaginary part of the energy of the metastable pseudo-ground state
$$\operatorname{Im} E(g) = C(-g)^{-\beta} \exp\left[-A(N)/(-g)^{1/(N-1)}\right] \qquad (8.84)$$

with
$$\beta = 1/[2(N-1)], \qquad C = \frac{2^{1/(N-1)}}{\sqrt{\pi}}.$$

Exercise 8.4
Classical diffusion equation: the Fokker–Planck equation.

Number of stochastic processes, such that random walk, thermal diffusion..., can be described by an equation of the type (see Section 5.5)
$$\frac{\partial P(q,t)}{\partial t} = \frac{\Omega}{2}\frac{\partial}{\partial q}\left(\frac{\partial P}{\partial q} + \frac{1}{\Omega}\frac{\partial E}{\partial q}P\right),$$

where $P(q,t)$ can be considered as a probability distribution and $\Omega > 0$ is a diffusion constant. Conservation of probabilities follows directly from the form of the equation:
$$\int dq\, P(q,t) = 1.$$

Moreover, one verifies that the equation has a stationary solution
$$P_0(q,t) = \exp(-E(q)/\Omega)$$

which, if it is *normalizable*, is proportional to the limiting distribution for $t \to +\infty$.

Below, one identifies the stochastic process with the diffusion of a particle on an axis $q \in (-\infty, +\infty)$. In Section 5.5, it has been proved that the probability for a particle that is at point x_0 at time $t=0$ to be at point x at later time $\tau > 0$, which we denote more explicitly by $P(x,\tau; x_0, 0)$, is given by the path integral
$$P(x,\tau; x_0, 0) = \int [dq] \exp[-\mathcal{S}(q)/\Omega]$$

with
$$\mathcal{S}(q) = \frac{1}{2}\int_0^\tau dt\left[\left(\dot{q} + \tfrac{1}{2}E'(q(t))\right)^2 - \tfrac{1}{2}\Omega E''(q(t))\right]$$

and the boundary conditions

$$q(0) = x_0, \quad q(\tau) = x.$$

1. One now considers the function

$$E(q) = q^2 - \tfrac{2}{3}q^3. \tag{8.85}$$

Show that in the limit $\Omega \to 0$ (low diffusivity) the path integral has the form of an integral corresponding to a potential with degenerate minima. Calculate for the two wells, the ground state energy in the gaussian approximation. Note that in the limit $\Omega \to 0$, one can neglect the contribution of order Ω in the action to determine the minima or the saddle points.

2. For the same function $E(q)$, still for $\Omega \to 0$ and for $\tau \to +\infty$, one notices that the saddle point equations correspond to a classical motion in a potential with degenerate minima. Such equations admit instanton solutions. What is their interpretation here?

Solutions.

1. Expanding the square in the action, one verifies that one term can be integrated exactly and, thus,

$$S(q) = \frac{1}{2}E(x) - \frac{1}{2}E(x_0) + \frac{1}{2}\int_0^\tau dt \left[\dot{q}^2 + q^2(1-q)^2 - \tfrac{1}{2}\Omega(1-2q) \right].$$

In the limit $\Omega \to 0$, the two wells corresponding to $q = 0$ and $q = 1$ are degenerate. In the gaussian approximation, the ground state energy ϵ of the harmonic oscillator is in both cases $\epsilon = \tfrac{1}{2}$. However, the classical value of the second derivative contributes at the same order in such a way that

$$\epsilon = \tfrac{1}{2} - \tfrac{1}{2} = 0 \quad \text{for } q = 0,$$
$$\epsilon = \tfrac{1}{2} + \tfrac{1}{2} = 1 \quad \text{for } q = 1.$$

Thus, beyond the classical approximation, the minima are no longer degenerate, and the expansion of the ground state energy is given by the perturbative expansion at $q = 0$. This result is quite reasonable if one considers the initial problem: it is hard for a particle to diffuse out of the minimum of the potential $E(q)$ at $q = 0$, but not to leave the maximum at $q = 1$.

2. For $\Omega \to 0$ and $\tau \to \infty$, the instantons that minimize the action satisfy

$$\dot{q} \pm \tfrac{1}{2}E'(q) = 0.$$

In the special case the solutions are

$$q_c(t) = \frac{1}{1 + e^{\pm(t-t_0)}}.$$

8.7 Barrier penetration: semi-classical approximation

The value of the classical action is

$$\mathcal{S}(q_c) = \tfrac{1}{6}(1 \mp 1)$$

The vanishing result corresponds to the solution with a particle leaving $q = 1$ at $t = -\infty$ and reaching $q = 0$ at $t = +\infty$, a process that clearly has a finite probability even for $\Omega \to 0$. The positive result corresponds to a particle that leaves $q = 0$ at $t = -\infty$ and reaches $q = 1$ at $t = +\infty$. This process is mathematically analogous to barrier penetration and the result characterizes the probability to escape the well around $q = 0$. For $\Omega \to 0$, it is of the order of $e^{-1/3\Omega}$.

9 QUANTUM EVOLUTION AND SCATTERING MATRIX

It is not our purpose, in this chapter, to present a detailed discussion of scattering theory in quantum mechanics and, in particular, of the calculation of physical observables such that cross-sections. The interested reader will find the necessary background material in classical quantum mechanics textbooks. Our goal, here, is to show how scattering problems are formulated in the framework of path integrals.

In quantum mechanics, the state of an isolated system evolves under the action of a unitary operator, as a consequence of the conservation of probabilities and, thus, of the norm of vectors in Hilbert space. We denote by $U(t'',t')$ the operator corresponding to the evolution between times t' and t'':

$$U(t'',t')U^\dagger(t'',t') = \mathbf{1}, \ U(t',t') = \mathbf{1}.$$

Moreover, quantum evolution of an isolated system is assumed to be markovian, a property expressed by the relation (2.1):

$$U(t_3,t_2)U(t_2,t_1) = U(t_3,t_1).$$

If $U(t,t')$ is differentiable, it satisfies the Schrödinger equation

$$i\hbar \frac{\partial U}{\partial t}(t,t') = H(t)U(t,t'), \qquad (9.1)$$

where $H(t)$ is a hermitian operator, the hamiltonian.

When H is time-independent, the evolution operator takes the special form $U(t'',t') \equiv U(t''-t') = e^{-iH(t''-t')/\hbar}$. It belongs to a representation of the abelian translation group and $H/\hbar$ is the generator of time translations.

In terms of the evolution operator, one defines a scattering matrix or S-matrix which describes how, asymptotically in time, evolution in presence of interactions differs from free evolution. We show in this chapter how the S-matrix can be obtained from the path integral representation of the evolution operator and how the path integral formalism leads to a simple derivation of some semi-classical approximations.

9.1 Evolution of the free particle and S-matrix

Even the evolution of a free quantum particle is slightly non-trivial; in general, one observes a spreading of wave packets. Scattering is then characterized by the asymptotic deviations at infinite time from this free evolution and this leads to the definition of a scattering or S-matrix.

9.1.1 Evolution of the free particle

For a free particle, the hamiltonian reduces to the kinetic term:
$$H_0 = \mathbf{p}^2/2m.$$

The matrix elements of the evolution operator $U_0 = e^{-iH_0(t''-t')}$ can be derived, for example, from an analytic continuation of expression (2.9). They can also be inferred more directly from their Fourier representation. In d space dimensions,

$$\langle \mathbf{q}''| U_0(t'',t') |\mathbf{q}'\rangle = \frac{1}{(2\pi)^d} \int d^d p \, e^{i[\mathbf{p}\cdot(\mathbf{q}''-\mathbf{q}')-\mathbf{p}^2(t''-t')/2m]/\hbar} \qquad (9.2a)$$

$$= \left(\frac{m}{2i\pi\hbar(t''-t')}\right)^{d/2} \exp\left[\frac{i}{\hbar}\frac{m(\mathbf{q}''-\mathbf{q}')^2}{2(t''-t')}\right]. \qquad (9.2b)$$

In particular, this evolution leads to the spreading of a wave packet. To exhibit this phenomenon, we define a wave function at initial time $t = 0$ by its Fourier representation $\tilde\psi(\mathbf{p})$ (its representation in the momentum basis), which we assume to have a support localized around a value $\mathbf{p} = \mathbf{p}_0$. At time t, one finds

$$\psi(\mathbf{q},t) = \langle q | e^{-iH_0 t/\hbar} |\psi\rangle = \int \frac{d^d p}{(2\pi)^d} \tilde\psi(\mathbf{p}) \exp\left[i\left(\mathbf{p}\cdot\mathbf{q} - t\frac{\mathbf{p}^2}{2m}\right)/\hbar\right]. \qquad (9.3)$$

When $t \to \infty$, the phase in the integral (9.3) varies rapidly and thus the integral is dominated by the points where the phase is stationary:

$$\frac{\partial}{\partial \mathbf{p}}\left(\mathbf{p}\cdot\mathbf{q} - t\frac{\mathbf{p}^2}{2m}\right) = 0 \implies \mathbf{q} = t\frac{\mathbf{p}}{m}. \qquad (9.4)$$

The integral (9.3) is thus equivalent to

$$\psi(\mathbf{q},t) \underset{|t|\to\infty}{\sim} \tilde\psi(\mathbf{p})\frac{1}{(2\pi)^{d/2}}\left(\frac{m\hbar}{|t|}\right)^{d/2} \exp\left(\frac{i\pi}{4}\operatorname{sgn} t + it\frac{\mathbf{p}^2}{2m\hbar}\right), \qquad (9.5)$$

with
$$\mathbf{p} = \frac{m}{t}\mathbf{q}.$$

Since the initial function has its support concentrated around $\mathbf{p} = \mathbf{p}_0$, the wave function $\psi(\mathbf{q},t)$ has its support concentrated around $\mathbf{q} = t\mathbf{p}_0/m$, that is around the classical trajectory. But, simultaneously the factor $|t|^{-d/2}$ shows the spreading of the wave packet. This spreading disappears only in the limit in which the wave function becomes a plane wave $e^{i\mathbf{p}\cdot\mathbf{q}}/\hbar$, but this limit is singular because a plane wave is not normalizable. Nevertheless, this is the limit in which it is convenient to discuss quantum scattering. Simply, one has to keep in mind that this limit may lead to difficulties. In this case, one has to explicitly introduce normalizable wave packets.

9.1.2 Particle in a potential and S-matrix

One now considers a particle in a potential (still in $\mathbb{R}^d$), with the hamiltonian

$$H = \mathbf{p}^2/2m + V(\mathbf{q}, t).$$

One assumes that the properties of the potential lead to classical scattering. This implies, in particular, that the potential $V(\mathbf{q}(t), t)$ decreases fast enough along the classical trajectory for $|t| \to \infty$, in such a way that asymptotically the trajectory converges toward the free motion corresponding to the hamiltonian

$$H_0 = \mathbf{p}^2/2m.$$

The scattering or S-matrix is then obtained by comparing quantum evolution in a potential, at large times $t \to \pm\infty$, with free evolution. More precisely, the S-matrix is defined as the limit of the evolution operator in the so-called interaction representation (see equation (9.51)):

$$S = \lim_{\substack{t' \to -\infty \\ t'' \to +\infty}} e^{iH_0 t''/\hbar} U(t'', t') e^{-iH_0 t'/\hbar}. \tag{9.6}$$

The multiplication on both sides by the free evolution operator is necessary for the existence of a large time limit. Indeed, even in the absence of a scattering potential, the operator $e^{-iH_0(t''-t')}$ depends on time and has no limit even though, when it acts on eigenfunctions of the momentum operator, it changes only their phase, as we have discussed above. The factor on the right corresponds to a time-reversed free evolution from time t' to time 0, while the factor on the left corresponds to a time-reversed free evolution from time 0 to time t''. With this definition, in the absence of a potential, the S-matrix is the identity. Moreover, one can show that if the potential decreases fast enough for $|t| \to \infty$, the limit exists and an S-matrix can be defined.

Note that in the basis where the momentum operator is diagonal, the relation (9.6) between S-matrix and evolution operator becomes simply

$$\langle \mathbf{p}''| S |\mathbf{p}'\rangle = \lim_{\substack{t' \to -\infty \\ t'' \to +\infty}} e^{iE'' t''/\hbar} \langle \mathbf{p}''|U(t'', t')|\mathbf{p}'\rangle e^{-iE' t'/\hbar}, \tag{9.7}$$

where

$$E' = E(\mathbf{p}'), \quad E'' = E(\mathbf{p}''), \quad E(\mathbf{p}) \equiv \mathbf{p}^2/2m. \tag{9.8}$$

However, the limit must be understood in the mathematical sense of distributions (it must be probed with test functions, which here are wave packets).

Finally, the elements of the S-matrix are generally parametrized in terms of the scattering matrix $\mathcal{T}$:

$$S = 1 - i\mathcal{T}, \Rightarrow \langle \mathbf{p}''|S|\mathbf{p}'\rangle = (2\pi\hbar)^d \delta^{(d)}(\mathbf{p}'' - \mathbf{p}') - i\langle \mathbf{p}''|\mathcal{T}|\mathbf{p}'\rangle. \tag{9.9}$$

When the potential is time-independent, energy is conserved and one sets

$$\langle \mathbf{p}''|\mathcal{T}|\mathbf{p}'\rangle = -2\pi\delta(E''-E')T(\mathbf{p}'',\mathbf{p}'). \tag{9.10}$$

In equation (9.9), the term proportional to the function $\delta^{(d)}(\mathbf{p}''-\mathbf{p}')$ corresponds to the unscattered contribution, and in most physical situations is not observed. Then, the scattering process can be entirely described by the $\mathcal{T}$-matrix. Finally, note that from this viewpoint the one-dimensional case is singular since conservation of energy also implies conservation of momentum up to the sign.

Differential cross-sections are then proportional to $|T(\mathbf{p}'',\mathbf{p}')|^2$, the proportionality factor being kinematical.

Path integral. To calculate the matrix elements of the evolution operator in the case, for example, of a hamiltonian of the form

$$H = \mathbf{p}^2/2m + V(\mathbf{q},t),$$

one can proceed by analytic continuation, replacing all time variables t par $t\,\mathrm{e}^{i\varphi}$ (except in the potential) in the expressions of Chapter 2 and rotating in the positive direction in the complex t plane from $\varphi = 0$ to $\varphi = \pi/2$.

The solution of equation (9.1) in terms of matrix elements in the position basis can then be written as

$$\langle \mathbf{q}''|U(t'',t')|\mathbf{q}'\rangle = \int_{\mathbf{q}(t')=\mathbf{q}'}^{\mathbf{q}(t'')=\mathbf{q}''} [d\mathbf{q}(t)]\exp\left[i\mathcal{A}(\mathbf{q})/\hbar\right]. \tag{9.11}$$

The function $\mathcal{A}(\mathbf{q})$ is now the usual action, integral of the classical lagrangian:

$$\mathcal{A}(\mathbf{q}) = \int_{t'}^{t''} dt\left[\tfrac{1}{2}m\dot{\mathbf{q}}^2 - V(\mathbf{q},t)\right]. \tag{9.12}$$

Expression (9.11) establishes a remarkable relation between quantum and classical physics. In quantum mechanics, all paths contribute to evolution but they are weighted by the phase factor $\mathrm{e}^{i\mathcal{A}/\hbar}$. The dominant contributions to the path integral are found in the neighbourhood of the paths for which the action is stationary, that is, paths solution of the equations of the classical motion. In particular, if the value of the classical action for classical paths is large relative to $\hbar$, the contributions to the path integral are completely localized near classical paths.

9.2 Perturbative expansion of the S-matrix

We first show how to derive an expansion of the evolution operator in powers of the potential from its path integral representation. The perturbative expansion of the S-matrix follows.

The path integral formalism then organizes the perturbative expansion in the same way as the operator formalism that is recalled in Section 9.7. For simplicity reasons, we assume that the potential is time-independent.

9.2.1 Perturbative expansion

We consider the hamiltonian
$$H = p^2/2m + V(q). \tag{9.13}$$

The classical actions that correspond to the free hamiltonian H_0 and to H are, respectively,
$$\mathcal{A}_0(q) = \int_{t'}^{t''} \tfrac{1}{2}m\dot{q}^2(t)\mathrm{d}t, \quad \mathcal{A}(q) = \int_{t'}^{t''} \left[\tfrac{1}{2}m\dot{q}^2(t) - V(q(t))\right]\mathrm{d}t. \tag{9.14}$$

The path integral (9.11), expanded in powers of V, then takes the form (setting for convenience $\hbar = 1$)
$$\langle q''|U(t'',t')|q'\rangle = \int_{q(t')=q'}^{q(t'')=q''} [\mathrm{d}q(t)]\exp[i\mathcal{A}(q)] = \sum_{\ell=0} \langle q''|U^{(\ell)}(t'',t')|q'\rangle$$

with
$$\langle q''|U^{(\ell)}(t'',t')|q'\rangle = \frac{(-i)^\ell}{\ell!} \int_{q(t')=q'}^{q(t'')=q''} [\mathrm{d}q(t)]\, e^{i\mathcal{A}_0(q)} \left[\int_{t'}^{t''} V(q(t))\mathrm{d}t\right]^\ell. \tag{9.15}$$

Potentials that lead to scattering must vanish at large distances, and this excludes the polynomial potentials that we have considered so far in this work. Instead, it makes sense to assume that the relevant potentials have a Fourier representation:
$$V(q) = (2\pi)^{-d} \int \mathrm{d}^d k\, e^{ikq}\, \tilde{V}(k). \tag{9.16}$$

Once the potential is replaced by its Fourier representation, the successive terms of the perturbative expansion are expressed in terms of simple gaussian path integrals. The ℓth term becomes
$$\langle q''|U^{(\ell)}(t'',t')|q'\rangle = \frac{(-i)^\ell}{\ell!} \int_{t'}^{t''} \prod_j \mathrm{d}\tau_j \int \prod_{j=1}^\ell \tilde{V}(k_j) \frac{\mathrm{d}^d k_j}{(2\pi)^d}$$
$$\times \int_{q(t')=q'}^{q(t'')=q''} [\mathrm{d}q(t)] \exp i\left[\int_{t'}^{t''} \tfrac{1}{2}m\dot{q}^2(t)\mathrm{d}t + \sum_j k_j q(\tau_j)\right]. \tag{9.17}$$

The integrand in expression (9.17) is symmetric in the times $\tau_1,\ldots,\tau_\ell$. We order times $t'' \geq \tau_\ell \geq \tau_{\ell-1}\cdots \geq \tau_1 \geq t'$ and simultaneously suppress the factor $1/\ell!$.

The gaussian integration over the path $q(t)$ is given, up to a normalization, by replacing $q(t)$ by the solution of the classical equation
$$-m\ddot{q} + \sum_j k_j \delta(t-\tau_j) = 0 \Rightarrow \dot{q}(\tau_{j+}) - \dot{q}(\tau_{j-}) = k_j/m.$$

The path integral thus gives a physical interpretation to the terms in the perturbative expansion: a path contributing to the ℓth order is a succession of free motions, where at times $\tau_1, \ldots, \tau_\ell$, the momentum changes by the amounts $k_1, \ldots, k_\ell$. Finally, the corresponding contributions must be averaged over all times and all momenta with weight $\tilde{V}(k)$.

To obtain matrix elements of the operator in the momentum basis, it is still necessary to Fourier transform with respect q' and q''. We call p' and p'' the corresponding momenta.

Remark. The dependence in q'' is given by the free evolution between τ_ℓ and t''. The Fourier transform thus involves the gaussian integral

$$\int d^d q'' \, e^{-ip''q'' + im(q'' - q(\tau_\ell))^2 / 2(t'' - \tau_\ell)}.$$

The result of the integration is obtained (up to a normalization) by replacing q'' by the minimum of the argument of the exponential:

$$q'' = q(\tau_\ell) + p''(t'' - \tau_\ell)/m.$$

One then notes that, for $t'' \to +\infty$, this amounts to calculating a path integral with classical scattering boundary conditions. Moreover, the complete result of the integration is

$$e^{-ip'' q(\tau_\ell)} e^{-ip''^2 (t'' - \tau_\ell)/2m},$$

generating a factor $e^{-ip''^2 t''/2m}$ that cancels in the S-matrix the factor arising from the free motion.

The same argument applies to the integration over q', which yields

$$\int d^d q' \, e^{ip'q' + im(q(\tau_1) - q')^2 / 2(\tau_1 - t')} \propto e^{ip' q(\tau_1)} e^{-ip'^2(\tau_1 - t')/2m}$$

and, thus,

$$q' = q(\tau_1) + (t' - \tau_1) p'/m.$$

9.2.2 Explicit calculation

The term of order zero in V yields $(2\pi)^d \delta(p'' - p')$. We now calculate explicitly the first order. We can write the evolution operator as the product of two free evolutions from t' to τ and from τ to t'', by introducing the representation

$$e^{ik_1 q(\tau_1)} = \int d^d q_1 \, \delta(q_1 - q(\tau_1)) e^{ik_1 q_1}.$$

Then, after Fourier transformation with respect to q' and q'', one finds

$$\langle p'' | U^{(1)}(t'', t') | p' \rangle = -i \int \frac{d^d k}{(2\pi)^d} \tilde{V}(k) \int_{t'}^{t''} d\tau \int d^d q \, e^{-ip''^2(t'' - \tau)/2m}$$
$$\times \langle p'' | q \rangle e^{ikq} \langle q | p' \rangle e^{-ip'^2(\tau - t')/2m}. \tag{9.18}$$

The integration over the variable q yields $(2\pi)^d \delta(k + p' - p'')$. After cancellation of the factors of the free motion, one obtains the first order contribution to the S-matrix:

$$\langle p''| S^{(1)} |p'\rangle = \lim_{\substack{t' \to -\infty \\ t'' \to +\infty}} -i\tilde{V}(p'' - p') \int_{t'}^{t''} d\tau \, e^{i\tau(p''^2 - p'^2)/2m} .$$

It is clear that the integral has a limit only in the sense of distributions:

$$\lim_{\substack{t' \to -\infty \\ t'' \to +\infty}} \int_{t'}^{t''} d\tau \, e^{i\tau s} = \int_{-\infty}^{+\infty} d\tau \, e^{i\tau s} = 2\pi\delta(s).$$

The integration thus generates a δ function of energy conservation:

$$\langle p''| S^{(1)} |p'\rangle = -i 2\pi \delta(E'' - E') \tilde{V}(p'' - p'), \tag{9.19}$$

with $E' = p'^2/2m$, $E'' = p''^2/2m$.

Higher order terms. We now divide the interval $[t', t'']$ into sub-intervals. We use the representation

$$\exp[ik_j q(\tau_j)] = \int d^d q_j \, \delta(q_j - q(\tau_j)) \exp(ik_j q_j) .$$

In each sub-interval, the propagation is free. We write the matrix elements of free evolution as Fourier transforms (equation (9.2a)). After Fourier transformation with respect to q' and q'', one then finds

$$\langle p''| U^{(\ell)}(t'', t') |p'\rangle = (-i)^\ell \int \prod_{j=1}^{\ell} d\tau_j \tilde{V}(k_j) \frac{d^d k_j}{(2\pi)^d} d^d q_j \prod_{j=2}^{\ell} \frac{d^d p_j}{(2\pi)^d}$$

$$\times \exp\left[\sum_{j=1}^{\ell+1} -ip_j^2 (\tau_j - \tau_{j-1})/2m + ip_j(q_j - q_{j-1}) + ik_j q_j\right]$$

with the conventions

$$\tau_0 = t', \quad \tau_{\ell+1} = t'', \quad p_{\ell+1} = p'', \quad p_1 = p', \quad q_0 = q_{\ell+1} = 0 .$$

The integrals over the variables q_j yield δ functions that determine the momenta $k_j = p_{j+1} - p_j$. After factorization of the two factors of the free motion, one can take the limits $t'' \to +\infty$, $t' \to -\infty$. One obtains

$$\langle p''| S^{(\ell)} |p'\rangle = (-i)^\ell \int \prod_{j=1}^{\ell} d\tau_j \prod_{j=2}^{\ell} \frac{d^d p_j}{(2\pi)^d} e^{ip''^2 \tau_\ell/2m} \tilde{V}(p'' - p_\ell) e^{ip_\ell^2(\tau_{\ell-1} - \tau_\ell)/2m}$$

$$\cdots \times e^{ip_2^2(\tau_1 - \tau_2)/2m} \tilde{V}(p_2 - p') e^{-ip'^2 \tau_1/2m} . \tag{9.20}$$

One still has to integrate over the variables τ_j, taking into account that these variables are ordered. One sets

$$\tau_{j+1} = \tau_j + u_j, \quad u_j \geq 0.$$

The remaining integral over τ_1, which is a consequence of time-translation invariance, yields a factor $2\pi\delta(E'' - E')$ of energy conservation. The integrals over the variables u_j on the real positive semi-axis yield (in the sense of distributions)

$$\int_0^{+\infty} du_j \, e^{i(E'' - E(p_j))u_j} = \frac{i}{E'' - E(p_j) + i0}, \quad E(p) \equiv p^2/2m,$$

that is distributions where the symbol $i0$ indicates the way to avoid the pole at $p_j^2 = p''^2$. The complete result thus is

$$\langle p''| S^{(\ell)} |p'\rangle = -2i\pi\delta(E'' - E') \int \tilde{V}(p'' - p_\ell) \prod_j \frac{d^d p_j}{(2\pi)^d} \frac{\tilde{V}(p_j - p_{j-1})}{E'' + i0 - E(p_j)}.$$

Lippman–Schwinger's equation. One notes that the perturbation series is a simple geometric series. Its sum is the solution of an integral equation, called Lippman–Schwinger's equation. In terms of the operator $T(E)$, where E is complex, solution of

$$T(E) = V - VG_0(E)T(E) \quad \text{with} \quad G_0(E) = (H_0 - E)^{-1}, \tag{9.21}$$

the quantity $T(p'', p')$ that appears in equation (9.10) is then given by

$$T(p'', p') = \langle p''| T(E + i0) |p'\rangle \quad \text{for } E = p'^2/2m = p''^2/2m.$$

9.2.3 Other method

As an exercise in path integral calculation, we give another derivation of the preceding results. We look for the limit of the expression

$$\langle p''| S^{(\ell)}(t'', t') |p'\rangle$$

$$= (-i)^\ell \, e^{iE''t'' - iE't'} \int d^d q'' d^d q' \, e^{-ip''q'' + ip'q'} \int_{t'}^{t''} \prod_j d\tau_j \int \prod_{j=1}^\ell \tilde{V}(k_j) \frac{d^d k_j}{(2\pi)^d}$$

$$\times \int_{q(t')=q'}^{q(t'')=q''} [dq(t)] \exp i \left[\int_{t'}^{t''} \tfrac{1}{2} m \dot{q}^2(t) dt + \sum_j k_j q(\tau_j) \right].$$

We translate $q(t) \mapsto q(t) + q'$ in such a way that $q(t') = 0$, $q(t'') = q'' - q'$. We then impose the condition $q(t'') = q'' - q'$ by a δ function:

$$\delta\bigl(q(t'') - q'' + q\bigr) = \frac{1}{(2\pi)^d} \int d^d \lambda \, e^{i\lambda(q'' - q' - q(t''))}. \tag{9.22}$$

After these transformations, all paths satisfying $q(t') = 0$, without any constraint over $q(t'')$, contribute to the integral. The dependence on q' and q'' is now explicit and the integrations can be performed:

$$\int d^d q' d^d q''\, e^{i(p'q' - p''q'' + \sum_j k_j q' + \lambda(q'' - q'))} = (2\pi)^{2d}\delta(\lambda - p'')\delta\left(p'' - p' - \sum k_j\right). \quad (9.23)$$

We then change variables $k_j \mapsto p_{j+1}$ with

$$k_j = p_{j+1} - p_j, \quad p_1 = p'.$$

The jacobian is unity. The second δ function in (9.23) implies

$$p_{\ell+1} = p''.$$

Then,

$$\sum_{j=1}^{\ell} k_j q(\tau_j) = \sum_{j=1}^{\ell}(p_{j+1} - p_j)q(\tau_j)$$

$$= -\sum_{j=1}^{\ell} p_j\big(q(\tau_j) - q(\tau_{j-1})\big) + p''q(\tau_\ell) - p'q(\tau_0). \quad (9.24)$$

The choice $\tau_0 = t'$ implies $q(\tau_0) = 0$.

We then express the dependence on $q(t)$ in terms of its derivative $\dot q(t)$. For example,

$$q(\tau_j) - q(\tau_{j-1}) = \int_{t'}^{t''} \theta_j(t)\dot q(t)dt,$$

where $\theta_j(t)$ is the characteristic function of the interval (τ_{j-1}, τ_j): it is 1 in the interval (τ_{j-1}, τ_j) and vanishes outside.

The coefficient of p'' comes from the contribution in (9.22) ($\lambda = p''$) and in (9.24):

$$p''\big(q(\tau_\ell) - q(t'')\big) = -p'' \int_{t'}^{t''} \theta_{\ell+1}(t)\dot q(t)dt, \quad \tau_{\ell+1} \equiv t''.$$

The path integral becomes

$$\int_{q(t')=0} [dq(t)] \exp i\left[\int_{t'}^{t''} dt\, \big(\tfrac{1}{2}m\dot q^2(t) - b(t)\dot q(t)\big)\right]$$

with

$$b(t) = \sum_{j=1}^{\ell+1} p_j \theta_j(t).$$

We change variables $q(t) \mapsto r(t)$ with

$$\dot q(t) = b(t)/m + \dot r(t)\,, \quad q(t') = r(t') = 0\,.$$

We see that p_j is the momentum inside the interval (τ_{j-1}, τ_j). The integral over $r(t)$ corresponds to a free motion. It is given by expression (9.2b) with $q' = 0$ and by integrating over q''. The result is 1.

It remains to calculate the explicit contribution generated by the change of variables

$$-\frac{1}{2m}\int_{t'}^{t''} b^2(t)\mathrm{d}t = -\frac{1}{2m}\sum_{j=1}^{\ell+1}(\tau_j - \tau_{j-1})p_j^2\,,$$

where we have used the property $\theta_j(t)\theta_k(t) = \delta_{jk}\theta_j(t)$. Taking into account the factors related to the free motion, one recovers expression (9.20).

9.3 S-matrix: bosons and fermions

We first discuss scattering within the holomorphic formalism of Chapter 6. This formalism is mainly useful when the asymptotic states are eigenstates of the harmonic oscillator, a situation that one encounters in the case of boson systems in the second quantization formulation as described in Section 6.6 and, as a consequence, in quantum field theory (both relativistic and non-relativistic).

We then discuss scattering within the Grassmann formalism of Chapter 7 a problem relevant to fermion scattering in the formulation of second quantization, and thus also in quantum field theory.

9.3.1 Holomorphic formalism and bosons

We first consider a hamiltonian of the form (6.51). We rewrite the path integral (6.54) and expression (6.55) in the case of real time evolution. The path integral representation of the evolution operator is given formally by a rotation $t \mapsto it$. Then,

$$\langle z''|U(t'', t')|\bar z'\rangle = \int \left[\frac{\mathrm{d}\bar z(t)\mathrm{d}z(t)}{2i\pi}\right] \mathrm{e}^{z(t')\bar z(t')} \exp\left[i\mathcal{A}(z, \bar z)\right],$$

$$\mathcal{A}(z, \bar z) = \int_{t'}^{t''} \mathrm{d}t\left[-i\bar z(t)\dot z(t) - H\bigl(z(t), \bar z(t); t\bigr)\right]$$

with the boundary conditions $z(t'') = z''$, $\bar z(t') = \bar z'$.

We assume that the corresponding asymptotic free hamiltonian H_0 has the form (6.20):

$$H_0 = \omega z\frac{\mathrm{d}}{\mathrm{d}z}\,, \quad \omega > 0\,.$$

The matrix elements of the free evolution operator $U_0(t) = \mathrm{e}^{-itH_0}$ can then be inferred from expression (6.29):

$$\langle z|U_0(t)|\bar z\rangle = \mathrm{e}^{z\bar z\,\mathrm{e}^{-i\omega t}}\,. \tag{9.25}$$

S-matrix. From the evolution operator, one derives the S-matrix. Defining the S-matrix by expression (9.6), one finds

$$S(z,\bar z) = \lim_{\substack{t'\to-\infty\\ t''\to+\infty}} \int \frac{\mathrm{d}z''\mathrm{d}\bar z''}{2i\pi}\frac{\mathrm{d}z'\mathrm{d}\bar z'}{2i\pi}\,\mathrm{e}^{-z''\bar z''}\,\mathrm{e}^{-z'\bar z'}\exp\left(z\bar z''\,\mathrm{e}^{i\omega t''}\right)$$

$$\times \langle z''|U(t'',t')|\bar z'\rangle \exp\left(z'\bar z\,\mathrm{e}^{-i\omega t'}\right).$$

Using equation (6.30), or integrating directly, one infers

$$S(z,\bar z) = \lim_{\substack{t'\to-\infty\\ t''\to+\infty}} \langle z\,\mathrm{e}^{i\omega t''}|U(t'',t')|\bar z\,\mathrm{e}^{-i\omega t'}\rangle. \tag{9.26}$$

As in the case of the integral over paths in real space, in the holomorphic path integral the configurations contributing to the S-matrix are, for large times, asymptotic to solutions of the equation of the classical motion. For the harmonic oscillator H_0, this implies

$$z(t'') \underset{t''\to+\infty}{\sim} z\,\mathrm{e}^{i\omega t''},\quad \bar z(t') \underset{t'\to-\infty}{\sim} \bar z\,\mathrm{e}^{-i\omega t'}.$$

A simple quantum application is the evaluation of transition probabilities between eigenstates of the harmonic oscillator induced by a time-dependent perturbation that vanishes at times $\pm\infty$. In the example of the hamiltonian (6.37) where we assume that $b(t), \bar b(t)$ vanish at $t\to\pm\infty$, expression (9.26) leads to

$$S(z,\bar z) = \exp\left[z\bar z + i\int_{-\infty}^{+\infty}\mathrm{d}t\,\left(\bar b(t)\,\mathrm{e}^{i\omega t}\,z + \bar z\,\mathrm{e}^{-i\omega t}\,b(t)\right)\right.$$

$$\left. - \int_{-\infty}^{+\infty}\mathrm{d}t\mathrm{d}\tau\,\bar b(t)\theta(\tau-t)\,\mathrm{e}^{-i\omega(\tau-t)}\,b(\tau)\right]. \tag{9.27}$$

The coefficients of the expansion of $S(z,\bar z)$ in powers of z and $\bar z$ yield the transition amplitudes S_{mn} between eigenstates of the corresponding harmonic oscillator, induced by a time-dependent potential linearly coupled to position and momentum:

$$S(z,\bar z) = \sum_{m,n} S_{mn}\frac{z^m}{\sqrt{m!}}\frac{\bar z^n}{\sqrt{n!}}.$$

9.3.2 Fermion S-matrix

One now considers a fermion hamiltonian represented by a differential operator $h(\boldsymbol\theta,\partial/\partial\boldsymbol\theta)$, acting on functions of the θ_i, written in normal form. The matrix elements of the corresponding evolution operator can be derived from equation (7.81) by substituting $t\mapsto it$:

$$\langle \boldsymbol\theta''|U(t'',t')|\bar{\boldsymbol\theta}'\rangle = \int_{\bar{\boldsymbol\theta}(t')=\bar{\boldsymbol\theta}'}^{\boldsymbol\theta(t'')=\boldsymbol\theta''}[\mathrm{d}\boldsymbol\theta(t)\mathrm{d}\bar{\boldsymbol\theta}(t)]\,\mathrm{e}^{-\bar{\boldsymbol\theta}(t')\cdot\boldsymbol\theta(t')}\exp\left[i\mathcal{A}(\boldsymbol\theta,\bar{\boldsymbol\theta})\right] \tag{9.28}$$

with
$$\mathcal{A}(\boldsymbol{\theta},\bar{\boldsymbol{\theta}}) = \int_{t'}^{t''} \mathrm{d}t \left\{ i\bar{\boldsymbol{\theta}}(t) \cdot \dot{\boldsymbol{\theta}}(t) - H\left[\boldsymbol{\theta}(t),\bar{\boldsymbol{\theta}}(t)\right] \right\}. \tag{9.29}$$

One then decomposes the hamiltonian into a sum of a free quadratic asymptotic term and an interaction:
$$H(\boldsymbol{\theta},\bar{\boldsymbol{\theta}}) = -\sum_i \omega_i \bar{\theta}_i \theta_i + H_{\mathrm{I}}(\boldsymbol{\theta},\bar{\boldsymbol{\theta}}).$$

The matrix elements of the free evolution operator $U_0(t) = \mathrm{e}^{-itH_0}$ are obtained from expression (7.56) by continuation in time. One finds
$$\langle \boldsymbol{\theta} | U_0(t) | \bar{\boldsymbol{\theta}} \rangle = \exp\left[-\sum_i \bar{\theta}_i \theta_i \,\mathrm{e}^{-i\omega_i t}\right].$$

The S-matrix defined by equation (9.6), is given by
$$S(\boldsymbol{\theta},\bar{\boldsymbol{\theta}}) = \lim_{\substack{t' \to -\infty \\ t'' \to +\infty}} \int \prod_i \mathrm{d}\theta_i'' \mathrm{d}\bar{\theta}_i'' \mathrm{d}\theta_i' \mathrm{d}\bar{\theta}_i' \, \mathrm{e}^{\bar{\boldsymbol{\theta}}'' \cdot \boldsymbol{\theta}''} \, \mathrm{e}^{\bar{\boldsymbol{\theta}}' \cdot \boldsymbol{\theta}'} \, \langle \boldsymbol{\theta} | U_0(-t'') | \bar{\boldsymbol{\theta}}'' \rangle$$
$$\times \langle \boldsymbol{\theta}'' | U(t'',t') | \bar{\boldsymbol{\theta}}' \rangle \langle \boldsymbol{\theta}' | U_0(-t') | \bar{\boldsymbol{\theta}} \rangle.$$

Using the result (7.58), one infers
$$S(\boldsymbol{\theta},\bar{\boldsymbol{\theta}}) = \lim_{\substack{t' \to -\infty \\ t'' \to +\infty}} \left\langle \theta_i \,\mathrm{e}^{i\omega_i t''} \middle| U(t'',t') \middle| \bar{\theta}_i \,\mathrm{e}^{i\omega_i t'} \right\rangle.$$

A simple example. We assume that the evolution is free except between times $-T/2$ and $T/2$, where the hamiltonian takes the form
$$H = \sum_{i,j} \Omega_{ij} \theta_i \frac{\partial}{\partial \theta_j}, \quad \Omega = \Omega^\dagger.$$

In the interval $[-T/2, T/2]$, the evolution operator then reads
$$\langle \boldsymbol{\theta} | U(t'',t') | \bar{\boldsymbol{\theta}} \rangle = \exp\left[\sum_{i,j} \theta_i [\mathrm{e}^{-i\boldsymbol{\Omega}(t''-t')}]_{ij} \bar{\theta}_j\right].$$

Multiplying $U(T/2, -T/2)$ with free evolution then yields
$$S(\boldsymbol{\theta},\bar{\boldsymbol{\theta}}) = \left\langle \theta_i \,\mathrm{e}^{iT\omega_i/2} \middle| U(T/2,-T/2) \middle| \bar{\theta}_i \,\mathrm{e}^{iT\omega_i/2} \right\rangle$$
$$= \exp\left[\sum_{i,j} \theta_i \,\mathrm{e}^{iT\omega_i/2} [\mathrm{e}^{-i\boldsymbol{\Omega} T}]_{ij} \,\mathrm{e}^{iT\omega_j/2} \bar{\theta}_j\right].$$

The elements of the matrix $\mathrm{e}^{iT\omega_i/2}[\mathrm{e}^{-i\boldsymbol{\Omega} T}]_{ij}\, \mathrm{e}^{iT\omega_j/2}$ yield the transition amplitudes between the different possible fermion states.

9.4 S-matrix in the semi-classical limit

We now show that the path integral representation of the evolution operator (equation (9.11)) leads to a representation of the elements of the S-matrix that is specially well adapted to a study of the semi-classical limit.

We calculate the elements of the S-matrix between two wave packets:

$$\langle\psi_2|\,S\,|\psi_1\rangle = \lim_{\substack{t'\to-\infty\\t''\to+\infty}} \int \mathrm{d}\mathbf{q}'\mathrm{d}\mathbf{q}''\,\langle\psi_2|\,\mathrm{e}^{iH_0 t''/\hbar}\,|\mathbf{q}''\rangle\,\langle\mathbf{q}''|\,U(t'',t')\,|\mathbf{q}'\rangle\,\langle\mathbf{q}'|\,\mathrm{e}^{-iH_0 t'/\hbar}\,|\psi_1\rangle. \quad (9.30)$$

Introducing two wave functions $\tilde\psi_1(\mathbf{p})$ and $\tilde\psi_2(\mathbf{p})$, in the momentum basis, associated to the vectors $|\psi_1\rangle$ et $|\psi_2\rangle$, we define

$$\psi_1(\mathbf{q},t) = \langle\mathbf{q}|\,\mathrm{e}^{-iH_0 t/\hbar}\,|\psi_1\rangle = \int \frac{\mathrm{d}^d p}{(2\pi)^d}\tilde\psi_1(\mathbf{p})\exp\left[i\left(\mathbf{p}\cdot\mathbf{q}-t\frac{\mathbf{p}^2}{2m}\right)\Big/\hbar\right], \quad (9.31)$$

and a similar expression for ψ_2.

We have shown that when $t\to\infty$, the integral is dominated by $\mathbf{p}=m\mathbf{q}/t$ (equations (9.3-9.5)). We then change variables in the integral (9.30), setting

$$\mathbf{q}' = \frac{t'}{m}\mathbf{p}',\ \mathbf{q}'' = \frac{t''}{m}\mathbf{p}'', \quad (9.32)$$

and obtain

$$\langle\psi_2|\,S\,|\psi_1\rangle \propto \lim_{\substack{t'\to-\infty\\t''\to+\infty}} \int \mathrm{d}\mathbf{p}'\mathrm{d}\mathbf{p}''\,\tilde\psi_2^*(\mathbf{p}'')\tilde\psi_1(\mathbf{p}')\exp\left[\frac{i}{\hbar}\left(t''\frac{\mathbf{p}''^2}{2m}-t'\frac{\mathbf{p}'^2}{2m}\right)\right]$$
$$\times \langle t''\mathbf{p}''/m|\,U(t'',t')\,|t'\mathbf{p}'/m\rangle. \quad (9.33)$$

In this equation, we can then replace the evolution operator by its expression in terms of the path integral (9.11):

$$\langle t''\mathbf{p}''/m|\,U(t'',t')\,|t'\mathbf{p}'/m\rangle = \int_{\mathbf{q}(t')=t'\mathbf{p}'/m}^{\mathbf{q}(t'')=t''\mathbf{p}''/m} [\mathrm{d}\mathbf{q}(t)]\exp(i\mathcal{A}(\mathbf{q})/\hbar).$$

We conclude that the S-matrix is obtained by calculating a path integral with boundary conditions corresponding to classical scattering, that is, by summing over paths asymptotic at large times to free motion. In particular, if we know how to solve the equations for the classical motion with such boundary conditions, we can calculate the evolution operator and, thus, the S-matrix for $\hbar\to 0$. This leads to semi-classical approximations for the S-matrix. Two calculations in this spirit are presented in this chapter, a semi-classical calculation in one space dimension and the eikonal approximation. Note, however, that the one-dimensional example is, from the scattering viewpoint, somewhat singular in the sense that obstacles cannot be bypassed.

9.5 Semi-classical approximation: one dimension

We consider the hamiltonian
$$H = p^2/2m + V(x).$$
We assume an analytic potential V, decreasing fast enough at large distances for the S-matrix to exist. We infer the elements of the S-matrix from the calculation of the path integral giving the evolution operator. As we have pointed out earlier, in the limit $\hbar \to 0$ the path integral is dominated by classical paths for which the action and, thus, the integrand are stationary. We thus determine first the relevant classical trajectories.

9.5.1 Forward scattering

We first consider a situation in which forward classical scattering is possible. This implies that the scattering energy is larger than the maximum value of the potential.

After a first integration of the equation of the classical motion, one finds
$$\tfrac{1}{2}m\dot{x}^2(\tau) + V(x) = \kappa^2/2m, \quad \kappa > 0,$$
with the boundary conditions
$$x(\tau') = x', \quad x(\tau'') = x''.$$
This equation can be integrated in the form
$$\tau = \pm m \int^x \frac{\mathrm{d}y}{\sqrt{\kappa^2 - 2mV(y)}}.$$
It is convenient to set $X = x'' - x'$, $T = \tau'' - \tau'$. Within the framework of perturbative calculations, we have pointed out that the relevant paths are those that satisfy classical scattering boundary conditions. A non-perturbative analysis has confirmed this result. This implies here that $X/T = k/m$ remains finite when $\tau' \to -\infty$, $\tau'' \to \infty$. In this regime, the equation
$$T = m \int_{x'}^{x''} \frac{\mathrm{d}x}{\sqrt{\kappa^2 - 2mV(x)}} \tag{9.34}$$
implies
$$\kappa = k + \frac{m}{T} \int_{-\infty}^{+\infty} \mathrm{d}x \left(\frac{k}{\sqrt{k^2 - 2mV(x)}} - 1 \right) + O\left(T^{-2}\right).$$
The action of the trajectory is then
$$\mathcal{A}_c = T\frac{\kappa^2}{2m} - 2 \int_{t'}^{t''} V(x(\tau))\,\mathrm{d}\tau$$
$$= T\frac{\kappa^2}{2m} - 2m \int_{x'}^{x''} \frac{V(x)\,\mathrm{d}x}{\sqrt{\kappa^2 - 2mV(x)}}$$
$$= T\frac{k^2}{2m} + \int_{-\infty}^{+\infty} \mathrm{d}x \left(\sqrt{k^2 - 2mV(x)} - k \right) + O\left(T^{-1}\right).$$

We now Fourier transform with respect to x' and x''. Since the result depends only on $x'' - x'$, we obtain a factor $\delta(k'' - k')$ that expresses, in this one-dimensional example, energy conservation. Indeed, in these special kinematic conditions, the energy conservation condition $k''^2 = k'^2$ implies momentum conservation. In the limit $\hbar \to 0$, the remaining integral over X can be estimated by the steepest descent method. At leading order for $T \to \infty$, only the terms of order T in the saddle point equation are relevant. One finds

$$X = k'T/m \;\Rightarrow\; k' = k\,.$$

The terms proportional to T then cancel the factors coming from the free motion in expression (9.7) of the S-matrix. The resulting semi-classical expression is then

$$\ln S_+(k) = \frac{i}{\hbar} \int_{-\infty}^{+\infty} \mathrm{d}x \left(\sqrt{k^2 - 2mV(x)} - k \right). \tag{9.35}$$

The result is a pure phase, $|S_+(k)| = 1$, $S_+(k) = e^{2i\delta_+(k)}$. Scattering yields only a phase shift in this semi-classical limit, because reflection is absent.

9.5.2 Backward scattering

We now assume, instead, that the energy is smaller than the maximum value of the potential in such a way that, classically, one observes total reflection. We also assume that the scattered particle comes from $-\infty$. Then, a similar calculation yields the action of the classical trajectory

$$\mathcal{A}_c = \frac{m}{2T}(x' + x'' - 2x_0)^2 + 2\int_{-\infty}^{x_0} \mathrm{d}x \left(\sqrt{k^2 - 2mV(x)} - k \right) + O\left(T^{-1}\right),$$

where x_0 is the point where reflection occurs, defined by $k^2 = 2mV(x_0)$, and $k = m(2x_0 - x' - x'')/T$.

Since the result does not depend on the combination $x' + x''$, after Fourier transformation a factor $\delta(k'' + k')$ appears that expresses energy conservation. The remaining integral over $X = x' + x''$, calculated by the steepest descent method, yields

$$2x_0 - X = k'T/m\,, \quad k' = k$$

and, thus,

$$\ln S_-(k) = \frac{2i}{\hbar} \int_{-\infty}^{x_0} \mathrm{d}x \left(\sqrt{k^2 - 2mV(x)} - k \right) + \frac{2ikx_0}{\hbar}\,.$$

Here again, since classically one observes total reflection, $S_-(k)$ is a pure phase.

9.5.3 Forbidden region

One might think that semi-classical approximations are only available in situations where classical scattering is allowed, though quantum scattering through a potential barrier is possible due to tunneling effects. However, we note that non-trivial, physically sensible, expressions are obtained by analytic continuation in the energy variable, the sign ambiguity in the analytic continuation being fixed by the condition $|S_\pm| < 1$.

In the case of forward scattering, if $k^2/2m$ is smaller than the maximum value of V, $i \ln S_+$ has a real part corresponding to a real trajectory, and an imaginary part that corresponds formally to a trajectory in the classically forbidden region. If, for example, the forbidden region in which $V(x) \geq k^2/2m$, extends from x_- to x_+, then expression (9.35) can be written as

$$\ln S_+(k) = \frac{i}{\hbar} \int_{-\infty}^{x_-} \mathrm{d}x \left(\sqrt{k^2 - 2mV(x)} - k \right) - \frac{ik(x_+ - x_-)}{\hbar}$$
$$+ \frac{i}{\hbar} \int_{x_+}^{+\infty} \mathrm{d}x \left(\sqrt{k^2 - 2mV(x)} - k \right) - \frac{1}{\hbar} \int_{x_-}^{x_+} \mathrm{d}x \sqrt{2mV(x) - k^2}.$$

This result is similar to the result obtained by a WKB calculation of the corresponding wave function. It is consistent with the evaluation of barrier penetration coefficients of Chapter 8. It has a physically reasonable dependence in the energy. Finally, one may wonder how a non-vanishing result is consistent with the unitarity of the S-matrix since S_- has already modulus one. The answer is that S_+ behaves like $e^{-\mathrm{const.}/\hbar}$ for $\hbar \to 0$, and gives contribution to the unitarity relation invisible to all orders in an expansion in powers of $\hbar$.

Barrier penetration and evolution in imaginary time. Returning to the classical equations, one notices that barrier penetration corresponds to an imaginary time in the integral (9.34). Moreover, the imaginary part of the action, that determines the probability of barrier penetration, is entirely related to the trajectory in imaginary time. We thus find an additional application of the path integral with imaginary or euclidean time, which we have introduced to study statistical problems: it allows evaluating barrier penetration amplitudes. In particular, it can be used to calculate the lifetimes of metastable states that decay through tunneling, as we have shown in Chapter 8.

9.6 Eikonal approximation

From the path integral representation of the S-matrix, one can derive an approximation for a scattering amplitude valid at high energy and small momentum transfer, called the eikonal approximation. From the viewpoint of the path integral, this corresponds to a situation where the kinetic term is large with respect to the potential, a situation quite analogous to the one discussed in Section 3.2.1.

9.6.1 Eikonal approximation

In the case of the free motion, the evolution operator is given by a gaussian path integral, which can be calculated by first solving the equations of the classical motion. The solution that satisfies the boundary conditions of the representation (9.11) and that corresponds to the free hamiltonian

$$H_0 = \mathbf{p}^2/2m \text{ with } \mathbf{p} \in \mathbb{R}^d,$$

is

$$\mathbf{q}(t) = \mathbf{q}' + (\mathbf{q}'' - \mathbf{q}')\frac{t - t'}{t'' - t'}. \tag{9.36}$$

Translating the integration variables $\mathbf{q}(t)$ by the classical solution (9.36), one then reduces the calculation to a normalization integral that can be determined by comparing it to the exact result ($\hbar = 1$)

$$\langle \mathbf{q}'' | U(t'', t') | \mathbf{q}' \rangle = \left(\frac{m}{2i\pi(t'' - t')}\right)^{d/2} \exp\left[i\frac{m}{2}\frac{(\mathbf{q}'' - \mathbf{q}')^2}{(t'' - t')}\right]. \tag{9.37}$$

The eikonal approximation corresponds to the kinematic regime

$$\mathbf{p}' = \mathbf{p} - \mathbf{k}/2, \quad \mathbf{p}'' = \mathbf{p} + \mathbf{k}/2, \quad \mathbf{p}^2 \to \infty, \quad \mathbf{p}^2 \gg \mathbf{k}^2.$$

In this situation, in the action the kinetic term dominates the potential. We thus calculate the path integral as in the free case, expanding paths around classical trajectories of the free motion (9.36). The calculation of the evolution operator at leading order is simple. Setting $\mathbf{q}'' - \mathbf{q}' = \mathbf{s}$, $(\mathbf{q}'' + \mathbf{q}')/2 = \mathbf{x}$, one obtains

$$\langle \mathbf{p} + \mathbf{k}/2 | U(t'', t') | \mathbf{p} - \mathbf{k}/2 \rangle \propto \int d^d s\, d^d x\, \exp[-i(\mathbf{p} \cdot \mathbf{s} + \mathbf{k} \cdot \mathbf{x}) + i\mathcal{A}(\mathbf{s}, \mathbf{x})], \tag{9.38}$$

where the classical action now is

$$\mathcal{A}(\mathbf{s}, \mathbf{x}) = \frac{im}{2}\frac{\mathbf{s}^2}{t'' - t'} - i\int_{t'}^{t''} dt\, V\left(\mathbf{x} - \frac{\mathbf{s}}{2} + \frac{t - t'}{t'' - t'}\mathbf{s}\right), \tag{9.39}$$

and the normalization in equation (9.38) is determined by comparing with the result (9.37) of the free motion.

In the large time limit, if the contribution of the potential is neglected, the integral over $\mathbf{s}$ is dominated by the saddle point

$$\mathbf{s} = (t'' - t')\mathbf{p}/m. \tag{9.40}$$

After the substitution (9.40) and the change of variables $t - (t' + t'')/2 \mapsto t$ and, thus, $|t| \leq (t'' - t')/2$, the argument in the potential becomes $\mathbf{x} + t\mathbf{p}/m$. We assume that the potential decreases fast enough for the integral in (9.39) to have a large

time limit, that is $t'' - t' \to \infty$. The contribution of the potential to the action then takes the form
$$-i \int_{-\infty}^{+\infty} dt\, V(\mathbf{x} + t\mathbf{p}/m).$$

Once the limit is taken, we can translate the integration variable t without changing the domain of integration. This amounts to translating the vector $\mathbf{x}$, in the argument of the potential V, by a vector proportional to $\mathbf{p}$. We choose this vector in such a way that V depends only on

$$\mathbf{b} = \mathbf{x} - \mathbf{p}\,(\mathbf{x}\cdot\mathbf{p}/\mathbf{p}^2), \tag{9.41}$$

the component of $\mathbf{x}$ orthogonal to $\mathbf{p}$. The integral over the component of $\mathbf{x}$ along $\mathbf{p}$ can then be performed and implies $\mathbf{p}\cdot\mathbf{k} = 0$. In the domain of validity of the eikonal approximation, this equation expresses energy conservation:

$$0 = \mathbf{p}''^2 - \mathbf{p}'^2 = (\mathbf{p} + \mathbf{k}/2)^2 - (\mathbf{p} - \mathbf{k}/2)^2 \sim 2\mathbf{p}\cdot\mathbf{k}.$$

One obtains

$$\langle \mathbf{p} + \mathbf{k}/2|\, U(t'',t')\, |\mathbf{p} - \mathbf{k}/2\rangle$$
$$\simeq \delta(\mathbf{p}\cdot\mathbf{k})\mathcal{N}(\mathbf{p}) \int d^{d-1}b\, e^{-i\mathbf{k}\cdot\mathbf{b}} \exp\left[-i\int_{-\infty}^{+\infty} dt\, V\left(\frac{\mathbf{p}t}{m} + \mathbf{b}\right)\right] \tag{9.42}$$

with

$$\mathcal{N}(\mathbf{p}) \sim \exp\left[i(t'' - t')\frac{\mathbf{p}^2}{2m}\right]. \tag{9.43}$$

Equation (9.42) yields, after Fourier transformation, the matrix elements of the scattering operator T in the momentum basis (defined by equation (9.9)):

$$\langle \mathbf{p} + \mathbf{k}/2|\, T\, |\mathbf{p} - \mathbf{k}/2\rangle$$
$$\simeq \frac{i|\mathbf{p}|}{m} \int \frac{d^{d-1}b}{(2\pi)^d}\, e^{-i\mathbf{k}\cdot\mathbf{b}} \left\{\exp\left[-i\int_{-\infty}^{+\infty} dt\, V\left(\frac{\mathbf{p}t}{m} + \mathbf{b}\right)\right] - 1\right\}. \tag{9.44}$$

In terms of the Fourier representation

$$V(\mathbf{q}) = \int d^d s\, e^{i\mathbf{q}\cdot\mathbf{s}}\, \tilde{V}(\mathbf{s}),$$

the integral over the potential can be written as

$$\int_{-\infty}^{+\infty} dt\, V\left(\frac{\mathbf{p}t}{m} + \mathbf{b}\right) = \int d^d s\, \tilde{V}(\mathbf{s}) \int_{-\infty}^{+\infty} dt\, e^{i(\mathbf{p}t/m + \mathbf{b})\cdot\mathbf{s}}.$$

The integral over time can now be performed and reduces the integral over $\mathbf{s}$ to an integral over the components of $\mathbf{s}$ orthogonal to $\mathbf{p}$:

$$\int_{-\infty}^{+\infty} dt\, V\left(\frac{\mathbf{p}t}{m} + \mathbf{b}\right) = 2\pi \frac{m}{|\mathbf{p}|} \int d^d s\, \delta(\mathbf{s}\cdot\hat{\mathbf{p}})\, e^{i\mathbf{b}\cdot\mathbf{s}}\, \tilde{V}(\mathbf{s}).$$

This expression involves the Fourier transform of $\tilde{V}$ but in the $(d-1)$-dimensional subspace orthogonal to $\mathbf{p}$.

Finally note that, at leading order in the potential V, the eikonal approximation reduces to
$$\langle \mathbf{p} + \mathbf{k}/2 | T | \mathbf{p} - \mathbf{k}/2 \rangle = \tilde{V}(\mathbf{k}),$$
which is the exact Born approximation (see Section 9.2).

9.6.2 Application to the Coulomb potential

We now apply the eikonal approximation to the evaluation of the scattering amplitude for a potential in $1/q$ of Coulomb type. The integral over the potential in this case has no limit for infinite time because th potential decreases too slowly at large distance. It is necessary to first integrate over a finite time interval. We parametrize the potential as
$$V(\mathbf{q}) = \frac{\alpha}{|\mathbf{q}|}. \tag{9.45}$$

Then,
$$\int_{(t'-t'')/2}^{(t''-t')/2} dt\, V\left(\frac{\mathbf{p}t}{m} + \mathbf{x}\right) \simeq \frac{2\alpha m}{p} \ln\bigl((t''-t')p/mb\bigr). \tag{9.46}$$

This infinite phase has the following origin: because the Coulomb potential decreases too slowly at large distance, the classical trajectory approaches too slowly the free motion, which, with our definition of the S-matrix, has been taken as the reference motion. In a Coulomb potential, amplitudes are not defined, only cross-sections are.

Factorizing the infinite phase, one can complete the calculation of the scattering amplitude. Integrating over the vector $\mathbf{b}$, one obtains
$$T(\mathbf{p}+\mathbf{k}/2, \mathbf{p}-\mathbf{k}/2) \simeq \frac{i\pi^{(d-1)/2}}{(2\pi)^d} \frac{p}{m} \exp\left[-i\frac{2\alpha m}{p} \ln\bigl((t''-t')p/mb\bigr)\right]$$
$$\times \frac{\Gamma\left[\frac{1}{2}(d-1)-\theta\right]}{\Gamma(\theta)} \left(\frac{k^2}{4}\right)^{[\theta+(1-d)/2]} \tag{9.47}$$

with
$$\theta = -i\alpha m/p. \tag{9.48}$$

In the physical dimension 3, expression (9.47) coincides with the exact result. For $\alpha < 0$ (the attractive case), it thus gives the energies E_n of bound states in a Coulomb potential, which appear as poles of the scattering amplitude:
$$\theta = \tfrac{1}{2}(d-1) + n \Rightarrow E_n = \frac{\mathbf{p}^2}{2m} = -\frac{2\alpha^2 m}{(d-1+2n)^2}. \tag{9.49}$$

One recognizes the well-known Coulomb spectrum.

Finally, the eikonal approximation has a relativistic generalization, which again yields quite interesting approximations to the energy spectrum in quantum electrodynamics. In the latter framework, it is based on an approximate summation of ladder and crossed ladder Feynman diagrams.

9.7 Perturbation theory and operators

To illustrate the correspondence between path integral and operator formulations of quantum mechanics, we recall here the basis of perturbation theory in the operator framework, in the case of time-independent hamiltonians.

The expansion of the S-matrix can be derived, for example, from the expansion of the operator (see equation (9.6))

$$\Omega(t) = e^{iH_0 t} e^{-iHt}, \tag{9.50}$$

where H_0 is the non-perturbed hamiltonian and

$$V = H - H_0$$

the perturbation.

Introducing the perturbation operator in the interaction representation

$$V_I(t) = e^{iH_0 t} V e^{-iH_0 t}, \tag{9.51}$$

one verifies that the operator $\Omega(t)$ satisfies the equation

$$\dot{\Omega}(t) = -iV_I(t)\Omega(t) \tag{9.52}$$

with the boundary condition

$$\Omega(0) = 1.$$

Equation (9.52) can be solved as a power series in the perturbation V_I. A straightforward calculation leads to the formal expansion

$$\Omega(t) = \sum_{n=0}^{\infty} (-i)^n \int dt_1\, dt_2 \ldots dt_n\, V_I(t_n)V_I(t_{n-1})\ldots V_I(t_2)V_I(t_1), \tag{9.53}$$

the integration domain in the r.h.s. being

$$0 \leq t_1 \leq t_2 \leq \cdots \leq t_n \leq t.$$

If one then introduces the time-ordered product defined in Section 4.6, one can rewrite the expansion as

$$\Omega(t) = \sum_{n=0}^{\infty} (-i)^n \int dt_1\, dt_2 \ldots dt_n\, T[V_I(t_n)V_I(t_{n-1})\ldots V_I(t_2)V_I(t_1)].$$

The product of factors that appears in the r.h.s. is now a symmetric function of all time arguments. The integration domain can then be symmetrized provided the integral is divided by a counting factor $n!$ and one obtains

$$\Omega(t) = \sum_{n=0}^{\infty} \frac{(-i)^n}{n!} \int_{\substack{0 \leq t_i \leq t \\ 1 \leq i < n}} dt_1\, dt_2 \ldots dt_n\, T\left[V_I(t_1)V_I(t_2)\ldots V_I(t_n)\right]. \tag{9.54}$$

The formal sum of this expansion can be written as

$$\Omega(t) = \mathrm{T}\left[\exp\left(-i\int_0^t \mathrm{d}t'\, V_\mathrm{I}(t')\right)\right]. \tag{9.55}$$

In particular, this result can be applied to a hamiltonian H perturbed by a term linear in the position operator $\hat{q}$. To the complete hamiltonian, corresponds also a path integral representation of the partition function. Comparing the expansion of the path integral in powers of the perturbation with expression (9.55), one recovers the relation between correlation functions and T-products established in Section 4.6 (equation (4.35)).

Exercises

Exercise 9.1

S-matrix calculation: the pseudo-potential $\delta(x)$. Sum the perturbative expansion of the S-matrix in the example of the one-dimensional potential

$$V(x) = \lambda \delta(x),$$

where the constant λ parametrizes the potential.

Solution. The potential has a Fourier representation:

$$\delta(x) = \frac{1}{2\pi}\int \mathrm{d}p\, e^{ipx},$$

which shows that $\tilde{V}$ is a constant. Expression (9.17) then simplifies drastically. In fact, in this example the Fourier representation is hardly useful; the integral over k_j yields the initial potential and implies $x(\tau_j) = 0$. One then uses expression (9.2b) (pour $\hbar = m = d = 1$) and finds

$$\langle x''|U^{(\ell)}(t'',t')|x'\rangle$$
$$= (-i\lambda)^\ell \int_{t'}^{t''}\left(\prod_{j=1}^\ell \mathrm{d}\tau_j\right) e^{ix''^2/2(t''-\tau_\ell)}\, e^{ix'^2/2(\tau_1-t')} \prod_{j=1}^{\ell+1}\frac{1}{\sqrt{2i\pi(\tau_j-\tau_{j-1})}},$$

where the times τ_j are ordered and $\tau_0 = t'$, $\tau_{\ell+1} = t'$. The result of the Fourier transformation is

$$\langle p''|U^{(\ell)}(t'',t')|p'\rangle$$
$$= (-i\lambda)^\ell \int_{t'}^{t''}\left(\prod_{j=1}^\ell \mathrm{d}\tau_j\right) e^{-ip''^2(t''-\tau_\ell)/2}\, e^{ip'^2(t'-\tau_1)/2} \prod_{j=2}^\ell \frac{1}{\sqrt{2i\pi(\tau_j-\tau_{j-1})}}.$$

This leads to the contribution

$$\langle p''|S^{(\ell)}|p'\rangle = (-i\lambda)^\ell \int \left(\prod_{j=1}^\ell \mathrm{d}\tau_j\right) e^{ip''^2\tau_\ell/2 - ip'^2\tau_1/2} \prod_{j=2}^\ell \frac{1}{\sqrt{2i\pi(\tau_j-\tau_{j-1})}}$$

to the S-matrix. One then changes variables:
$$\tau_{j+1} = \tau_j + u_j, \quad u_j \geq 0.$$

The remaining integral over τ_1 yields a δ function of energy conservation:
$$E(p') \equiv p'^2/2 = E(p'') \equiv p''^2/2.$$

The integrals over the u_i's take all the form
$$\int_0^\infty \frac{du}{\sqrt{2i\pi u}} e^{ip''^2 u/2} = \frac{1}{\sqrt{2E}}.$$

Thus,
$$\langle p''| S^{(\ell)} |p'\rangle = 2\pi\delta\big(E(p'') - E(p')\big)(-i\lambda)^\ell (2E)^{-(\ell-1)/2}.$$

The perturbative contributions form a geometric series that can be summed:
$$\langle p''| S |p'\rangle = 2\pi\delta(p' - p'') - 2\pi \frac{i\lambda}{1 + i\lambda/\sqrt{2E}} \delta\big(E(p'') - E(p')\big).$$

In one dimension, energy conservation implies $p'' = \pm p'$, since
$$\delta\big(E(p'') - E(p')\big) = \frac{1}{\sqrt{2E}} \left[\delta(p'' - p') + \delta(p'' + p')\right],$$

the two cases corresponding to transmission and reflection. Calling $S_\pm$ the corresponding elements of the S-matrix, one finds
$$S_+(E) = 1 - i\frac{\lambda/\sqrt{2E}}{1 + i\lambda/\sqrt{2E}}, \quad S_-(E) = -i\frac{\lambda/\sqrt{2E}}{1 + i\lambda/\sqrt{2E}}.$$

The result is consistent with conservation of probabilities since
$$|S_+|^2 + |S_-|^2 = 1.$$

Exercise 9.2

Semi-classical limit. Use expression (9.33) to derive a semi-classical approximation to the S-matrix for the one-dimensional potential $V(q) = \lambda/\cosh^2 q$. Compare with the exact results that can be found, for example, in R.G. Newton, *Scattering theory Waves and Particles* (McGraw-Hill, New York 1966)). The elements of the S-matrix, S_+ and S_- that correspond to transmission and reflection, respectively, at energy $E = k^2/2$, are
$$S_+ = \frac{\Gamma(d+\alpha)\Gamma(\alpha+1-d)}{\Gamma(1+\alpha)\Gamma(\alpha)}, \quad S_- = \frac{\sin \pi d}{\sin \pi \alpha} \frac{\Gamma(d+\alpha)\Gamma(\alpha+1-d)}{\Gamma(1+\alpha)\Gamma(\alpha)}$$

with
$$d = \tfrac{1}{2}(1 - \sqrt{1 - 8\lambda}) \quad \alpha = -ik, \quad k > 0.$$

10 PATH INTEGRALS IN PHASE SPACE

In this chapter, we generalize the construction of the path integral of Chapter 2 to hamiltonians general functions of the phase space variables, position and momentum. This leads to a path integral in which the action takes its hamiltonian form and one integrates over trajectories in phase space with a generalized Liouville measure. Let us point out here that this formalism has many features in common with the holomorphic formalism of Chapter 6.

We examine more specifically the important example of hamiltonians quadratic in the momentum variables. We first verify that in the simplest situations already discussed in Chapters 2 and 5, after an explicit integration over momenta $p(t)$ one recovers the usual path integral.

More general hamiltonians are often met, for example, in the quantization of the motion on riemannian manifolds. We illustrate the analysis with the quantization of free motion on a sphere (or hypersphere) S_{N-1}.

However, before entering in such a discussion, we recall a few relevant elements of classical mechanics

10.1 A few elements of classical mechanics

For the discussion of path integrals in this more general context, it is useful to recall a few elements of classical mechanics. In this whole section time is real. We consider only the situation where the equations of the classical motion derive from an action principle. The action is the time integral of a lagrangian:

$$\mathcal{A}(q) = \int dt\, \mathcal{L}(q, \dot{q}; t), \tag{10.1}$$

where the variables $q_i(t)$ characterize, for example, the position of a particle at time t and $\dot{q}_i$ is its time derivative. The equations of the classical motion are obtained by expressing that the action is stationary with respect to variations of the trajectory $q(t)$:

$$\frac{\delta \mathcal{A}}{\delta q_i(t)} = 0 \Leftrightarrow \frac{d}{dt}\frac{\partial \mathcal{L}}{\partial \dot{q}_i} = \frac{\partial \mathcal{L}}{\partial q_i}. \tag{10.2}$$

10.1.1 Symmetries. Conservation laws

Continuous symmetries (i.e. symmetries associated with Lie groups) of the action lead to quantities conserved in the classical motion. A general strategy to derive these conservation laws is to make time-dependent group transformations and to express that the action is stationary at the classical motion.

Let us illustrate this strategy with the example of the rotation group, the $SO(N)$ group in N space dimensions. We thus assume that a lagrangian is rotation invariant, that is invariant under all transformations of the form

$$q_i(t) \mapsto \sum_j R_{ij} q_j(t),$$

where R is an orthogonal matrix $R^T R = 1$, with determinant 1:

$$\mathcal{L}(q, \dot{q}; t) = \mathcal{L}(Rq, R\dot{q}; t).$$

One then expresses that if $q_i(t)$ is a classical solution, the action is stationary with respect to any variation of $q_i(t)$ and thus, in particular, with respect to a variation that takes the form of an infinitesimal time-dependent rotation

$$R_{ij}(t) = \delta_{ij} + \tau_{ij}(t).$$

Since the matrix R_{ij} is orthogonal, τ_{ij} is antisymmetric: $\tau_{ji} = -\tau_{ij}$. The corresponding variations of the trajectory are

$$\delta q_i(t) = \sum_j \tau_{ij}(t) q_j(t) \Rightarrow \delta \dot{q}_i(t) = \sum_j \left[\tau_{ij}(t) \dot{q}_j(t) + \dot{\tau}_{ij}(t) q_j(t) \right]. \qquad (10.3)$$

If τ is time-independent, the symmetry assumption implies that the variation of the action vanishes. The variation of the action at first order in τ thus can come only from the derivative of τ, that is from $\dot{q}$:

$$\delta \mathcal{A}(q) = \int dt \sum_{i,j} \dot{\tau}_{ij}(t) q_j(t) \frac{\partial \mathcal{L}}{\partial \dot{q}_i(t)} .$$

The variation of the action vanishes only if the derivative of the coefficient of $\dot{\tau}$ vanishes and, thus, if the coefficient of $\dot{\tau}$ is a constant. Since the matrix τ is antisymmetric, only the antisymmetric part in ij of its coefficient must vanish. We define

$$L_{ij} = q_i \frac{\partial \mathcal{L}}{\partial \dot{q}_j} - q_j \frac{\partial \mathcal{L}}{\partial \dot{q}_i} . \qquad (10.4)$$

Then,

$$\dot{L}_{ij} = 0.$$

To the rotation symmetry are associated conserved quantities L_{ij} in direct correspondence with the generators of the rotation group.

In dimension 3, the antisymmetric representation of the $SO(3)$ group is isomorphic to the vector representation. The conserved quantities correspond to the angular momentum vector

$$L_i = \frac{1}{2} \sum_{j,k} \epsilon_{ijk} L_{jk},$$

where ϵ_{ijk} is the completely antisymmetric symbol with $\epsilon_{123} = 1$ and, thus, in vector notation

$$\mathbf{L} = \mathbf{q} \times \frac{\partial \mathcal{L}}{\partial \dot{\mathbf{q}}} .$$

10.1.2 Time-translation invariance. Hamiltonian formalism

Another (slightly singular) example is provided by lagrangians that do not depend *explicitly* on time. We assume that $q_c(t)$ is a classical trajectory, and we calculate the action corresponding to $q_c(t + \epsilon(t))$, that is, resulting from a general change of parametrization of time. In an expansion in powers of $\epsilon(t)$, the first order in ϵ vanishes as a consequence of the equations of motion. Then,

$$\frac{\mathrm{d}}{\mathrm{d}t} q_c(t + \epsilon(t)) = (1 + \dot\epsilon) \dot q_c(t + \epsilon(t)).$$

In the action, we change variables

$$t \mapsto t' = t + \epsilon(t) \;\Rightarrow\; \mathrm{d}t' = (1 + \dot\epsilon)\mathrm{d}t\,.$$

After the change of variables, the variations of the action come from the derivative of the measure and the explicit dependence of $\mathcal{L}$ in the time variable:

$$\delta \mathcal{A} = \int \mathrm{d}t \left[\dot\epsilon \left(\sum_i \dot q_i \frac{\partial \mathcal{L}}{\partial \dot q_i} - \mathcal{L} \right) - \epsilon \frac{\partial \mathcal{L}}{\partial t} \right].$$

We then integrate by parts and express that the coefficient of $\epsilon(t)$ vanishes:

$$\frac{\mathrm{d}}{\mathrm{d}t} \left(\sum_i \dot q_i \frac{\partial \mathcal{L}}{\partial \dot q_i} - \mathcal{L} \right) = -\frac{\partial \mathcal{L}}{\partial t}\,.$$

One concludes that if a system is invariant under time translation, which implies that $\partial \mathcal{L}/\partial t$ vanishes, the quantity

$$E = \sum_i \dot q_i \frac{\partial \mathcal{L}}{\partial \dot q_i} - \mathcal{L}$$

is a conserved quantity, which is the energy.

Hamiltonian and Legendre transformation. More generally, even when the lagrangian is time-dependent, it is useful to introduce the hamiltonian

$$H = \sum_i \dot q_i \frac{\partial \mathcal{L}}{\partial \dot q_i} - \mathcal{L}(q, \dot q; t)$$

and to express it in terms of phase space variables, the position q_i and

$$p_i(t) = \frac{\partial \mathcal{L}}{\partial \dot q_i(t)} \tag{10.5}$$

called the momentum conjugate to q_i. The hamiltonian then is given by

$$H(p, q; t) = \sum_i p_i(t) \dot q_i(t) - \mathcal{L}(\dot q, q; t), \tag{10.6}$$

which, as a consequence of equation (10.5), is the Legendre transform of the lagrangian. Equation (10.5) expresses that the r.h.s. of equation (10.6) is stationary with respect to $\dot q$, at p, q fixed. The Legendre transformation is involutive: $\mathcal{L}$ and $\dot q$ play a role symmetric to H and p. Indeed, varying expression (10.6) with respect to p (at q fixed), $\dot q$ being considered as a function of p through (10.5), one obtains

$$\dot q_i(t) = \frac{\partial H}{\partial p_i},$$

which is one of the two equations of motion in the hamiltonian formalism. Moreover, it follows from the stationarity conditions, that the derivative of $H + \mathcal{L}$ with respect to any argument that does not participate in the Legendre transformation vanishes. In particular, differentiating equation (10.6) with respect to q at p fixed yields

$$\left.\frac{\partial H}{\partial q_i}\right|_p = -\left.\frac{\partial \mathcal{L}}{\partial q_i}\right|_{\dot q} + \frac{\partial \dot q}{\partial q}\left.\frac{\partial}{\partial \dot q}\right|_{p,q}[\mathbf{p}\cdot\dot{\mathbf{q}} - \mathcal{L}(\dot q, q; t)],$$

and, thus,

$$\frac{\partial H}{\partial q_i} + \frac{\partial \mathcal{L}}{\partial q_i} = 0.$$

Combining this equation with the equation of motion (10.2), one obtains the second equation of motion in the hamiltonian formalism. One thus finds

$$\dot q_i(t) = \frac{\partial H}{\partial p_i}, \quad \dot p_i(t) = -\frac{\partial H}{\partial q_i}, \tag{10.7}$$

equations that can also be considered as equations of motion in phase space.

These equations can also be derived from an action principle, by expressing that the action, expressed in terms of the hamiltonian,

$$\mathcal{A}(p,q) = \int \mathrm{d}t \left[\sum_i p_i(t)\dot q_i(t) - H(p(t), q(t); t)\right] \tag{10.8}$$

is stationary with respect to the variations both of $p(t)$ and $q(t)$.

In this form the property of energy conservation, when H does not depend explicitly on time, is immediate.

Let us also point out that the term

$$\int \mathrm{d}t \sum_i p_i(t)\dot q_i(t) = \oint \sum_i p_i \mathrm{d}q_i \tag{10.9}$$

depends on the geometric trajectory in phase space but not on the motion on the trajectory. More precisely, it is equal to a sum of areas between the trajectory and the axes. It can be antisymmetrized in p and q. In the more mathematical language of differential geometry, it is the integral of a two-form, the so-called symplectic form $\omega = \sum_i \mathrm{d}p_i \wedge \mathrm{d}q_i$. This viewpoint is specially useful when phase space has a non-trivial topology (like in the problem of spin quantization).

The conserved quantities (10.4) associated to the rotation symmetry (in N dimensional space the group $SO(N)$) in the phase space formalism then read

$$L_{ij} = q_i p_j - p_i q_j. \tag{10.10}$$

10.1.3 Canonical transformations

Canonical transformations are transformations in phase space $\{q_i, p_i\} \mapsto \{Q_i, P_i\}$ that leave the symplectic form invariant. They have a group structure. One set of trivial transformations corresponds to adding to p_i a gradient $\partial \Pi(q)/\partial q_i$. This transformation is induced by the addition to a total time derivative to the lagrangian. It is then easy to characterize the infinitesimal transformations that do not have this form. One finds

$$Q_i = q_i + \varepsilon \frac{\partial T(p,q)}{\partial p_i} + O(\varepsilon^2), \quad P_i = p_i - \varepsilon \frac{\partial T(p,q)}{\partial q_i} + O(\varepsilon^2), \tag{10.11}$$

where $T(p,q)$ is arbitrary. If T has the form of a hamiltonian, one recognizes the equations of motion integrated between time t and $t+\varepsilon$. To the set of all hamiltonians is associated a set of canonical transformations: the mapping that associates the position in phase space at time t to the position at initial time is a canonical transformation.

This observation indicates how to determine the finite form of canonical transformations. One introduces a generating function $\mathcal{S}(q, Q)$ (the classical action of a trajectory that goes from q to Q) and one sets

$$p_i = \frac{\partial \mathcal{S}}{\partial q_i} \quad P_i = -\frac{\partial \mathcal{S}}{\partial Q_i}. \tag{10.12}$$

One directly verifies the invariance of the form by changing variables in two steps: $\{q_i, p_i\} \mapsto \{q_i, Q_i\}$ then $\{q_i, Q_i\} \mapsto \{Q_i, P_i\}$. Along the same lines, one verifies that the transformation also leaves invariant the Liouville measure $\prod_i dq_i dp_i$. In particular, these results contribute to the justification of the choice of the thermal measure

$$\prod_i dq_i\, dp_i\, e^{-\beta H(p,q)}$$

in classical statistical mechanics.

10.1.4 Poisson brackets

It is convenient to introduce Poisson brackets, semi-classical limit of the commutators of quantum mechanics. If $A(\hat{p}, \hat{q}), B(\hat{p}, \hat{q})$ are two hermitian operators, one defines

$$\{A(p,q), B(p,q)\} = \lim_{\hbar \to 0} \frac{1}{i\hbar} [A(\hat{p}, \hat{q}), B(\hat{p}, \hat{q})]\bigg|_{\hat{p} \mapsto p, \hat{q} \mapsto q}$$

$$= \sum_i \frac{\partial A}{\partial q_i} \frac{\partial B}{\partial p_i} - \frac{\partial B}{\partial q_i} \frac{\partial A}{\partial p_i}. \tag{10.13}$$

With this definition $\{q_i, p_j\} = \delta_{ij}$ and an infinitesimal canonical transformation (10.11) acting on a function $A(p,q)$ reads

$$A(P,Q) = A(p,q) + \varepsilon\{A, T\}. \tag{10.14}$$

Since canonical transformations form a group, Poisson brackets induce a Lie algebra structure. In particular, they satisfy the Jacobi identity

$$\{A, \{B, C\}\} + \{B, \{C, A\}\} + \{C, \{A, B\}\} = 0. \tag{10.15}$$

It is straightforward to verify that the time derivative of any function of p, q takes the form

$$d_t F(p, q) = \{F, H\}. \tag{10.16}$$

The canonical invariance of Poisson brackets $\{q_i(t), p_j(t)\}$ follows.

Moreover, conserved quantities generate a Lie algebra. Indeed, one verifies that if $A(p, q)$ and $B(p, q)$ are two conserved quantities, then

$$\{A, H\} = \{B, H\} = 0, \Rightarrow \{\{A, B\}, H\} = 0$$

and, thus, $\{A, B\}$ is also conserved. In particular, the Lie algebra corresponding to a continuous symmetry of the action is represented in terms of Poisson brackets.

For example, the square of the angular momentum (10.10),

$$\mathbf{L}^2 = \tfrac{1}{2} \sum_{i,j} L_{ij}^2 = \mathbf{p}^2 \mathbf{q}^2 - (\mathbf{p} \cdot \mathbf{q})^2, \tag{10.17}$$

commutes, in the sense of Poisson brackets, with all generators L_{ij} and corresponds to the Casimir of the algebra.

10.2 The path integral in phase space

From now on, we return to the euclidean formalism unless stated otherwise.

If a quantum hamiltonian is known explicitly and if it is quadratic in the conjugate momenta p, the strategy of Section 2.2 can be generalized and the matrix elements of the quantum statistical operator can be calculated for infinitesimal time intervals by solving the Schrödinger equation. In Section 10.2.1 we indicate another method for more general hamiltonians. However, often the problem has a different formulation: only the classical hamiltonian is known and it remains necessary to guess the corresponding quantum operator. Of course, the solution of the problem is in general not unique since the quantum hamiltonian depends on the ordering of operators in products. In real situations, the choice of a quantization is constrained by the symmetries of the classical hamiltonian that one wants to also implement in the quantum theory.

10.2.1 Path integral

The starting point is the same as in Section 2.1. We consider a bounded operator in Hilbert space, $U(t, t')$, $t \geq t'$, which describes evolution from time t' to time t and which satisfies a Markov property in time:

$$U(t, t'')U(t'', t') = U(t, t') \text{ for } t \geq t'' \geq t' \text{ and } U(t', t') = \mathbf{1}. \tag{10.18}$$

Moreover, we assume that $U(t, t')$ is differentiable with a continuous derivative. We set
$$\left. \frac{\partial U(t, t')}{\partial t} \right|_{t=t'} = -\hat{H}(t)/\hbar,$$
where $\hbar$ is Planck's constant. We then differentiate equation (10.18) with respect to t, take the $t'' = t$ limit and find
$$\hbar \frac{\partial U}{\partial t}(t, t') = -\hat{H}(t) U(t, t'). \tag{10.19}$$

Quite generally, the Markov property (10.18) allows writing $U(t'', t')$ as a time-ordered product of n operators corresponding to time intervals $\varepsilon = (t'' - t')/n$ that can be chosen arbitrarily small by increasing n:
$$U(t'', t') = \prod_{m=1}^{n} U[t' + m\varepsilon, t' + (m-1)\varepsilon], \quad n\varepsilon = t'' - t'. \tag{10.20}$$

The basic idea now is to introduce matrix elements in a mixed position–momentum representation, somewhat analogous to the holomorphic formalism of Section 6.4 (see also Section 5.5.2). In what follows we use the letters p and q to represent the eigenvectors of momentum and position operators, respectively. With this convention, the basic identities are
$$\langle q | p \rangle = \langle q | \mathbf{1} | p \rangle = e^{ipq/\hbar} \Rightarrow \langle p | \mathbf{1} | q \rangle = e^{-ipq/\hbar}. \tag{10.21}$$

We assume that the hamiltonian $\hat{H}$ is a function of the momentum and position operators $\hat{p}$ and $\hat{q}$ and we write it in normal-ordered form with all operators $\hat{p}$ at the left of all operators $\hat{q}$
$$\hat{H} = \sum_{m,n} H_{mn} \hat{p}^m \hat{q}^n \Rightarrow \hat{H}^\dagger = \sum_{m,n} \bar{H}_{mn} \hat{q}^n \hat{p}^m.$$

When the hamiltonian is hermitian, as we assume below, the two forms are equivalent.

We now write the matrix elements of $U(t+\varepsilon, t)$ in the position basis as
$$\langle q_2 | U(t+\varepsilon, t) | q_1 \rangle = \int \frac{d^d p_2}{(2\pi\hbar)^d} \langle q_2 | U(t+\varepsilon, t+\varepsilon/2) | p_2 \rangle \langle p_2 | U(t+\varepsilon/2, t) | q_1 \rangle.$$

We then use alternatively the two forms of the hamiltonian to calculate the matrix elements at order ε:
$$\langle p | U(t+\varepsilon, t) | q \rangle \underset{\varepsilon \to 0}{=} e^{-ipq/\hbar} e^{-\varepsilon H_{\text{n.o.}}(p,q;t)/\hbar} + O(\varepsilon^2), \tag{10.22}$$
$$\underset{\varepsilon \to 0}{=} e^{ipq/\hbar} e^{-\varepsilon H_{\text{a.n.o.}}(p,q;t+\varepsilon)/\hbar} + O(\varepsilon^2), \tag{10.23}$$

where $H_\text{n.o.}$ and $H_\text{a.n.o.}$ correspond to the quantum operator $\hat{H}$ written in normal and anti-normal ordered form, in which the operators $\hat{p}, \hat{q}$ have been replaced by the corresponding classical variables. In particular, the two functions are complex conjugate:
$$H_\text{a.n.o.}(p,q;t) = \bar{H}_\text{n.o.}(p,q;t).$$

Neglecting corrections of order ε^2, one infers
$$\langle q_2 | U(t+\varepsilon, t) | q_1 \rangle \underset{\varepsilon \to 0}{=} \int \frac{\mathrm{d}^d p_2}{(2\pi\hbar)^d} e^{i(q_2-q_1)p_2/\hbar}$$
$$\times \exp\left\{-\varepsilon[H_\text{n.o.}(p_2,q_2;t+\varepsilon/2) + \bar{H}_\text{n.o.}(p_2,q_1;t+\varepsilon/2)]/2\hbar\right\}, \quad (10.24)$$

an expression that defines explicitly a hermitian operator because the hamiltonian contribution is symmetric in $q_1 \leftrightarrow q_2$.

One then uses this form to write equation (10.20) in terms of the matrix elements of U in the position basis (Section 2.1.2) as

$$\langle q'' | U(t'',t') | q' \rangle = \int \prod_{k=1}^n \frac{\mathrm{d}^d p_k \mathrm{d}^d q_k}{(2\pi\hbar)^d} \delta(q_n - q'') \prod_{k=1}^n \langle q_k | U(t_k, t_{k-1}) | q_{k-1} \rangle$$
$$= \lim_{n \to \infty} \int \prod_{k=1}^n \frac{\mathrm{d}^d p_k \mathrm{d}^d q_k}{(2\pi\hbar)^d} \delta(q_n - q'') \exp\left[-\mathcal{S}_\varepsilon(p,q)/\hbar\right] \quad (10.25)$$

with

$$\mathcal{S}_\varepsilon(p,q) = -\sum_{k=1}^{n-1} i p_k (q_k - q_{k-1}) + \frac{\varepsilon}{2} \sum_{k=1}^n \left[H_\text{n.o.}(p_k, q_k; t_k) + \bar{H}_\text{n.o.}(p_k, q_{k-1}; t_k)\right]$$
$$(10.26)$$

and the conventions
$$t_k = t' + (k-1/2)\varepsilon, \quad t'' = t' + n\varepsilon, \quad q_0 = q', \quad q_n = q''.$$

Introducing a trajectory in phase space $\{p(t), q(t)\}$ that interpolates between discrete times:
$$p(t_k) = p_k, \quad q(t_k) = q_k,$$
one can take a formal continuum limit $\varepsilon \to 0$, $n \to \infty$. Setting
$$H(p,q;t) = \tfrac{1}{2}[H_\text{n.o.}(p,q;t) + \bar{H}_\text{n.o.}(p,q;t)] = \operatorname{Re} H_\text{n.o.}(p,q;t),$$
one then obtains a path integral in phase space of the form
$$\langle q'' | U(t'',t') | q' \rangle = \int_{q(t')=q'}^{q(t'')=q''} [\mathrm{d}p(t)\mathrm{d}q(t)] \exp\left[-\mathcal{S}(p,q)/\hbar\right] \quad (10.27)$$

with
$$S(p,q) = \int_{t'}^{t''} dt \left[-ip(t)\dot{q}(t) + H(p(t), q(t), t)\right]. \tag{10.28}$$

The function $S(p,q)$ has the form of a classical action in euclidean or imaginary time, as written in the hamiltonian formalism, but $H(p,q;t)$ may differ from the classical hamiltonian $H_{\text{cl.}}(p,q;t)$ by quantum corrections. In some sense

$$H_{\text{cl.}}(p,q;t) = \lim_{\hbar \to 0} H(p,q;t).$$

Partition function. For a time-independent hamiltonian, the partition function is then given by the path integral (10.27) with periodic boundary conditions:

$$\mathcal{Z}(\tau/\hbar) = \text{tr}\, U(\tau/2, -\tau/2) = \int [\mathrm{d}p(t)\mathrm{d}q(t)] \exp\left[-\mathcal{S}(p,q)/\hbar\right] \tag{10.29}$$

with $q(\tau/2) = q(-\tau/2)$. In this form the partition function is simply related to the holomorphic integral (6.57), after a change of variables of the form (6.63).

Quantum evolution. In the case of quantum evolution, which corresponds to the operator $e^{-itH/\hbar}$ when the hamiltonian is time-independent, the representation (10.27) is replaced by

$$\langle q'' | U(t'', t') | q' \rangle = \int [\mathrm{d}p(t)\mathrm{d}q(t)] \exp\left[i\mathcal{A}(p,q)/\hbar\right]. \tag{10.30}$$

The euclidean (or generalized euclidean) action $\mathcal{S}$ in the path integral is replaced by $\mathcal{A}(p,q)$, the classical action (10.8) in the hamiltonian formalism (with possible quantum corrections):

$$\mathcal{A}(p,q) = \int_{t'}^{t''} [p(t)\dot{q}(t) - H(p,q,t)]\, \mathrm{d}t. \tag{10.31}$$

Even in this general situation, quantum evolution is obtained by summing over all paths with a complex weight $e^{i\mathcal{A}/\hbar}$. Therefore, for $\hbar \to 0$, the paths close to the extrema of the action, which are the classical paths, still give the leading contributions to the path integral.

10.2.2 Discussion

The expressions (10.27, 10.31) are quite aesthetic since they involve only a generalized (euclidean or real) classical action and the invariant Liouville measure in phase space. In particular, they are formally invariant under the canonical transformations (10.12) defined in Section 10.1.

Note that although we have calculated an euclidean operator, the measure is not real. The imaginary part originates from the symplectic form. As we have pointed out, its contribution depends only on the geometric trajectory, but not on the motion along it. The contribution, thus, is the same, irrespective whether time is real or imaginary.

Let us also point out that if phase space is compact (as in the quantization of spin variables), the integral of the symplectic form is defined modulo the total area, and this implies quantization properties: indeed, the total area must be a multiple of $2\pi\hbar$ in order for the path integral to be defined and non-vanishing.

Integration space. When one tries to characterize more precisely the space of trajectories that contribute to the path integral (10.27), one encounters the first difficulty. The term that connects the different times in expression (10.26) is now $ip_k(q_k - q_{k-1})$. It generates oscillations in the integral (10.25) that suppress trajectories that are not regular enough. The typical size of the difference $(q_k - q_{k-1})$ for trajectories that contribute is given by the typical values of p_k. For example if, as in Section 2.2, the hamiltonian is quadratic in p, the typical values of p_k in the integral (10.25) are of order $1/\sqrt{\varepsilon}$ and one recovers that $(q_k - q_{k-1})$ is of order $\sqrt{\varepsilon}$, a result consistent with the analysis of Section 2.2. In the same way, if one rewrites expression (10.26) to transform the term $ip_k(q_k - q_{k-1})$ into $iq_k(p_k - p_{k-1})$ ('integration by parts'), one finds the regularity conditions satisfied by contributing functions $p(t)$. It is related to the typical values of q_k in integral (10.25). Again, considering the example of a hamiltonian of the form $p^2/2m + V(q)$, we notice that if, for example, $V(q)$ increases as q^{2N} for $|q| \to \infty$, the typical values of q_k are such that εq^{2N} is of order 1 and, thus, the difference $(p_k - p_{k-1})$ must be of order $\varepsilon^{1/2N}$. For a general hamiltonian, the same discussion clearly may become quite involved.

Remark. The formal canonical invariance of the path integral can really be true only for a very limited class of transformations. Indeed, one can show that in the one-dimensional case, a hamiltonian H with only one degree of freedom can always be transformed into a free hamiltonian:

$$p\dot{q} - H \longmapsto P\dot{Q} - \frac{1}{2m}P^2.$$

One would then be tempted to conclude that the semi-classical approximation is exact. It is easy to produce counter-examples. The discrete form (10.26) shows the origin of this difficulty. A variable p_k is associated with a pair (q_k, q_{k-1}), and the discrete form is, in general, not invariant.

The path integral in phase space thus is more difficult to handle than the simpler integral in position space defined before, and has found so far fewer practical applications. For each non standard example, it is necessary to return to the discrete form (10.25) and to perform a separate analysis.

Again, the extension to an arbitrary number of degrees of freedom is immediate. Expressed in terms of the classical action and the Liouville measure, the path integral representation of the statistical operator has the same form as in equation (10.27).

10.3 Harmonic oscillator. Perturbative calculations

We consider perturbations of the hamiltonian of the harmonic oscillator

$$H_0 = \tfrac{1}{2}(\hat{p}^2 + \omega^2 \hat{q}^2), \tag{10.32}$$

and we restrict calculations to the partition function, that is, to path integrals with periodic boundary conditions, setting $\hbar = 1$. First, adding two terms linearly coupled to p and q, one can determine the various two-point functions by solving the two equations of motion derived from the action

$$\mathcal{S}(p,q) = \int_{-\tau/2}^{\tau/2} dt \left[-ip(t)\dot{q}(t) + \tfrac{1}{2}p^2(t) + \tfrac{1}{2}\omega^2 q^2(t) - c(t)p(t) - b(t)q(t) \right]. \tag{10.33}$$

The classical equations, obtained by varying p and q, then are

$$-i\dot{q}(t) + p(t) = c(t),$$
$$\dot{p}(t) + \omega^2 q(t) = b(t).$$

Straightforward calculations lead to the classical action, from which by functional differentiation one derives

$$\langle q(t)q(u)\rangle = \frac{1}{2\omega\sinh(\omega\tau/2)} \cosh\bigl(\omega(\tau/2 - |t-u|)\bigr) \tag{10.34a}$$

$$\langle p(t)p(u)\rangle = \omega^2 \langle q(t)q(u)\rangle \tag{10.34b}$$

$$\langle q(t)p(u)\rangle = \frac{i\,\mathrm{sgn}(t-u)}{2\sinh(\omega\tau/2)} \sinh\bigl(\omega(\tau/2 - |t-u|)\bigr). \tag{10.34c}$$

These expressions can also be derived from the form (6.43) of the $\langle \bar{z}(t)z(0)\rangle$ function, after the change of variables

$$z = \frac{1}{\sqrt{2\omega}}(\omega q - ip), \quad \bar{z} = \frac{1}{\sqrt{2\omega}}(\omega q + ip).$$

In the $q(t)$ two-point function one recognizes expression (2.52), as expected.

Then, for a general hamiltonian $H(p,q) = H_0(p,q) + H_\mathrm{I}(p,q)$, where H_I is a polynomial in p and q, the partition function can be calculated by expanding in powers of H_I and evaluating all contributions using Wick's theorem.

It follows immediately that if one adds to $\mathcal{S}(p,q)$ a perturbation linear in p of the form of a magnetic term $pA(q)$, one needs the equal-time two-point function $\langle q(t)p(u)\rangle$. One then faces the problem already encountered in Section 5.3.1, that is, the ambiguity of $\mathrm{sgn}(0)$. It is a direct reflection of the problem of ordering of quantum operators in products. Since we have adopted in this chapter a construction almost symmetric in time, the convenient prescription $\mathrm{sgn}(0) = 0$ can be adopted.

10.4 Lagrangians quadratic in the velocities

We have discussed, in Section 10.2, some of the difficulties one faces when one tries to define a path integral in phase space. To show that expression (10.28) has, at least, some heuristic value, we now examine the special example of general hamiltonians quadratic in the momentum variables.

10.4.1 Verifications

We first verify that in the case of hamiltonians of the form

$$H = \frac{p^2}{2m} + V(q),$$

one recovers by integrating over $p(t)$ in expression (10.27) the result (2.22).

The classical action is

$$S(p,q) = \int_{t'}^{t''} dt \left[-ip\dot{q} + \frac{p^2}{2m} + V(q) \right]. \tag{10.35}$$

In expression (10.27), the integral over momentum variables p is gaussian. As usual, we change variables to eliminate the term linear in p, setting

$$p(t) = im\dot{q}(t) + r(t). \tag{10.36}$$

The action becomes

$$S(q) = \int_{t''}^{t'} dt \left[\frac{1}{2m} r^2(t) + \frac{1}{2} m\dot{q}^2 + V(q) \right]. \tag{10.37}$$

The path integral thus factorizes into an integral over $r(t)$:

$$\mathcal{N}(t', t'') = \int [dr(t)] \exp\left[-\frac{1}{\hbar} \int_{t'}^{t''} \frac{r^2(t)}{2m} dt \right], \tag{10.38}$$

which does not depend on the potential $V(q)$ anymore and yields simply a normalization $\mathcal{N}$ function of t' and t'', and an integral over $q(t)$:

$$\langle q''| U(t'',t') | q' \rangle = \mathcal{N} \int_{q(t')=q'}^{q(t'')=q''} [dq(t)] \exp\left[-\frac{1}{\hbar} \int_{t'}^{t''} dt \left(\tfrac{1}{2} m\dot{q}^2 + V(q) \right) \right]. \tag{10.39}$$

The normalization factor can be calculated from expression (10.25):

$$\mathcal{N} = \left(\frac{m}{2\pi\hbar\varepsilon} \right)^{n/2}. \tag{10.40}$$

Thus, we have explicitly verified that expression (10.27) is consistent with expression (2.22).

Magnetic field. We now consider a particle in a magnetic field. The hamiltonian can be written as (equation (5.12))

$$H = \frac{1}{2m}[\hat{\mathbf{p}} + e\mathbf{A}(\hat{\mathbf{q}})]^2 + V(\hat{\mathbf{q}}). \tag{10.41}$$

Normal and anti-normal ordering generate two complex conjugate contributions proportional to $ie\nabla \cdot \mathbf{A}(\hat{\mathbf{q}})$, which cancel in the sum. The path integral thus involves the simple classical hamiltonian (5.11).

To eliminate the term linear in $\mathbf{p}$ in the action, one changes variables, setting

$$\mathbf{p}(t) = im\dot{\mathbf{q}}(t) - e\mathbf{A}(\mathbf{q}(t)) + \mathbf{r}(t). \tag{10.42}$$

After integration over $\mathbf{r}(t)$, one obtains an integral over paths $\mathbf{q}(t)$ with the action

$$\mathcal{S}(\mathbf{q}) = \int_{t'}^{t''} dt \left[\tfrac{1}{2} m \dot{\mathbf{q}}^2 + ie\mathbf{A}(\mathbf{q}) \cdot \dot{\mathbf{q}} + V(\mathbf{q}) \right],$$

which is identical to the euclidean classical action (5.13).

10.4.2 General quadratic lagrangian

In the general situation, the quadratic form in the conjugate momenta $\mathbf{p}$ of the hamiltonian depends on the position variables. Since the new problems are generated by this quadratic part, we restrict below the discussion to purely quadratic hamiltonians.

Notation. In this part, it is convenient to use a compact notation, based on the convention of implicit summation over repeated upper and lower indices. For example,

$$x^\alpha y_\alpha \equiv \sum_\alpha x^\alpha y_\alpha .$$

The most general hamiltonian quadratic in the momenta can be derived from a general lagrangian quadratic in the velocities. Actually, in most quantization problems of this type, the initial data are classical lagrangians.

We thus consider the real time lagrangian

$$\mathcal{L}(\dot{\mathbf{q}}, \mathbf{q}) = \tfrac{1}{2} \dot{q}^\alpha g_{\alpha\beta}(\mathbf{q}) \dot{q}^\beta, \tag{10.43}$$

where $g_{\alpha\beta}(\mathbf{q})$ is a positive matrix.

We use also the traditional notation $g^{\alpha\beta}$ for the matrix inverse of $g_{\alpha\beta}$:

$$g_{\alpha\gamma}(\mathbf{q}) g^{\gamma\beta}(\mathbf{q}) = \delta_\alpha^\beta ,$$

where δ^β_α is the Kronecker δ, because the most interesting examples correspond to quantum mechanics on riemannian manifolds. The tensor $g_{\alpha\beta}(\mathbf{q})$ then is the metric tensor.

The corresponding classical hamiltonian is obtained by a Legendre transformation. The conjugate momentum is

$$p_\alpha = \frac{\partial \mathcal{L}(\dot{\mathbf{q}}, \mathbf{q})}{\partial \dot{q}^\alpha} = g_{\alpha\beta}(\mathbf{q})\dot{q}^\beta.$$

The hamiltonian is then

$$H(\mathbf{p}, \mathbf{q}) = p_\alpha \dot{q}^\alpha - \mathcal{L}(\dot{\mathbf{q}}, \mathbf{q}) = \tfrac{1}{2} p_\alpha g^{\alpha\beta}(\mathbf{q}) p_\beta. \tag{10.44}$$

If we simply use the classical hamiltonian in the phase space path integral, the integration over $\mathbf{p}(t)$ is gaussian and can be performed. However, a difficulty arises with the evaluation of the determinant resulting from the integration and, thus, we integrate over $\mathbf{p}(t)$ first in the discretized form (10.24).

$$\langle \mathbf{q} | U(t, t-\varepsilon) | \mathbf{q}' \rangle \underset{\varepsilon \to 0}{=} \int \frac{\mathrm{d}^d p}{(2\pi\hbar)^d} e^{i(q-q')^\alpha p_\alpha/\hbar}$$
$$\times \exp\left\{-\varepsilon [p_\alpha g^{\alpha\beta}(\mathbf{q}) p_\beta + p_\alpha g^{\alpha\beta}(\mathbf{q}') p_\beta]/4\hbar\right\}. \tag{10.45}$$

With the notation

$$\bar{g}_{\alpha\beta} = \tfrac{1}{2}\left[g^{\alpha\beta}(\mathbf{q}) + g^{\alpha\beta}(\mathbf{q}')\right],$$

the integration over p_α leads to

$$\langle \mathbf{q} | U(t, t-\varepsilon) | \mathbf{q}' \rangle = [2\pi\hbar\varepsilon \det \bar{\mathbf{g}}]^{1/2} \exp\left[-\mathcal{S}(\mathbf{q}, \mathbf{q}'; \varepsilon)/\hbar\right]$$

with

$$\frac{\mathcal{S}}{\varepsilon} = \frac{1}{2}\frac{(q-q')^\alpha}{\varepsilon} \bar{g}_{\alpha\beta}(\mathbf{q},\mathbf{q}') \frac{(q-q')^\beta}{\varepsilon}, \tag{10.46}$$

and where $\bar{\mathbf{g}}$ is the matrix with elements $\bar{g}_{\alpha\beta}$.

Returning to expression (10.25) and taking the formal continuum limit, one verifies that the argument in the exponential again becomes the euclidean action $\mathcal{S}$, integral of the classical lagrangian (10.43) where now time is euclidean:

$$\mathcal{S}(\mathbf{q}) = \int_{t'}^{t''} \mathrm{d}t\, \tfrac{1}{2}\dot{q}^\alpha g_{\alpha\beta}(\mathbf{q})\dot{q}^\beta. \tag{10.47}$$

This is not surprising since, to integrate over momenta p_α, one first solves the classical equation of motion for p_α.

However, in contrast with the two preceding examples, the integration has generated a normalization $\mathcal{N}(\mathbf{q})$ that has a non-trivial dependence in $\mathbf{q}(t)$:

$$\langle \mathbf{q}'' | U(t''t') | \mathbf{q}' \rangle = \int [\mathrm{d}q(t)]\, \mathcal{N}(\mathbf{q}) \exp\left[-\mathcal{S}(\mathbf{q})/\hbar\right] \tag{10.48}$$

with

$$\mathcal{N}(\mathbf{q}) \underset{\varepsilon \to 0}{\sim} (2\pi\hbar\varepsilon)^{-n/2} \prod_{i=1}^{n} [\det \bar{\mathbf{g}}(\mathbf{q})]^{1/2}. \tag{10.49}$$

It has the form of an infinite quantum correction (it has no $1/\hbar$ factor) to the classical action:

$$\mathcal{N}(\mathbf{q}) \propto \exp\left[\frac{1}{2\varepsilon} \int_{t'}^{t''} \ln \det \mathbf{g}\left[\mathbf{q}(t)\right] \mathrm{d}t\right], \tag{10.50}$$

or, equivalently, using the identity $\ln \det \mathbf{g} = \operatorname{tr} \ln \mathbf{g}$,

$$\mathcal{N}(\mathbf{q}) \propto \exp\left[\frac{1}{2\varepsilon} \int_{t'}^{t''} \operatorname{tr} \ln \mathbf{g}\left[\mathbf{q}(t)\right] \mathrm{d}t\right]. \tag{10.51}$$

A formal calculation, starting from expression (10.27), yields a similar result with $1/\varepsilon$ replaced by $\delta(0)$ (δ being Dirac's function). The problem one faces here is directly related to the question of the ordering of quantum operators in products. In an expansion of the path integral in powers of $\hbar$ (semi-classical expansion), the contributions (quantum corrections) generated by the classical action alone are divergent starting with order $\hbar$. These divergences are cancelled by the contributions coming from the measure (10.51). However, in the continuum the cancellation is formal and the remaining finite part is not defined. It is necessary to return to the discretized form (10.25), which implies a choice of quantization, to determine it. Another direct way of understanding the ambiguity is the following. A change in the quantization rule in expression (10.49) symmetric in $\mathbf{q}, \mathbf{q}'$, does not affect the hermiticity of the discretized operator, but affects $\mathbf{g}$ at order $(\mathbf{q} - \mathbf{q}')^2$. Since the difference $|\mathbf{q} - \mathbf{q}'|$ is generically of order $\sqrt{\varepsilon}$, such a change modifies $\mathbf{g}$, as well as $\operatorname{tr} \ln \mathbf{g}$, by a quantity of order ε. The modification of $\mathcal{N}(\mathbf{q})$ then generates a *finite* quantum correction (of order $\hbar$) to the classical action, typical of the commutation of the momentum and position operators.

Once more, one finds a relation between the ordering of quantum operators in products and ambiguities of the continuum limit. In a generic situation (i.e. in the absence of symmetries), the different continuum limits differ here by an arbitrary potential, and thus by the values of an infinite number of parameters.

Remark. When $g_{\alpha\beta}$ is a metric tensor on a riemannian manifold, the factor (10.49) generates formally the covariant measure on the manifold. This can be shown by verifying the invariance of the form of the path integral in a change of coordinates in position space. One sets

$$q^\alpha = f^\alpha(\mathbf{q}'), \quad T^\alpha_\beta(\mathbf{q}') = \frac{\partial f^\alpha}{\partial q'^\beta}, \quad \det \mathbf{T} \neq 0.$$

Then,

$$\mathcal{L} = \tfrac{1}{2} \dot{q}'^\alpha g'_{\alpha\beta}(\mathbf{q}') \dot{q}'^\beta$$

with
$$g'_{\alpha\beta}(\mathbf{q}') = T^\gamma_\alpha g_{\gamma\delta}(\mathbf{q})T^\delta_\beta.$$

Simultaneously, the functional measure is multiplied by the jacobian of the change of variables:
$$dq = \det \mathbf{T}\, dq'.$$

One then notes
$$\sqrt{\det \mathbf{g}}\,\det \mathbf{T} = \sqrt{\det \mathbf{g}'}.$$

With these definitions, the form of the path integral remains unchanged.

10.5 Free motion on the sphere or rigid rotator

To illustrate the discussion of Section 10.4.2, we now quantify free motion on the sphere, or equivalently the rigid rotator with $O(N)$ symmetry (this defines also the $O(N)$ non-linear σ model in one space dimension). The example $N = 2$ has already been discussed in Section 5.6.

The lagrangian quadratic in the velocities that describes free motion on the sphere S_{N-1}, and thus is rotation invariant, has necessarily the form
$$\mathcal{L} = \tfrac{1}{2}R^2\,\dot{\mathbf{r}}^2(t),$$

where R is a constant, and $\mathbf{r}$ a vector with unit length:
$$\mathbf{r}^2(t) = 1. \tag{10.52}$$

To obtain the hamiltonian, one must either introduce a Lagrange multiplier to impose the constraint (10.52), or parametrize the sphere in terms of independent variables q^α. The lagrangian then takes the form
$$\mathcal{L} = \tfrac{1}{2}\dot{q}^\alpha g_{\alpha\beta}(\mathbf{q})\dot{q}^\beta,$$

where $g_{\alpha\beta}(\mathbf{q})$ is the metric tensor on the sphere.

10.5.1 Hamiltonian

The corresponding hamiltonian then reads
$$H = \tfrac{1}{2}g^{\alpha\beta}(\mathbf{q})p_\alpha p_\beta, \tag{10.53}$$

where $g^{\alpha\beta}$ is the inverse of the matrix $g_{\alpha\beta}$.

Following the discussion of Section 10.4, the matrix elements of $e^{-\beta H}$ are then given by the path integral
$$\langle \mathbf{q}''|e^{-\beta H}|\mathbf{q}'\rangle = \int \left[\sqrt{g[\mathbf{q}(t)]}d\mathbf{q}(t)\right]\exp\left[-\frac{1}{2}\int_0^\beta dt\,\dot{q}^\alpha(t)g_{\alpha\beta}(\mathbf{q}(t))\dot{q}^\beta(t)\right], \tag{10.54}$$

where $g(\mathbf{q})$, the determinant of the matrix $g_{\alpha\beta}$, generates the invariant measure on the sphere.

To parametrize the sphere, one can, for example, choose the $N-1$ first components q^α of the vector $\mathbf{r}$ as parameters. This parametrization is clearly singular, as any other global parametrization of the sphere for $N > 2$ (one needs at least two patches to cover the sphere). Locally,

$$\mathbf{r} \equiv (\mathbf{q}, \sqrt{1-\mathbf{q}^2}).$$

One infers

$$\dot{\mathbf{r}}^2 = \dot{\mathbf{q}}^2 + \left(d_t \sqrt{1-\mathbf{q}^2}\right)^2 = \dot{\mathbf{q}}^2 + (\mathbf{q} \cdot \dot{\mathbf{q}})^2/(1-\mathbf{q}^2).$$

The metric of the sphere $g_{\alpha\beta}$ in this parametrization reads

$$g_{\alpha\beta}(\mathbf{q})/R^2 = \delta_{\alpha\beta} + \frac{q_\alpha q_\beta}{1-\mathbf{q}^2}. \tag{10.55}$$

The hamiltonian is expressed in terms of the inverse matrix $g^{\alpha\beta}$:

$$R^2 g^{\alpha\beta} = \delta_{\alpha\beta} - q_\alpha q_\beta$$

and, thus,

$$H = \frac{1}{2R^2} \left[\mathbf{p}^2 - (\mathbf{p} \cdot \mathbf{q})^2\right].$$

In this form, it is possible to recognize the square of the angular momentum (10.17), since the position vector $\mathbf{r}$ has unit length, and one component of $\mathbf{r}$ is constrained and thus has no conjugate momentum. Indeed, the variations of q_α in an infinitesimal rotation (10.3) have the form

$$\delta q_\alpha = \sum_\beta \tau_{\alpha\beta} q_\beta + \tau_{\alpha N} \sqrt{1-\mathbf{q}^2}.$$

The conserved quantities (10.4, 10.10) are then

$$L_{\alpha\beta} = q_\alpha p_\beta - q_\beta p_\alpha, \quad L_{\alpha N} = \sqrt{1-\mathbf{q}^2}\, p_\alpha.$$

One infers

$$\mathbf{L}^2 = \sum_{\alpha<\beta} L_{\alpha\beta}^2 + \sum_\alpha L_{\alpha N}^2 = \mathbf{p}^2 - (\mathbf{p} \cdot \mathbf{q})^2,$$

Therefore, the hamiltonian corresponding to free motion on the sphere S_{N-1} can be written as

$$H = \frac{1}{2R^2} \mathbf{L}^2, \tag{10.56}$$

where the vector $\mathbf{L}$ is the angular momentum operator, which represents the set of the $N(N-1)/2$ generators of the Lie algebra of the group $SO(N)$ acting on Hilbert space.

The quantum hamiltonian of the rigid rotator $O(N)$ can also been obtained from the free hamiltonian in $\mathbb{R}^N$, by introducing radial and angular coordinates and by freezing the radial coordinate.

10.5.2 The spectrum of the rigid rotator: path integral

Path integral. To write a path integral, one still needs the determinant of the metric tensor (10.55):
$$\det \mathbf{g} = R^{2N}/(1-\mathbf{q}^2).$$

One then notes that the integration measure in the path integral can be rewritten in terms of the initial vector $\mathbf{r}$. Indeed,
$$\frac{\mathrm{d}^{N-1}q}{\sqrt{1-\mathbf{q}^2}} = \mathrm{d}^N r\, \delta(1-\mathbf{r}^2).$$

In this form, the formal rotation invariance of the path integral is exhibited explicitly. One can then rewrite the path integral (10.54) in terms of a vector $\mathbf{r}$ in $\mathbb{R}^N$ with unit length as
$$\langle \mathbf{r}''|\mathrm{e}^{-\beta H}|\mathbf{r}'\rangle = \int_{\mathbf{r}(0)=\mathbf{r}'}^{\mathbf{r}(\beta)=\mathbf{r}''} [\mathrm{d}\mathbf{r}(t)\delta\left(1-\mathbf{r}^2(t)\right)] \exp\left[-\frac{1}{2}R^2 \int_0^\beta \mathrm{d}t\,\dot{\mathbf{r}}^2(t)\right]. \qquad (10.57)$$

The spectrum: semi-classical calculation. We now describe an application of the formalism, which provides also another example of a semi-classical calculation.

We consider the hamiltonian (10.56) for $R=1$:
$$H = \tfrac{1}{2}\mathbf{L}^2. \qquad (10.58)$$

The path integral (10.57) then reduces to
$$\langle \mathbf{r}''|\mathrm{e}^{-\beta H}|\mathbf{r}'\rangle = \int_{\mathbf{r}(0)=\mathbf{r}'}^{\mathbf{r}(\beta)=\mathbf{r}''} [\mathrm{d}\mathbf{r}(t)\delta\left(1-\mathbf{r}^2(t)\right)] \exp\left[-\frac{1}{2}\int_0^\beta \mathrm{d}t\,\dot{\mathbf{r}}^2(t)\right]. \qquad (10.59)$$

We call θ the angle between $\mathbf{r}'$ and $\mathbf{r}''$:
$$\cos\theta = \mathbf{r}'\cdot\mathbf{r}'',\ 0\leq\theta\leq\pi. \qquad (10.60)$$

We now introduce the matrix $\mathbf{R}(t)$ that acts on $\mathbf{r}(t)$ and transforms, by a rotation in the $(\mathbf{r}',\mathbf{r}'')$ plane, $\mathbf{r}'$ into $\mathbf{r}''$ in a time β. Its restriction to the plane $(\mathbf{r}',\mathbf{r}'')$ reads
$$\begin{bmatrix} \cos(\theta t/\beta) & \sin(\theta t/\beta) \\ -\sin(\theta t/\beta) & \cos(\theta t/\beta) \end{bmatrix}. \qquad (10.61)$$

It is the identity matrix in the subspace orthogonal to the plane $(\mathbf{r}',\mathbf{r}'')$.

One then changes variables, setting
$$\mathbf{r}(t) = \mathbf{R}(t)\boldsymbol{\rho}(t). \qquad (10.62)$$

10.5 Path integrals in phase space

We call u and v the two components of ρ in the plane $(\mathbf{r}', \mathbf{r}'')$, u being the component along $\mathbf{r}'$, and ρ_T the component in the orthogonal subspace. With this notation, the path integral can be written as

$$\langle \mathbf{r}''| e^{-\beta H} |\mathbf{r}'\rangle = \int_{\rho(0)=\mathbf{r}'}^{\rho(\beta)=\mathbf{r}'} [d\rho(t)\delta\left(1 - \rho^2(t)\right)] \exp\left[-\mathcal{S}(\rho)\right] \tag{10.63}$$

with

$$\mathcal{S}(\rho) = \frac{1}{2}\int_0^\beta dt \left(\dot{\rho}_T^2 + \dot{u}^2 + \dot{v}^2 + \frac{\theta^2}{\beta^2}(u^2 + v^2) + 2\frac{\theta}{\beta}(\dot{v}u - \dot{u}v)\right). \tag{10.64}$$

The constraint

$$u^2 + v^2 + \rho_T^2 = 1, \tag{10.65}$$

can then be used to rewrite the action as

$$\mathcal{S}(\rho) = \frac{1}{2}\frac{\theta^2}{\beta} + \frac{1}{2}\int_0^\beta dt \left(\dot{\rho}_T^2 - \frac{\theta^2}{\beta^2}\rho_T^2 + \dot{u}^2 + \dot{v}^2 + 2\frac{\theta}{\beta}(\dot{v}u - \dot{u}v)\right). \tag{10.66}$$

Unlike the abelian case where the calculation is exact, here one is able only to expand for β small (high temperature). This also corresponds to a semi-classical or WKB limit, which is valid in the limit of large quantum numbers. In what follows, we thus neglect contributions that decrease exponentially in β^{-1}. At leading order, only the small fluctuations around the classical solution are relevant. One thus eliminates the variable u from the action (10.66), using equation (10.65):

$$u = \left(1 - v^2 - \rho_T^2\right)^{1/2}, \tag{10.67}$$

and expands the action in powers of ρ_T et v. At leading order, the result is given by the gaussian approximation and depends only on the quadratic terms. In particular, the term linear in θ in (10.66) does not contribute at this order, the integral over $v(t)$ thus is independent of θ and contributes only to the normalization. The last part of the calculation is then similar the last part of the calculation in Section 2.4, which leads to the result (2.35). The components of ρ_T become independent variables, and the integration over ρ_T yields the $(N-2)$th power of the integral over one component. Since each component satisfies the boundary conditions

$$\rho_i(0) = \rho_i(\beta) = 0,$$

the integration yields the result (2.39) for the harmonic oscillator for $q' = q'' = 0$, $\hbar = m = 1$ and $\omega = i\theta/\beta$ and, thus,

$$\langle \mathbf{r}''| e^{-\beta H} |\mathbf{r}'\rangle \sim K(\beta) \left(\frac{\theta}{2\pi\beta \sin\theta}\right)^{(N-2)/2} e^{-\theta^2/2\beta}, \tag{10.68}$$

where the normalization constant $K(\beta)$ is independent of θ:

$$K(\beta) = (2\pi\beta)^{-1/2}.$$

To extract the eigenvalues of H, one can project this expression onto the orthogonal polynomials $P_l^N(\cos\theta)$ associated to the group $SO(N)$:

$$\int_0^\pi d\theta\, (\sin\theta)^{N-2} P_l^N(\cos\theta) P_{l'}^N(\cos\theta) = \delta_{ll'}, \qquad (10.69)$$

which are proportional to the Gegenbauer polynomials $C_l^{(N-2)/2}$. For $\beta \to 0$, only the first two terms of the expansion of the polynomials P_l^N at $\theta = 0$ are needed:

$$P_l^N(\cos\theta) = P_l^N(1)\left(1 - \frac{l(l+N-2)}{2(N-1)}\theta^2 + O\left(\theta^4\right)\right). \qquad (10.70)$$

If one assumes that to each value of l corresponds only one eigenvalue E_l of H, with a degeneracy

$$\delta_l(N) = \frac{\Gamma(l+N-2)(N+2l-2)}{\Gamma(N-1)\Gamma(l+1)},$$

one obtains

$$e^{-\beta E_l} \sim \frac{1}{\delta_l(N)} \frac{1}{(2\pi\beta)^{(N-2)/2}} \int_0^\pi d\theta\, P_l^N(\cos\theta)(\theta\sin\theta)^{(N-2)/2} e^{-\theta^2/(2\beta)}$$

$$= e^{-\beta E_0}\left(1 - \tfrac{1}{2}l(l+N-2)\beta + O\left(\beta^2\right)\right) \qquad (10.71)$$

and, thus,

$$E_l = E_0 + \tfrac{1}{2}l(l+N-2) + O(\beta). \qquad (10.72)$$

Since E_l is independent of β, we infer from the calculation the exact result, up to an additive constant E_0. The corrections vanishing for $\beta \to 0$ must vanish identically.

Concerning this result a comment now is in order: we have shown in Section 10.4 that the path integral (10.54, 10.59) is ill-defined because the measure factor yields formally divergent contributions. We have explained that these divergences are cancelled by other divergences that appear in the perturbative expansion. Therefore, the resulting expressions are finite but ambiguous and these ambiguities reflect the problem of operator ordering in the quantization of the classical hamiltonian. However, we obtain here a well-defined result. The reason is that at each step in the calculation we have maintained explicitly the $SO(N)$ symmetry. Therefore, implicitly among all possible quantizations (and all possible definitions of the path integral) we have chosen an element of the subclass that corresponds $O(N)$ symmetric quantizations. A general study of the $O(N)$ non-linear σ model (the quantum field generalization of the rigid rotator) shows that such a hamiltonian is determined up to an additive constant. The remaining quantization ambiguities are thus entirely contained in the eigenvalue E_0.

Exercises

Exercise 10.1

A hamiltonian quartic in the momentum. The results obtained in this chapter can be illustrated by considering a quantum hamiltonian of degree 4 in p:

$$\hat{H} = \tfrac{1}{2}(\hat{p}^2 + \omega^2 \hat{q}^2) + \tfrac{1}{4}\lambda(\hat{p}^2 + \omega^2 \hat{q}^2)^2,$$

with $\lambda > 0$.

The exact spectrum is simply related to the spectrum of the harmonic oscillator:

$$E_k = \omega(k + \tfrac{1}{2}) + \lambda\omega^2(k + \tfrac{1}{2})^2.$$

Calculate the energy spectrum at first order in λ, starting from the path integral representation in phase space of the partition function and using Wick's theorem. In the path integral one will use the classical hamiltonian. Some identities of Section 3.1 are useful.

Solution. The partition function is given by

$$\mathcal{Z}(\beta) = \int [dp\, dq]\exp\int_{-\beta/2}^{\beta/2} dt\, [ip(t)\dot{q}(t) - H(p(t), q(t))], \qquad (10.73)$$

where paths are periodic and H is the classical hamiltonian.

At order λ, the expansion of the partition function reads

$$\mathcal{Z}(\beta)/\mathcal{Z}_0(\beta) = 1 - \tfrac{1}{4}\lambda \int_{-\beta/2}^{\beta/2} dt\, \left\langle \left(p^2(t) + \omega^2 q^2(t)\right)^2 \right\rangle_0 + O(\lambda^2),$$

where $\langle \bullet \rangle_0$ means gaussian expectation value. The needed expressions are given in equations (10.34). With the symmetric convention $\mathrm{sgn}(0) = 0$, one obtains

$$\mathcal{Z}(\beta)/\mathcal{Z}_0(\beta) = 1 - 2\lambda\beta\omega^4 \left(\langle q^2(0) \rangle\right)^2 + O(\lambda^2)$$
$$= 1 - \tfrac{1}{2}\lambda\beta\omega^2 \cosh^2(\omega\beta/2)/\sinh^2(\omega\beta/2) + O(\lambda^2).$$

Comparing with expression (3.2), one concludes

$$E_k = \omega(k + \tfrac{1}{2}) + \lambda\omega^2(k^2 + k + \tfrac{1}{2}) + O(\lambda^2).$$

Therefore, by using the classical hamiltonian, one finds a global shift of the spectrum with respect to the exact result by $\lambda\omega^2/4$. Such a shift corresponds to the difference between two symmetric quantizations.

Exercise 10.2

One now changes the sign in front of the quadratic terms:
$$\hat{H} = -\tfrac{1}{2}(\hat{p}^2 + \omega^2\hat{q}^2) + \tfrac{1}{4}\lambda(\hat{p}^2 + \omega^2\hat{q}^2)^2.$$
The spectrum is still simply related to the spectrum of the harmonic oscillator:
$$E_k = -\omega(k + \tfrac{1}{2}) + \lambda\omega^2(k + \tfrac{1}{2})^2.$$
Calculate the partition function from the path integral in phase space for $\lambda \to 0$, still using the classical hamiltonian. Infer the spectrum.

Solution. One method is the following: one changes variables $p, q \mapsto \rho, \theta$ in the path integral:
$$p(t) = \sqrt{\rho(t)} \cos\theta(t), \quad q(t) = \sqrt{\rho(t)} \sin\theta(t)/\omega.$$
The Liouville measure becomes
$$dpdq = d\rho d\theta/2\omega.$$
After an integration by parts,
$$\int p\dot{q}\, dt = \frac{1}{2\omega} \int \rho\dot{\theta}\, dt,$$
the action can be written as
$$S(\rho, \theta) = \int dt \left[-i\rho\dot{\theta}/2\omega - \tfrac{1}{2}\rho + \tfrac{1}{4}\lambda\rho^2\right].$$
The integral over ρ is gaussian but with a constraint on the domain of integration: $\rho \geq 0$. However, in the limit $\lambda \to 0$, the integral is dominated by a saddle point up to corrections of order $e^{-1/4\lambda}$. Neglecting these corrections, one can integrate over ρ and one finds the action for the angular variable θ:
$$S(\theta) = -\frac{1}{4\lambda} \int dt \left(1 + i\dot{\theta}/\omega\right)^2.$$
One can then adapt the method used for the $O(2)$ model (Section 5.6). One sums over the contributions of periodic trajectories that wind around the circle n times:
$$\mathcal{Z}(\beta) \propto \sum_n \exp\left[\frac{\beta}{4\lambda}(1 + 2ni\pi/\omega\beta)^2\right].$$
One then uses the identity
$$\sum_n e^{-an^2 + in\varphi} = \sqrt{\pi/a} \sum_l e^{(\varphi - 2\pi l)^2/4a}.$$
The spectrum follows:
$$E_l = \lambda\omega^2 l^2 - l\omega.$$
This result is valid a priori only for $E_l \ll 0$, that is for l around the minimum $l \sim 1/2\omega\lambda$ where E_l is of order $-1/\lambda$. In particular, it is valid for large quantum numbers.

A comparison with the exact result shows that $k + \tfrac{1}{2}$ has been replaced by l. The shift of $1/2$ is not surprising because the terms linear in k depend on the ordering in operator products.

APPENDIX A: QUANTUM MECHANICS: MINIMAL BACKGROUND

A major difference between quantum and classical mechanics is the superposition principle: physical states are associated with vectors of a complex vector space, and physical predictions are related to expectation values of operators in these states.

A1 Hilbert space and operators

Hilbert space. A possible construction of quantum mechanics is based on the concept of complex Hilbert space that we briefly recall.

One considers the vector space of complex sequences $\{\varphi_n\}$ endowed with a scalar product, straightforward generalization of the scalar product of complex vectors in $\mathbb{C}^N$. The scalar product (φ, ψ) of two sequences $\{\varphi_n\}$ et $\{\psi_n\}$ is defined by

$$(\varphi, \psi) = (\psi, \varphi)^* = \sum_{n=1}^{\infty} \varphi_n^* \psi_n,$$

where φ_n^* is the complex conjugate of φ_n.

The scalar product implies a norm, and the Hilbert space is the complex vector space $\mathcal{H}$ of the vectors with finite norm:

$$\sum_{n=1}^{\infty} |\varphi_n|^2 < \infty.$$

In this vector space, one can introduce a basis of vectors $v^{(N)}$, $N \in \mathbb{N}$ defined by the sequences

$$v_N^{(N)} = 1, \ v_n^{(N)} = 0, \ \forall \, n \neq N.$$

All elements ψ of $\mathcal{H}$ can be written as

$$\psi = \sum_{n=1}^{\infty} \psi_n v^{(n)}.$$

One can give a physical interpretation to these vectors as the normalized eigenvectors of the quantum harmonic oscillator (see Section 6.2).

Notation: bras and kets. Dirac has introduced a convenient notation: vectors in Hilbert space are denoted by $|\psi\rangle$, the complex conjugate vectors by $\langle\psi|$ and the scalar product by

$$(\varphi, \psi) \equiv \langle\varphi|\psi\rangle = (\langle\psi|\varphi\rangle)^*.$$

For simplicity, we then denote by $|n\rangle$ the basis vector $v^{(n)}$. The vector complex conjugate $\langle n|$ correspond to the same sequence. The relation

$$\langle n|m\rangle = \delta_{mn}.$$

then expresses that the vectors form an orthonormal basis.

Operators. On the vectors in Hilbert space act operators, which correspond to linear internal mappings. The representation of an operator O acting on a sequence $\varphi = \{\varphi_n\}$ generalizes matrices acting on vectors:

$$(O\varphi)_m = \sum_n O_{mn}\varphi_n .$$

In the bra and ket notation, the action of an operator O on a vector $|\psi\rangle$ is denoted by $O|\psi\rangle$.

The hermitian conjugate $O^\dagger$ can be defined by

$$(\varphi, O\psi) = (O^\dagger \varphi, \psi).$$

One then verifies that $O^\dagger$ is represented by the elements O^*_{nm}, and that in the bra, ket notation

$$\langle \varphi| O |\psi\rangle = \left(\langle\psi| O^\dagger |\varphi\rangle\right)^* .$$

Physical interpretation. In quantum mechanics, the state of a system is characterized by a vector in Hilbert space. If $|\psi\rangle$ is a vector with unit norm,

$$\langle\psi|\psi\rangle = \sum_n |\psi_n|^2 = 1,$$

the quantities $|\psi_n|^2$ have a probabilistic interpretation: the probability to find, in a measure, the quantum system in the state n.

More generally, physical observables are associated to hermitian operators

$$O = O^\dagger.$$

A hermitian operator is diagonalizable, has real eigenvalues and orthogonal eigenvectors. We assume, for simplicity, that its spectrum is discrete and denote by $|\varphi_N\rangle$ its eigenvectors and O_N the corresponding eigenvalues. The operator can then be written as

$$O = \sum_{N=1}^{\infty} O_N |\varphi_N\rangle \langle\varphi_N| .$$

For a physical system in a state characterized by a normalized vector $|\psi\rangle$ ($\langle\psi|\psi\rangle = 1$), the quantities $|\langle\varphi_N|\psi\rangle|^2$ again have a probabilistic interpretation: the probability to find the value O_N in a measure of the physical observable associated to O. Quantum mechanics thus predicts only expectation values. For an operator O and a normalized state vector ψ, the quantity

$$\langle\psi|O|\psi\rangle , \tag{A1}$$

is the expectation value of the measures of the physical observable associated to O. The condition of hermiticity quite generally ensures that the expectation value is real.

A2 Quantum evolution, symmetries and density matrix

Quantum evolution. Evolution of states in an isolated quantum system is determined by the action of an evolution operator. The state vector at time t'' is inferred from the vector at time t' by

$$|\psi(t'')\rangle = U(t'',t')|\psi(t')\rangle \quad \text{and thus} \quad U(t',t') = \mathbf{1}.$$

Since the norm of vectors has a probabilistic interpretation, it must be conserved during time evolution: the total probability must remain equal to 1. Therefore, the transformation $U(t'',t')$ must conserve norms and, as a consequence, scalar products of vectors and, thus, must be unitary:

$$U(t'',t')U^\dagger(t'',t') = \mathbf{1}.$$

Moreover, the evolution of an isolated quantum system is assumed to be markovian, a property defined by the multiplication rule

$$U(t_3,t_2)U(t_2,t_1) = U(t_3,t_1).$$

If the operator $U(t,t')$ is differentiable, it satisfies the equation

$$i\hbar \frac{\partial U}{\partial t}(t,t') = \hat{H}(t)U(t,t'), \qquad (A2)$$

where the hermitian operator $\hat{H}(t)$ is the hamiltonian. The Planck constant $\hbar$ appears in the equation for dimensional reasons.

Another form of the equation is

$$i\hbar \frac{\partial U}{\partial t'}(t,t') = -U(t,t')\hat{H}(t'). \qquad (A3)$$

When $\hat{H}$ is time-independent, $U(t'',t') = e^{-i\hat{H}(t''-t')/\hbar}$ and the evolution operators belong to a representation of the translation group (time-translations), which has $\hat{H}/\hbar$ as a generator. Moreover, the expectation value of $\hat{H}$ in any state is time-independent, a property that corresponds, in quantum mechanics, to energy conservation.

Let us point out here that the evolution operator is bounded, but not necessarily the hamiltonian, which is defined, in general, only in a dense subspace of the Hilbert space. Equation ($A2$) is thus only meaningful in the subspace of finite energy states.

Wave packet reduction. After the measure of an observable, a vector $|\psi\rangle$ is projected onto the eigenvector corresponding to the eigenvalue of the associated operator found in the measure, an operation called wave packet reduction. This evolution, obviously, is not unitary, but this is not contradictory with unitary evolution because, during the measure, the quantum system can no longer be considered as isolated: the wave packet reduction results from the interaction between a macroscopic apparatus (which has essentially an infinite number of degrees of freedom) and the measured quantum system.

Heisenberg representation. As a function of time, the expectation value of an operator O can thus be written as

$$\langle\psi(t)|O|\psi(t)\rangle = \langle\psi(t')|U^\dagger(t,t')OU(t,t')|\psi(t')\rangle.$$

Introducing the Heisenberg representation of the operator O,

$$O(t,t') = U^\dagger(t,t')OU(t,t'), \qquad (A4)$$

one can transfer the evolution in time from states to operators:

$$\langle\psi(t)|O|\psi(t)\rangle = \langle\psi(t')|O(t,t')|\psi(t')\rangle.$$

The operator $O(t,t')$ then satisfies the evolution equation

$$i\hbar\frac{\partial O(t,t')}{\partial t} = [O(t,t'), U^\dagger(t,t')\hat{H}(t)U(t,t')],$$

or

$$i\hbar\frac{\partial O(t,t')}{\partial t'} = [\hat{H}(t'), O(t,t')].$$

Symmetries. Conservation of the norm implies that symmetries of physical systems are represented in Hilbert space by unitary (or anti-unitary, i.e. unitary followed by a complex conjugation) transformations.

A unitary transformation is implemented by the action of a unitary operator S, and the symmetry condition implies that the operator commutes with the hamiltonian:

$$[S,\hat{H}] = 0.$$

When symmetries belong to continuous groups, the corresponding unitary operators form a group representation. The symmetries then lead, as in classical mechanics (see Section 10.1.1), to quantities conserved in time: they are associated with the expectation values of the generators (chosen hermitian) of the group representation.

Density matrix. Quantum statistical mechanics still requires the introduction of the density matrix. If one has only a partial information about a quantum system, a state in Hilbert space is replaced by a statistical mixture that can be represented by a density matrix ρ. A density matrix is hermitian and positive, with unit trace in order to ensure that total probability is equal to 1:

$$\rho = \rho^\dagger, \quad \rho > 0, \quad \text{tr}\,\rho = 1.$$

These conditions can simply be understood by diagonalizing the density matrix, which then can be written as

$$\rho = \sum_n \rho_n |\psi_n\rangle \langle\psi_n|,$$

$\{|\psi_n\rangle\}$ being its eigenvectors and $\{\rho_n\}$ the corresponding eigenvalues. The eigenvalues ρ_n have an interpretation as usual statistical (and not quantum) probabilities to be in a state n. Therefore, they must be positive and sum to 1.

The expectation value of an observable associated with an operator O then becomes

$$\langle O \rangle = \operatorname{tr} \rho O.$$

Expression (A1) corresponds to a situation with perfect information, where all eigenvalues of ρ vanish but one–one then speaks of a pure state–and ρ is a projector: $\rho^2 = \rho$.

Time evolution of the density matrix then can be derived from the evolution of state vectors. One finds

$$\rho(t) = U(t,t')\rho(t')U^\dagger(t,t') \;\Rightarrow\; i\hbar \frac{\partial \rho(t)}{\partial t} = [\hat{H}(t), \rho(t)].$$

For a system at thermal equilibrium at a temperature $k_B T = 1/\beta$ (k_B is Boltzmann's constant that is set to 1 in this work), the density matrix must commute with the hamiltonian. Various arguments then lead to the conclusion

$$\rho = \mathrm{e}^{-\beta \hat{H}} / \mathcal{Z}(\beta) \text{ with } \mathcal{Z}(\beta) = \operatorname{tr} \mathrm{e}^{-\beta \hat{H}}.$$

Operator algebra. In the case of a pure state, the density matrix remains unchanged when the state vector is multiplied by a phase factor and thus the relation between physical state and density matrix is biunique. This observation leads naturally to a formulation of quantum mechanics entirely based on a complex algebra of bounded operators with a hermitian conjugation and a trace operation. Such a description of quantum mechanics is formally more satisfactory, but not very intuitive.

A3 Position and momentum. Schrödinger equation

Position and momentum operators. Two hermitian operators play an important role: the position operator $\hat{\mathbf{q}} = (\hat{q}_1, \hat{q}_2, \ldots, \hat{q}_d)$ (in d-dimensional space) and the conjugate momentum operator $\hat{\mathbf{p}} = (\hat{p}_1, \hat{p}_2, \ldots, \hat{p}_d)$, which in the simplest situation corresponds to the usual particle momentum (see Sections 5.1 and 10.1 for details), whose expectation values in a state thus characterize the average position and momentum, respectively. The commutation relations between these operators are

$$[\hat{q}_\alpha, \hat{q}_\beta] = [\hat{p}_\alpha, \hat{p}_\beta] = 0, \; [\hat{q}_\alpha, \hat{p}_\beta] = i\hbar \delta_{\alpha\beta}. \tag{A5}$$

Many observables can be expressed in terms of these two operators (in particular, in the case of particles without spin or other quantum numbers, a situation we assume here for simplicity).

Again, these operators are not bounded and thus not defined in the whole Hilbert space. Therefore, a more careful construction is based on the elements of the unitary groups they generate: the operators

$$T(\mathbf{a}) = e^{i\hat{\mathbf{p}}\cdot\mathbf{a}/\hbar} \text{ and } V(\mathbf{k}) = e^{-i\hat{\mathbf{q}}\cdot\mathbf{k}/\hbar},$$

where $\mathbf{a}$ and $\mathbf{k}$ are two constant vectors, form a representation of the abelian groups of translations and momentum changes, respectively.

The commutation relations become

$$T(\mathbf{a})V(\mathbf{k}) = e^{-i\mathbf{a}\cdot\mathbf{k}/\hbar} V(\mathbf{k})T(\mathbf{a}).$$

It is then convenient to introduce the eigenvectors of these operators. However, they do not belong to Hilbert space. For example, the eigenvectors of $V(\mathbf{k})$ correspond to states localized in space, which are singular limits of vectors in Hilbert space: a state strictly localized at a point $\mathbf{q}_0$ in $\mathbb{R}^d$ can be obtained as

$$\lim_{\varepsilon \to 0} \frac{1}{(\pi\varepsilon)^{d/2}} e^{-(\mathbf{q}-\mathbf{q}_0)^2/\varepsilon} = \delta(\mathbf{q} - \mathbf{q}_0),$$

which is a mathematical distribution, Dirac's δ function, with a divergent norm. Nevertheless, these vectors form an orthogonal basis,

$$\int d^d q \, \delta(\mathbf{q} - \mathbf{q}_0)\delta(\mathbf{q} - \mathbf{q}_1) = \delta(\mathbf{q}_1 - \mathbf{q}_0),$$

which is complete since

$$\int d^d q_0 \, \delta(\mathbf{q} - \mathbf{q}_0))\delta(\mathbf{q}' - \mathbf{q}_0) = \delta(\mathbf{q} - \mathbf{q}').$$

If we denote by $\delta(\mathbf{q} - \mathbf{q}_0) \mapsto |\mathbf{q}_0\rangle$ a vector in the basis, these relations become

$$\langle \mathbf{q}_0 | \mathbf{q}_1 \rangle = \delta(\mathbf{q}_0 - \mathbf{q}_1), \quad \int d^d q_0 \, |\mathbf{q}_0\rangle \langle \mathbf{q}_0| = 1.$$

In this basis the position operator, generator of momentum translations, is diagonal:

$$\hat{\mathbf{q}} |\mathbf{q}_0\rangle = \mathbf{q}_0 |\mathbf{q}_0\rangle,$$

Any vector $|\psi\rangle$ in Hilbert space can be decomposed on the basis:

$$\psi(\mathbf{q}) = \langle \mathbf{q}|\psi\rangle,$$

and the function $\psi(\mathbf{q})$, the set of all components, is called wave function. Wave functions, corresponding to vectors in Hilbert space, form a Hilbert space of square integrable functions since

$$\int d^d q\, |\psi(\mathbf{q})|^2 = \int d^d q\, \langle \psi|\mathbf{q}\rangle \langle \mathbf{q}|\psi\rangle = \langle \psi|\psi\rangle < \infty.$$

The scalar product of two vectors $|\psi\rangle$ and $|\varphi\rangle$ then reads

$$\langle \psi|\varphi\rangle = \int d^d q\, \psi^*(\mathbf{q})\varphi(\mathbf{q}).$$

The position operator acts on wave functions by multiplication since

$$\langle \mathbf{q}|\,\hat{\mathbf{q}}\,|\psi\rangle = \mathbf{q}\,\langle \mathbf{q}|\psi\rangle = \mathbf{q}\,\psi(\mathbf{q}).$$

The two operators multiplication by $\mathbf{q}$ and differentiation $-i\hbar\nabla_\mathbf{q}$ satisfy the commutation relations ($A5$) since

$$\left[q_\alpha, \frac{\hbar}{i}\frac{\partial}{\partial q_\beta}\right] = i\hbar\delta_{\alpha\beta}.$$

It follows that the momentum operator acting on wave functions can be represented by

$$\langle \mathbf{q}|\,\hat{\mathbf{p}}\,|\psi\rangle = \frac{\hbar}{i}\nabla_\mathbf{q}\psi(\mathbf{q}),$$

if necessary after a gauge transformation (5.5).

Finally, another generalized basis is provided by plane waves

$$\psi_\mathbf{p}(\mathbf{q}) = e^{i\mathbf{p}\cdot\mathbf{q}/\hbar},$$

which are eigenvectors of the momentum operator:

$$\frac{\hbar}{i}\nabla_\mathbf{q}\psi_\mathbf{p}(\mathbf{q}) = \mathbf{p}\,\psi_\mathbf{p}(\mathbf{q}),$$

and correspond to states invariant under space translation. They also form an orthogonal and complete basis:

$$\int d^d q\, e^{-i\mathbf{p}_0\cdot\mathbf{q}/\hbar}\, e^{i\mathbf{p}_1\cdot\mathbf{q}/\hbar} = (2\pi\hbar)^d \delta(\mathbf{p}_0 - \mathbf{p}_1),$$

$$\int d^d p\, e^{-i\mathbf{p}\cdot\mathbf{q}/\hbar}\, e^{i\mathbf{p}\cdot\mathbf{q}'/\hbar} = (2\pi\hbar)^d \delta(\mathbf{q} - \mathbf{q}').$$

The representations of vectors in these two bases are related by a Fourier transformation. Defining, abstractly, the eigenvectors of $\hat{\mathbf{p}}$ by

$$\hat{\mathbf{p}}\widetilde{|\mathbf{p}\rangle} = \mathbf{p}\widetilde{|\mathbf{p}\rangle},$$

and, thus,
$$e^{i\mathbf{p}\cdot\mathbf{q}/\hbar} = \langle \mathbf{q} | \widetilde{\mathbf{p}} \rangle,$$

one finds
$$\int d^d q \, e^{i\mathbf{q}\cdot\mathbf{p}/\hbar} |\mathbf{q}\rangle = \widetilde{|\mathbf{p}\rangle}.$$

Then, defining
$$\widetilde{\langle \mathbf{p} |} \psi \rangle = \tilde{\psi}(\mathbf{p}),$$

one obtains
$$\psi(\mathbf{q}) = \langle \mathbf{q} | \psi \rangle = \frac{1}{(2\pi\hbar)^d} \int d^d p \, e^{i\mathbf{q}\cdot\mathbf{p}/\hbar} \tilde{\psi}(\mathbf{p}).$$

Free particle. The hamiltonian $\hat{H}_0$ of the free particle is a hamiltonian invariant under space translation
$$[\hat{H}_0, T(\mathbf{a})] = 0 \;\Rightarrow\; [\hat{H}_0, \hat{\mathbf{p}}] = 0.$$

The momentum then is a conserved quantity and the hamiltonian can be expressed only as a function of the generators $\hat{\mathbf{p}}$. For a system invariant under rotation, $\hat{H}_0$ is a function of $\hat{\mathbf{p}}^2$ and for small momentum it can be written as
$$\hat{H}_0 = \frac{1}{2m}\hat{\mathbf{p}}^2,$$

where m is the mass of a particle.

Particle with infinite mass. The hamiltonian of a particle with infinite mass is invariant under momentum change:
$$[\hat{H}_I, V(\mathbf{k})] = 0 \;\Rightarrow\; [\hat{H}_I, \hat{\mathbf{q}}] = 0.$$

The position of a particle is then a conserved quantity and the hamiltonian H_I is function only of the position operator
$$\hat{H}_I = V(\hat{\mathbf{q}}),$$

where V is called potential.

Schrödinger equation. A large class of hamiltonians can be written as the sum of these two terms:
$$\hat{H} = \frac{1}{2m}\hat{\mathbf{p}}^2 + V(\hat{\mathbf{q}}). \tag{A6}$$

The time evolution of a wave function then follows from a partial differential equation of the form
$$i\hbar \frac{\partial}{\partial t}\psi(\mathbf{q},t) = -\frac{\hbar^2}{2m}\nabla_{\mathbf{q}}^2 \psi(\mathbf{q},t) + V(\mathbf{q})\psi(\mathbf{q},t),$$

called Schrödinger equation.

A simple extension is provided by a particle in a magnetic field, which leads to a hamiltonian of the form (5.11):

$$H = \frac{1}{2m}[\hat{\mathbf{p}} + e\mathbf{A}(\hat{\mathbf{q}})]^2 + V(\hat{\mathbf{q}}), \qquad (A7)$$

where $\mathbf{A}(\mathbf{q})$ is the vector potential from which derives the magnetic field.

Positions and momenta in Heisenberg representation. We can now introduce the Heisenberg representation (A4) of the position operator

$$\hat{\mathbf{Q}}(t) = U^\dagger(t,0)\,\hat{\mathbf{q}}\,U(t,0).$$

In the case of the time-independent hamiltonian (A6), the time derivative of $\hat{\mathbf{Q}}(t)$ is given by

$$\frac{d}{dt}\hat{\mathbf{Q}}(t) = \frac{1}{i\hbar}[\hat{\mathbf{Q}}(t), \hat{H}] = \frac{1}{m}\hat{\mathbf{P}}(t),$$

where $\hat{\mathbf{P}}(t)$ is the momentum operator in Heisenberg representation. In the example of the magnetic hamiltonian (A7), instead

$$\frac{d}{dt}\hat{\mathbf{Q}}(t) = \hat{\mathbf{P}}(t) + e\mathbf{A}(\hat{\mathbf{Q}}(t)).$$

These examples show that the physical interpretation of the conjugate momentum operator depends on the hamiltonian, a property also true in classical mechanics (equation (10.5)).

Bibliography

[1] N. Wiener, *J. of Math. and Phys.* 2 (1923) 131, reprinted in *Selected Papers of N. Wiener*, p. 55, MIT Press (Cambridge 1964). See also N. Wiener, *Proc. London Math. Soc.* 22 (1924) 455.

[2] G. Wentzel, *Z. Phys.* 22 (1924) 193.

[3] P.A.M. Dirac, *Physik. Z. Sowjetunion* 3 (1933) 64, reprinted in *Collected Papers on Quantum Electrodynamics*, J. Schwinger ed., Dover (New-York 1958).

[4] R.P. Feynman, *Rev. Mod. Phys.* 20 (1948) 267; *Phys. Rev.* 80 (1950) 440, reprinted in *Collected Papers on Quantum Electrodynamics*, J. Schwinger ed., Dover (New-York 1958).

[5] M. Kac, *Trans. Amer. Math. Soc.* 65 (1949) 1. See also
M. Kac, chap. 4 *Probability and Related Topics in Physical Sciences, Lectures in Math. Phys.*, Interscience (New York 1959).

[6] One early review of the properties of path integrals, with many references to the corresponding mathematical literature is
I.M. Gelfand and A.M. Yaglom, *Fortsch. Phys.* 5 (1957) 517,
J. Math. Phys. 1 (1960) 48.

[7] E. Nelson, *J. Math. Phys.* 5 (1964) 332.

[8] In the article
J.R. Klauder, *Ann. Phys.* 11 (1960) 123,
integrals over complex phase space variables are introduced. The general formulation can be found in S.S. Schweber, *J. Math. Phys.* 3 (1962) 831,
where the relation with the holomorphic formalism introduced in
V. Bargmann, *Commun. Pure and Appl. Math.* 14 (1961) 187,
is explicitly shown.

[9] In the framework of quantum field theory, several authors have suggested introducing anti-commuting variables to represent fermion fields, for example,
P.T. Matthews and A. Salam, *Nuovo Cimento* 2 (1955) 120.
Differentiation in the case of Grassmann variables is defined in J.L. Martin, *Proc. Roy. Soc. (London)* A251 (1959) 543. The rules of integration and differentiation in Grassmann algebras have been clarified and formalized in
F.A. Berezin, *Dokl. Akad. Nauk SSSR* 137 (1961) 311.

[10] Functional methods to deal with boson and fermion systems are presented in
F.A. Berezin, *The Method of Second Quantization* (Academic Press, New York 1966).

[11] R.P. Feynman, *Phys. Rev.* 84 (1951) 108 (Appendix B); W. Tobocman, *Nuovo Cimento* 3 (1956) 1213; C. Garrod, *Rev. Mod. Phys.* 38 (1966) 483.

[12] L.D. Faddeev, in *Methods in Field Theory*, Les Houches School 1975, R. Balian and J. Zinn-Justin eds., (North-Holland, Amsterdam 1976).

[13] For a review of the different forms of path integrals see, for example,
F.A. Berezin, *Theor. Math. Phys.* 6 (1971) 141, and the reference [15].

[14] In the article
C. Morette, *Phys. Rev.* 81 (1951) 848
several results are derived: the calculation of path integrals by the stationary phase method for $\hbar \to 0$ and the relation with the WKB approximation derived from the Schrödinger equation, the determination of the non-trivial measure generated by the quantization of potentials quadratic in the velocities. This measure, there, is obtained from the condition of unitarity of the evolution operator at short times, a strategy analogous to the one used in
J.H. Van Vleck, *Proc. Natl. Acad. Sci. USA* 14 (1928) 178.

[15] The quantization of constrained systems is discussed in
L.D. Faddeev, *Theor. Math. Phys.* 1 (1969) 3.

[16] R.P. Feynman and A.R. Hibbs, *Quantum Mechanics and Path Integrals* (McGraw Hill, New York 1965).

[17] L. Cohen, *J. Math. Phys.* 11 (1970) 3296; J.S. Dowker, *J. Math. Phys.* 17 (1976) 1873.

[18] An early reference is P. Pechukas, *Phys. Rev.* 181 (1969) 166. See also reference [12].

[19] The semi-classical spectrum is derived, in a very general framework, from the semi-classical approximation to the evolution operator in
M.C. Gutzwiller, *J. Math. Phys.* 11 (1970) 1791.

[20] Path integrals can be used to calculate decay rate of metastable states as it has been noticed in
J.S. Langer, *Ann. Phys.* 41 (1967) 108.

[21] S. Coleman, The uses of instantons in *The Whys of Subnuclear Physics, Erice 1977*, A. Zichichi ed. (Plenum, New York 1979).

[22] In the framework of quantum field theory, collective coordinates have been discussed in
J.-L. Gervais et B. Sakita, *Phys. Rev.* D11 (1975) 2943; V.E. Korepin, P.P. Kulish and L.D. Faddeev, *JETP Lett.* 21 (1975) 138.

[23] C.S. Lam, *Nuovo Cimento* A55 (1968) 258; E. Brézin, J.C. Le Guillou and J. Zinn-Justin, *Phys. Rev.* D15 (1977) 1554 et 1558.

[24] L.R. Schulman, *Techniques and Applications of Path Integration* (Wiley, New York 1981).

H. Kleinert, *Path Integrals in Quantum Mechanics, Statistics and Polymer Physics* (World Scientific, Singapore 1995).

A recent book with many references is

C. Grosche and F. Steiner, *Handbook of Feynman path integrals* (Springer, Berlin, Heidelberg 1998).

Index

analytic functions:
   Hilbert space, 140.
   scalar product, 140.
angular momentum:
   $SO(2)$ group, 126.
   $SO(3)$ group, 280.
   $SO(N)$ group, 295.
anharmonic oscillator:
   quartic, 247.
annihilation and creation operators:
   fermions, 197, 199.
anti-periodic boundary conditions, 203, 205.
asymptotic expansion, 240.

Bargmann–Wigner's potential, 237, 251.
barrier penetration, 225.
berezinian, 212.
Bessel function, 15, 17.
Bohr–Sommerfeld quantization condition, 69.
Bohr–Sommerfeld's formula:
   generalization, 237.
Boltzmann's constant, 305.
Bose–Einstein condensation, 161.
Bose–Einstein condensation:
   free bosons, 163.
   harmonic potential, 161.
boson hamiltonian:
   kernel, 158.
bosons:
   equation of state, 160.
   hamiltonian, 158.
   partition function, 159.
   second quantization, 156.
bras and kets, 29, 92, 301.
brownian motion, 34, 122.
brownian paths, 105.

canonical transformations, 283.
central limit theorem, 123.
chemical potential, 159, 199.
circle:
   topology, 127.
Clifford algebra, 181.
cluster property, 96.
collective coordinates, 232, 234, 243.
complex gaussian integral, 136.
connected contributions, 8.
connected correlation functions, 11.
conservation laws, 279.
continuum limit, 99.
correlation functions, 41, 94.
   connected, 101.
   continuum limit, 100.
   gaussian, 46.
correlation length, 96, 98.
covariant derivative, 115.
covariant measure, 293.
creation and annihilation operators, 141.
cumulants, 10.

density matrix, 27, 304.
dimensional reduction, 101.
Dirac's δ function, 43.
   holomorphic formalism, 141.
divergent series, 239.
double-well potential:
   quartic, 226.

equation of state:
   fermions, 205.
equilibrium distribution, 125.
evolution in imaginary time, 225.
evolution operator, 303.
expansion in powers of $\hbar$, 54.

Faddeev–Popov's method, 233, 234.
Fermi energy, 206.
Fermi gas, 206.
   free, 206.
fermion creation–annihilation symmetry, 197.
fermions, 179.
   hamiltonian, 195.
   partition function, 202.
   path integrals, 200.
   state vectors, 197.
Feynman diagrams, 8.
field integral, 166.
   fermions, 207.
finite size effect, 99.
Fock space, 165.
Fokker Planck equation, 124, 253.
   dissipative, 125.
   examples, 132.
   hamiltonian, 122.
Fourier transformation:
   convention, 30.
free energy, 81.
free particle, 308.
functional derivative, 42.
functional integral, 166.

Gamma function, 14.
gauge field, 114.
gauge invariance, 114.
gauge symmetry, 113.
gauge transformations, 111.
gaussian expectation value, 4.
gaussian integral, 2.
   complex matrices, 20.
gaussian model, 97.
gaussian path integrals, 38.
Gegenbauer polynomials, 298.
generating function, 1, 10.
generating functional, 42, 101.
   bosons, 164.
   gaussian measure, 44.
grand canonical formalism, 159.

Grassmann algebra, 179.
   complex conjugation, 180.
   derivative, 181.
   gaussian integral, 184.
   integration, 182.
   scalar product, 191.
Grassmann analytic function, 191, 192.
Grassmann δ function, 192.
ground state:
   properties, 85.

hamiltonian formalism, 281.
harmonic oscillator, 39, 141, 143, 221, 267, 289.
   partition function, 48.
   perturbed, 52.
Heisenberg representation, 106, 304.
Hermite polynomials, 25, 221.
Hilbert space, 301.
Hölder condition, 36.
holomorphic formalism, 140, 142.
   operator kernels, 143.
   trace, 144.

instanton, 225, 228, 230, 241, 251.
   jacobian, 245.

Källen–Lehmann representation, 105.

lattice model, 91.
Legendre transformation, 281.
Lie algebra, 284.
Lippman–Schwinger's equation, 264.
locality, 29.

magnetic field, 113.
   constant, spectrum, 131.
magnetic hamiltonian, 291.
Markov property, 285.
markovian processes, 28, 284.
metastable states: decay, 238.
metric tensor, 292.

nearest-neighbour interaction, 92.

non-relativistic quantum field theory, 167, 207.
normal order, 143, 150, 286.

O(N) symmetry:
  limit $N \to \infty$, 76.
  quartic potential, 76.
  steepest descent method, 77, 80.
operators:
  determinants, 84.

particle number operator, 158.
partition function:
  classical, 92.
  Fermi gas, 207.
  fermions, 205.
  harmonic oscillator, 149.
  holomorphic, 152.
path integral and quantization, 111.
path integral, 27, 35.
  ambiguities, 117, 153.
  changes of variables, 129.
  gaussian, 38.
  gaussian, holomorphic formalism, 147.
  holomorphic, 146, 151.
  phase space, 284.
periodic boundary conditions, 37, 149.
perturbation theory, 6, 53, 84, 101, 289.
perturbative expansion, 152.
pfaffian, 210.
phase space:
  partition function, 287.
  path integral, 284.
  position–momentum, 155.
Poisson brackets, 283.
Poisson's formula, 128.
propagator, 45.
  holomorphic formalism, 148.
psi-function, 70.

quantization ambiguities, 125.
quantization and path integral, 114.
quantum evolution, 257, 287.

quantum partition function, 36.
random matrices:
  gaussian unitary ensemble (GUE), 220.
random walk, 122.
resolvent, 67.
Ricatti equation, 72.
rigid rotator:
  $O(2)$ symmetry, 126.
  $O(N)$ symmetry, 294.

saddle points:
  degenerate, 231.
scattering matrix T, 259.
Schrödinger equation, 32, 71, 309.
Schwinger–Dyson equation, 102.
second quantization, 156.
semi-classical approximation, 161, 225, 296.
semi-classical expansion:
  partition function, 55.
sign function sgn, 45, 119.
S-matrix, 259.
  bosons, 266.
  fermions, 267.
  perturbative expansion, 260.
  semi-classical approximation, 269.
spectrum:
  expansion in powers of $\hbar$, 66.
  perturbative calculation, 63.
  semi-classical expansion, 66.
spin:
  fermion representation, 219.
square-well potential:
  spectrum, 59.
statistical operator, 27.
steepest descent method, 12, 228.
step function θ, 69, 119, 149.
Stirling's formula, 15.
supertrace, 213.
symmetries, 279.
symmetry:
  spontaneous breaking, 96.

symplectic form, 282.

thermal equilibrium, 305.
thermal wave length, 163.
thermodynamic limit, 93, 95, 100.
theta step function, 148.
time translation invariance, 281.
time-ordered product, 106.
trace:
   operator, 37.
transfer matrix, 93.
translation invariance, 229.
translation operator, 306.
two-point function, 95.
   connected, 96.
   gaussian, holomorphic, 149.

perturbative calculation, 103.
spectral representation, 104.

universality, 99.

vacuum, 157, 195.
vacuum expectation value, 100.
virial theorem, 242.

wave function, 307.
wave packet reduction, 304.
wave packet spreading, 258.
Wick's theorem, 5, 47.
   complex, 139.
   fermions, 188, 211.
Wigner's semi-circle law, 223.
WKB approximation, 71.

zero mode, 231, 243.